# COLLECTED WORKS
OF
# COUNT RUMFORD

VOLUME I

# COLLECTED WORKS OF COUNT RUMFORD

EDITED BY SANBORN C. BROWN

## VOLUME I

## THE NATURE OF HEAT

THE BELKNAP PRESS OF
HARVARD UNIVERSITY PRESS
CAMBRIDGE, MASSACHUSETTS
1968

Distributed in Great Britain by Oxford University Press, London
Library of Congress Catalog Card Number 68-17633
Printed in the United States of America

# PREFACE

In July 1796 Count Rumford wrote to the Honorable John Adams, President of the American Academy of Arts and Sciences, announcing the gift of five thousand dollars to the Academy. In this letter Rumford stipulated "that the Interest . . . may be . . . given once every second year as a premium to the Author of the most Important Discovery or useful Improvement which shall be made and published by printing, or in any way made Known to the Public, in any part of the Continent of America, or in any of the American Islands, during the preceding two years, on HEAT or on LIGHT . . . If during any term of two years, reckoning from the last adjudication . . . no new Discovery or Improvement should be made in any part of America relative to either of the subjects in question (Heat or Light,) which, in the opinion of the Academy, shall be of sufficient Importance to deserve this premium, in that case, it is my desire that the premium *may not be given*, but that the Value of it may be reserved, and by laying out in the Purchase of Additional Stock in the American Funds may be employed to Augment the Capital of this premium."

The Academy was delighted to accept this gift, but it soon became apparent that the restrictions were too severe. Up until 1829 no Rumford premium had been given, and a committee of the Academy was formed to look into the matter. As a result of their study, the Academy applied to the Supreme Court of Massachusetts for relief from the conditions of the gift. The case was argued in May 1832, the Academy's point of view being presented by the President, Nathaniel Bowditch. The opinion of the Court relieved the Academy of the impractical features of the original gift while at the same time fully retaining the spirit of Count Rumford's donation. This opinion, delivered by Judge Shaw on May 18, 1832, stipulated the conditions under which the Rumford Fund still operates:

"It is, therefore, by the court ordered . . . that the plaintiffs be . . . empowered to make from the income of said funds, as it now exists, at any annual meeting of the Academy, instead of biennially, as directed by the said Benjamin Count Rumford, award of a gold and silver medal . . . as a premium to the author of any important discovery or useful improvement on heat or on light . . . in any part of the Continent of America or any of the American Islands . . .

"And it is further ordered . . . that the plaintiffs may appropriate from time to time, as the same can advantageously be done, the residue of the income of said fund, hereafter to be received . . . to the purchase of such books and papers and philosophical apparatus (to be the property of said Academy) and in making such publications or procuring such lectures, experiments, or investigations, as shall in their opinion best facilitate and encourage the making of discoveries and improvements which may merit the premium so as aforesaid to be by them awarded."

Historically the responsibility for operating the Rumford Fund has been assigned by the Council of the Academy to a standing committee called the

Rumford Committee, consisting of seven Fellows elected annually. It is the duty of this Committee "to use all proper means to make the Rumford Fund constantly active and useful so as to carry out the donor's intention in the manner defined by the decree of the Supreme Court in 1832, not only by investigating all applications and claims to the Rumford medals, but also by such other means as have been already indicated, and in general to see to the due and proper execution of the trust." The honoring of Count Rumford by sponsoring studies and publications about his life and scientific works has been considered one of the continuing responsibilities of the Rumford Committee since the 1830's.

The concept of collecting Rumford's works and publishing them under the sponsorship of the Academy was initiated by Dr. Morrill Wyman in 1862 but it was not until six years later that the project actually got under way: "The American Academy of Arts and Sciences, at a meeting held June 9, 1868, resolved to publish a complete edition of Count Rumford's Works, and made an appropriation of money for that purpose. In accordance with this resolution, the 'Rumford Committee' of the Academy undertook to collect the writings of Count Rumford, which were scattered through various scientific Journals and Transactions . . . The Committee first prepared as complete a list of Rumford's works as possible, arranging them in chronological order. In selecting the papers for publication it was decided to arrange them according to the order of the list, as far as this could be done without unduly separating papers which relate to the same subject. It has been thought best, after mature deliberation, to publish the papers in their original form, without reference to their relations either to the history or to the subsequent progress of the sciences of which they treat."

The first volume of these collected works, described in the paragraph just quoted from the Preface of Volume I, was published in 1870. Subsequent Rumford Committee members apparently did not agree with the publication of the *Complete Works* in chronological order, and four years later Volume III appeared, containing the three long essays on kitchen fireplaces. Volume II, published in 1875, which with Volume I had been assembled by Professor Joseph Winlock, the 1870 chairman of the Rumford Committee, continued to follow the chronological order. The 1874–1875 Rumford Committee, under the chairmanship of Josiah P. Cooke, Jr., put out Volumes III and IV largely ignoring chronological order and grouping papers according to their content. In 1871, in connection with the edition of Rumford's *Complete Works*, the Academy also published the *Memoir of Sir Benjamin Thompson, Count Rumford, with Notices of His Daughter*, by George E. Ellis.

Since the publication of this series, the five volumes have been in considerable demand, and they are essentially unobtainable today. Yet, as interest in the history of science has increased, the demand for this material in a readily available form has risen markedly. It is for this reason that the 1965–66 Rumford Committee voted to subsidize a reprinting of the collected works of Count Rumford, and by approving this request endorsed this new edition of the Rumford *Works*, to be published in collaboration with the Harvard University Press. Several major changes have been incorporated into this new edition of

Rumford's *Collected Works*, some reflecting trends in modern scholarship and some those in the newer techniques of reproduction. Insofar as is possible, this new edition arranges the papers by subject matter, retaining chronological order only within a particular subject area. Bibliographic techniques have changed considerably since the eighteenth century, and modern bibliographic form has been incorporated into this edition. One of the frustrations of using the previous edition was the lack of a suitable index and a concentrated attempt to produce a useful working index for the historian has been undertaken here. In the nineteenth-century edition, the figures were all relithographed, but not always with the care for detail that at times is essential to an understanding of the scientific operation of Rumford's apparatus. In this present edition we have photographically reproduced the original figures, with the attendant increase in faithfulness of reproduction.

The language problem is not inconsiderable in producing any collected works of Count Rumford. To quote from the Preface to the fourth volume (1875): "The Rumford Committee have spared no pains to make the edition complete in every detail, hoping that it might be accepted by scholars as a worthy memorial of the great services which Count Rumford rendered to mankind both in science and in philanthropy. They have sought, however, to avoid needless repetition; and where, as was the case in several instances, the same matter appeared in different publications, and even under a changed title, they have only reproduced those parts which seemed to be the more mature or the more complete. The selection, however, has not always been without difficulty, owing to the circumstance that Count Rumford published his papers in three different languages, and those originally published in one were generally subsequently translated into the other two, not unfrequently with emendations and additions by the Count himself. Hence it has sometimes been necessary in carrying out the proposed plan to reproduce different portions of the same paper from versions in different languages, but in every case the sources have been indicated, and, other things being equal, preference has always been given to the English version; for, although so long a resident both at Munich and at Paris, Count Rumford always wrote in English with greater clearness and skill than in either German or French . . . Both for the sake of uniformity and also in order to render the work more accessible to his own countrymen, the committee decided to print the whole in the Count's vernacular language. All the new translations from the French and German have been made . . . not without difficulty; for not only was the foreign text in many cases obscure, so different from the clear English style of the author, but, moreover, it was often evident that the German or French version was itself a translation from a draft written originally in English." In addition to these problems the translations were chosen and made by a nonphysicist, and study of the original texts shows that in some cases important physical ideas have been left out in this process of editing and reconstituting.

One publication in French, which was passed over by previous committees as being redundant, has been translated and included in the present volume because it contains in fact clearly different material.

Sanborn C. Brown

# CONTENTS

# COLLECTED WORKS
OF
# COUNT RUMFORD

VOLUME I

# AN EXPERIMENTAL INQUIRY

CONCERNING THE

# SOURCE OF THE HEAT WHICH IS EXCITED BY FRICTION.

## AN INQUIRY

CONCERNING THE

## SOURCE OF THE HEAT WHICH IS EXCITED BY FRICTION.

IT frequently happens that in the ordinary affairs and occupations of life, opportunities present themselves of contemplating some of the most curious operations of Nature; and very interesting philosophical experiments might often be made, almost without trouble or expence, by means of machinery contrived for the mere mechanical purposes of the arts and manufactures.

I have frequently had occasion to make this observation; and am persuaded that a habit of keeping the eyes open to everything that is going on in the ordinary course of the business of life has oftener led, as it were by accident, or in the playful excursions of the imagination, put into action by contemplating the most common appearances, to useful doubts and sensible schemes for investigation and improvement, than all the more intense meditations of philosophers in the hours expressly set apart for study.

It was by accident that I was led to make the experiments of which I am about to give an account; and, though they are not perhaps of sufficient importance to merit so formal an introduction, I cannot help flattering myself that they will be thought curious in several respects, and worthy of the honour of being made known to the Royal Society.

Being engaged lately in superintending the boring of cannon in the workshops of the military arsenal at Munich, I was struck with the very considerable degree of Heat which a brass gun acquires in a short time in being bored, and with the still more intense Heat (much greater than that of boiling water, as I found by experiment) of the metallic chips separated from it by the borer.

The more I meditated on these phænomena, the more they appeared to me to be curious and interesting. A thorough investigation of them seemed even to bid fair to give a farther insight into the hidden nature of Heat; and to enable us to form some reasonable conjectures respecting the existence, or non-existence, of an *igneous fluid*, — a subject on which the opinions of philosophers have in all ages been much divided.

In order that the Society may have clear and distinct ideas of the speculations and reasonings to which these appearances gave rise in my mind, and also of the specific objects of philosophical investigation they suggested to me, I must beg leave to state them at some length, and in such manner as I shall think best suited to answer this purpose.

From *whence comes* the Heat actually produced in the mechanical operation above mentioned?

Is it furnished by the metallic chips which are separated by the borer from the solid mass of metal?

If this were the case, then, according to the modern doctrines of latent Heat, and of caloric, the *capacity for Heat* of the parts of the metal, so reduced to chips, ought not only to be changed, but the change undergone by them should be sufficiently great to account for *all* the Heat produced.

But no such change had taken place; for I found, upon taking equal quantities, by weight, of these chips, and of thin slips of the same block of metal separated by means of a fine saw, and putting them at the same temperature (that of boiling water) into equal quantities of cold water (that is to say, at the temperature of $59\frac{1}{2}°$ F.), the portion of water into which the chips were put was not, to all appearance, heated either less or more than the other portion in which the slips of metal were put.

This experiment being repeated several times, the results were always so nearly the same that I could not determine whether any, or what change had been produced in the metal, *in regard to its capacity for Heat*, by being reduced to chips by the borer.*

From hence it is evident that the Heat produced could not possibly have been furnished at the expence of the

* As these experiments are important, it may perhaps be agreeable to the Society to be made acquainted with them in their details.

One of them was as follows: —

To 4590 grains of water, at the temperature of $59\frac{1}{2}°$ F. (an allowance as compensation, reckoned in water, for the capacity for Heat of the containing cylindrical tin vessel being included), were added $1016\frac{1}{8}$ grains of gun-metal in thin slips, separated from the gun by means of a fine saw, being at the temperature of 210° F. When they had remained together 1 minute, and had been well stirred about, by means of a small rod of light wood, the Heat of the mixture was found to be = 63°.

From this experiment the *specific Heat* of the metal, calculated according to the rule given by Dr. Crawford, turns out to be = 0.1100, that of water being = 1.0000.

An experiment was afterwards made with the metallic chips as follows: —

To the same quantity of water as was used in the experiment above mentioned, at the same temperature (*viz.* $59\frac{1}{2}°$), and in the same cylindrical tin vessel, were now put $1016\frac{1}{8}$ grains of metallic chips of gun-metal bored out of the same gun from which the slips used in the foregoing experiment were taken, and at the same temperature (210°). The Heat of the mixture at the end of 1 minute was just 63°, as before; consequently the specific Heat of these metallic chips was = 0.1100. Each of the above experiments was repeated three times, and always with nearly the same results.

latent Heat of the metallic chips. But, not being willing to rest satisfied with these trials, however conclusive they appeared to me to be, I had recourse to the following still more decisive experiment.

Taking a cannon (a brass six-pounder), cast solid, and rough as it came from the foundry (see Fig. 1, Tab. IV.), and fixing it (horizontally) in the machine used for boring, and at the same time finishing the outside of the cannon by turning (see Fig. 2), I caused its extremity to be cut off, and, by turning down the metal in that part, a solid cylinder was formed, $7\frac{3}{4}$ inches in diameter, and $9\frac{8}{10}$ inches long, which, when finished, remained joined to the rest of the metal (that which, properly speaking, constituted the cannon) by a small cylindrical neck, only $2\frac{1}{5}$ inches in diameter, and $3\frac{8}{10}$ inches long.

This short cylinder, which was supported in its horizontal position and turned round its axis by means of the neck by which it remained united to the cannon, was now bored with the horizontal borer used in boring cannon; but its bore, which was 3.7 inches in diameter, instead of being continued through its whole length (9.8 inches) was only 7.2 inches in length; so that a solid bottom was left to this hollow cylinder, which bottom was 2.6 inches in thickness.

This cavity is represented by dotted lines in Fig. 2; as also in Fig. 3, where the cylinder is represented on an enlarged scale.

This cylinder being designed for the express purpose of generating Heat *by friction*, by having a blunt borer forced against its solid bottom at the same time that it should be turned round its axis by the force of horses, in order that the Heat accumulated in the cylinder might from time to time be measured, a small round hole (see *d*, *e*, Fig. 3), 0.37 of an inch only in diameter, and 4.2 inches in depth, for the purpose of introducing a small cylindrical mercurial thermometer, was made in it, on one side, in a direction perpendicular to the axis of the cylinder, and ending in the middle of the solid part of the metal which formed the bottom of its bore.

The solid contents of this hollow cylinder, exclusive of the cylindrical neck by which it remained united to the cannon, were $385\frac{3}{4}$ cubic inches, English measure, and it weighed 113.13 lb., avoirdupois ; as I found on weighing it at the end of the course of experiments made with it, and after it had been separated from the cannon with which, during the experiments, it remained connected.*

### *Experiment No.* 1.

This experiment was made in order to ascertain how much Heat was actually generated by friction, when a blunt steel borer being so forcibly shoved (by means of a strong screw) against the bottom of the bore of the cylinder, that the pressure against it was equal to the weight of about 10,000 lb., avoirdupois, the cylinder

* For fear I should be suspected of prodigality in the prosecution of my philosophical researches, I think it necessary to inform the Society that the cannon I made use of in this experiment was not sacrificed to it. The short hollow cylinder which was formed at the end of it was turned out of a cylindrical mass of metal, about 2 feet in length, projecting beyond the muzzle of the gun, called in the German language the *verlorner kopf* (the head of the cannon to be thrown away), and which is represented in Fig 1.

This original projection, which is cut off before the gun is bored, is always cast with it, in order that, by means of the pressure of its weight on the metal in the lower part of the mould during the time it is cooling, the gun may be the more compact in the neighbourhood of the muzzle, where, without this precaution, the metal would be apt to be porous, or full of honeycombs.

was turned round on its axis (by the force of horses) at the rate of about 32 times in a minute.

This machinery, as it was put together for the experiment, is represented by Fig. 2. W is a strong horizontal iron bar, connected with proper machinery carried round by horses, by means of which the cannon was made to turn round its axis.

To prevent, as far as possible, the loss of any part of the Heat that was generated in the experiment, the cylinder was well covered up with a fit coating of thick and warm flannel, which was carefully wrapped round it, and defended it on every side from the cold air of the atmosphere. This covering is not represented in the drawing of the apparatus, Fig. 2.

I ought to mention that the borer was a flat piece of hardened steel, 0.63 of an inch thick, 4 inches long, and nearly as wide as the cavity of the bore of the cylinder, namely, $3\frac{1}{2}$ inches. Its corners were rounded off at its end, so as to make it fit the hollow bottom of the bore; and it was firmly fastened to the iron bar (*m*) which kept it in its place. The area of the surface by which its end was in contact with the bottom of the bore of the cylinder was nearly $2\frac{1}{3}$ inches. This borer, which is distinguished by the letter *n*, is represented in most of the figures.

At the beginning of the experiment, the temperature of the air in the shade, as also that of the cylinder, was just 60° F.

At the end of 30 minutes, when the cylinder had made 960 revolutions about its axis, the horses being stopped, a cylindrical mercurial thermometer, whose bulb was $\frac{32}{100}$ of an inch in diameter, and $3\frac{1}{4}$ inches in length, was introduced into the hole made to receive it, in the side of the cylinder, when the mercury rose almost instantly to 130°.

Though the Heat could not be supposed to be quite equally distributed in every part of the cylinder, yet, as the length of the bulb of the thermometer was such that it extended from the axis of the cylinder to near its surface, the Heat indicated by it could not be very different from that of the *mean temperature* of the cylinder; and it was on this account that a thermometer of that particular form was chosen for this experiment.

To see how fast the Heat escaped out of the cylinder (in order to be able to make a probable conjecture respecting the quantity given off by it during the time the Heat generated by the friction was accumulating), the machinery standing still, I suffered the thermometer to remain in its place near three quarters of an hour, observing and noting down, at small intervals of time, the height of the temperature indicated by it.

| | The Heat, as shown by the thermometer, was |
|---|---|
| Thus at the end of 4 minutes | 126° |
| after 5 minutes, always reckoning from the first observation | 125 |
| at the end of 7 minutes | 123 |
| 12 " | 120 |
| 14 " | 119 |
| 16 " | 118 |
| 20 " | 116 |
| 24 " | 115 |
| 28 " | 114 |
| 31 " | 113 |
| 34 " | 112 |
| 37½ " | 111 |
| and when 41 minutes had elapsed | 110 |

Having taken away the borer, I now removed the

metallic dust, or, rather, scaly matter, which had been detached from the bottom of the cylinder by the blunt steel borer, in this experiment; and, having carefully weighed it, I found its weight to be 837 grains, Troy.

Is it possible that the very considerable quantity of Heat that was produced in this experiment (a quantity which actually raised the temperature of above 113 lb. of gun-metal at least 70 degrees of Fahrenheit's thermometer, and which, of course, would have been capable of melting $6\frac{1}{2}$ lb. of ice, or of causing near 5 lb. of ice-cold water to boil) could have been furnished by so inconsiderable a quantity of metallic dust? and this merely in consequence of *a change* of its capacity for Heat?

As the weight of this dust (837 grains, Troy) amounted to no more than $\frac{1}{948}$th part of that of the cylinder, it must have lost no less than 948 degrees of Heat, to have been able to have raised the temperature of the cylinder 1 degree; and consequently it must have given off 66,360 degrees of Heat to have produced the effects which were actually found to have been produced in the experiment!

But without insisting on the improbability of this supposition, we have only to recollect, that from the results of actual and decisive experiments, made for the express purpose of ascertaining that fact, the capacity for Heat of the metal of which great guns are cast *is not sensibly changed* by being reduced to the form of metallic chips in the operation of boring cannon; and there does not seem to be any reason to think that it can be much changed, if it be changed at all, in being reduced to much smaller pieces by means of a borer that is less sharp.

If the Heat, or any considerable part of it, were produced in consequence of a change in the capacity for Heat of a part of the metal of the cylinder, as such change could only be *superficial*, the cylinder would by degrees be *exhausted;* or the quantities of Heat produced in any given short space of time would be found to diminish gradually in successive experiments. To find out if this really happened or not, I repeated the last-mentioned experiment several times with the utmost care; but I did not discover the smallest sign of exhaustion in the metal, notwithstanding the large quantities of Heat actually given off.

Finding so much reason to conclude that the Heat generated in these experiments, or *excited*, as I would rather choose to express it, was not furnished *at the expense of the latent Heat* or *combined caloric* of the metal, I pushed my inquiries a step farther, and endeavoured to find out whether the air did, or did not, contribute anything in the generation of it.

### *Experiment No.* 2.

As the bore of the cylinder was cylindrical, and as the iron bar (*m*), to the end of which the blunt steel borer was fixed, was square, the air had free access to the inside of the bore, and even to the bottom of it, where the friction took place by which the Heat was excited.

As neither the metallic chips produced in the ordinary course of the operation of boring brass cannon, nor the finer scaly particles produced in the last-mentioned experiments by the friction of the blunt borer, showed any signs of calcination, I did not see how the air could possibly have been the cause of the Heat that was produced; but, in an investigation of this kind, I thought that no pains should be spared to clear away the rubbish, and leave the subject as naked and open to inspection as possible.

In order, by one decisive experiment, to determine whether the air of the atmosphere had any part, or not, in the generation of the Heat, I contrived to repeat the experiment under circumstances in which *it was evidently impossible for it to produce any effect whatever.* By means of a piston exactly fitted to the mouth of the bore of the cylinder, through the middle of which piston the square iron bar, to the end of which the blunt steel borer was fixed, passed in a square hole made perfectly air-tight, the access of the external air to the inside of the bore of the cylinder was effectually prevented. (In Fig. 3, this piston (*p*) is seen in its place; it is likewise shown in Fig. 7 and 8.)

I did not find, however, by this experiment, that the exclusion of the air diminished, in the smallest degree, the quantity of Heat excited by the friction.

There still remained one doubt, which, though it appeared to me to be so slight as hardly to deserve any attention, I was however desirous to remove. The piston which closed the mouth of the bore of the cylinder, in order that it might be air-tight, was fitted into it with so much nicety, by means of its collars of leather, and pressed against it with so much force, that, notwithstanding its being oiled, it occasioned a considerable degree of friction when the hollow cylinder was turned round its axis. Was not the Heat produced, or at least some part of it, occasioned by this friction of the piston? and, as the external air had free access to the extremity of the bore, where it came in contact with the piston, is it not possible that this air may have had some share in the generation of the Heat produced?

## *Experiment No. 3.*

A quadrangular oblong deal box (see Fig. 4), water-tight, $11\frac{1}{2}$ English inches long, $9\frac{4}{10}$ inches wide, and $9\frac{6}{10}$ inches deep (measured in the clear), being provided with holes or slits in the middle of each of its ends, just large enough to receive, the one the square iron rod to the end of which the blunt steel borer was fastened, the other the small cylindrical neck which joined the hollow cylinder to the cannon; when this box (which was occasionally closed above by a wooden cover or lid moving on hinges) was put into its place, that is to say, when, by means of the two vertical openings or slits in its two ends (the upper parts of which openings were occasionally closed by means of narrow pieces of wood sliding in vertical grooves), the box (*g*, *h*, *i*, *k*, Fig. 3) was fixed to the machinery in such a manner that its bottom (*i*, *k*) being in the plane of the horizon, its axis coincided with the axis of the hollow metallic cylinder; it is evident, from the description, that the hollow metallic cylinder would occupy the middle of the box, without touching it on either side (as it is represented in Fig. 3); and that, on pouring water into the box, and filling it to the brim, the cylinder would be completely covered and surrounded on every side by that fluid. And farther, as the box was held fast by the strong square iron rod (*m*) which passed in a *square hole* in the center of one of its ends (*a*, Fig. 4), while the round or cylindrical neck, which joined the hollow cylinder to the end of the cannon, could turn round freely on its axis in the *round hole* in the center of the other end of it, it is evident that the machinery could be put in motion without the least danger of forcing the box out of its place, throwing the water out of it, or deranging any part of the apparatus.

Everything being ready, I proceeded to make the experiment I had projected in the following manner.

The hollow cylinder having been previously cleaned out, and the inside of its bore wiped with a clean towel till it was quite dry, the square iron bar, with the blunt steel borer fixed to the end of it, was put into its place; the mouth of the bore of the cylinder being closed at the same time by means of the circular piston, through the center of which the iron bar passed.

This being done, the box was put in its place, and the joinings of the iron rod and of the neck of the cylinder with the two ends of the box having been made water-tight by means of collars of oiled leather, the box was filled with cold water (*viz.* at the temperature of 60°), and the machine was put in motion.

The result of this beautiful experiment was very striking, and the pleasure it afforded me amply repaid me for all the trouble I had had in contriving and arranging the complicated machinery used in making it.

The cylinder, revolving at the rate of about 32 times in a minute, had been in motion but a short time, when I perceived, by putting my hand into the water and touching the outside of the cylinder, that Heat was generated; and it was not long before the water which surrounded the cylinder began to be sensibly warm.

At the end of 1 hour I found, by plunging a thermometer into the water in the box (the quantity of which fluid amounted to 18.77 lb., avoirdupois, or $2\frac{1}{4}$ wine gallons), that its temperature had been raised no less than 47 degrees; being now 107° of Fahrenheit's scale.

When 30 minutes more had elapsed, or 1 hour and 30 minutes after the machinery had been put in motion, the Heat of the water in the box was 142°.

At the end of 2 hours, reckoning from the beginning of the experiment, the temperature of the water was found to be raised to 178°.

At 2 hours 20 minutes it was at 200°; and at 2 hours 30 minutes it ACTUALLY BOILED!

It would be difficult to describe the surprise and astonishment expressed in the countenances of the bystanders, on seeing so large a quantity of cold water heated, and actually made to boil, without any fire.

Though there was, in fact, nothing that could justly be considered as surprising in this event, yet I acknowledge fairly that it afforded me a degree of childish pleasure, which, were I ambitious of the reputation of a *grave philosopher*, I ought most certainly rather to hide than to discover.

The quantity of Heat excited and accumulated in this experiment was very considerable; for, not only the water in the box, but also the box itself (which weighed 15¼ lb.), and the hollow metallic cylinder, and that part of the iron bar which, being situated within the cavity of the box, was immersed in the water, were heated 150 degrees of Fahrenheit's scale; *viz.* from 60° (which was the temperature of the water and of the machinery at the beginning of the experiment) to 210°, the Heat of boiling water at Munich.

The total quantity of Heat generated may be estimated with some considerable degree of precision as follows:—

| | Quantity of ice-cold water which, with the given quantity of Heat, might have been heated 180 degrees, or made to boil. In avoirdupois weight. |
|---|---|
| Of the Heat excited there appears to have been actually accumulated, — | |
| In the water contained in the wooden box, $18\frac{3}{4}$ lb., avoirdupois, heated 150 degrees, namely, from 60° to 210° F. . . . . . . . | lb. 15.2 |
| In 113.13 lb. of gun-metal (the hollow cylinder), heated 150 degrees; and, as the capacity for Heat of this metal is to that of water as 0.1100 to 1.0000, this quantity of Heat would have heated $12\frac{1}{2}$ lb. of water the same number of degrees . | 10.37 |
| In 36.75 cubic inches of iron (being that part of the iron bar to which the borer was fixed which entered the box), heated 150 degrees; which may be reckoned equal in capacity for Heat to 1.21 lb. of water . . . . . . | 1.01 |
| N. B. No estimate is here made of the Heat accumulated in the wooden box, nor of that dispersed during the experiment. | |
| Total quantity of ice-cold water which, with the Heat actually generated by friction, and accumulated in 2 hours and 30 minutes, might have been heated 180 degrees, or made to boil . . . . . . . . . . . | 26.58 |

From the knowledge of the *quantity* of Heat actually produced in the foregoing experiment, and of the *time* in which it was generated, we are enabled to ascertain *the velocity of its production*, and to determine how large a fire must have been, or how much fuel must have been consumed, in order that, in burning equably, it should have produced by combustion the same quantity of Heat in the same time.

In one of Dr. Crawford's experiments (see his Treatise on Heat, p. 321), 37 lb. 7 oz., Troy, = 181,920 grains of water, were heated $2\frac{1}{10}$ degrees of Fahrenheit's thermometer with the Heat generated in the combustion of 26 grains of wax. This gives 382,032 grains of water heated 1 degree with 26 grains of wax, or $14,693\frac{14}{26}$ grains of water heated 1 degree, or $\frac{14693}{180} = 81.631$

grains heated 180 degrees, with the Heat generated in the combustion of 1 grain of wax.

The quantity of ice-cold water which might have been heated 180 degrees with the Heat generated by friction in the before-mentioned experiment was found to be 26.58 lb., avoirdupois, = 188,060 grains; and, as 81.631 grains of ice-cold water require the Heat generated in the combustion of 1 grain of wax to heat it 180 degrees, the former quantity of ice-cold water, namely, 188,060 grains, would require the combustion of no less than 2303.8 grains (= $4\frac{8}{10}$ oz., Troy) of wax to heat it 180 degrees.

As the experiment (No. 3) in which the given quantity of Heat was generated by friction lasted 2 hours and 30 minutes, = 150 minutes, it is necessary, for the purpose of ascertaining how many wax candles of any given size must burn together, in order that in the combustion of them the given quantity of Heat may be generated in the given time, and consequently *with the same celerity* as that with which the Heat was generated by friction in the experiment, that the size of the candles should be determined, and the quantity of wax consumed in a given time by each candle in burning equably should be known.

Now I found, by an experiment made on purpose to finish these computations, that when a good wax candle, of a moderate size, $\frac{3}{4}$ of an inch in diameter, burns with a clear flame, just 49 grains of wax are consumed in 30 minutes. Hence it appears that 245 grains of wax would be consumed by such a candle in 150 minutes; and that, to burn the quantity of wax (= 2303.8 grains) necessary to produce the quantity of Heat actually obtained by friction in the experiment in question, and in the given time (150 minutes), *nine candles*, burning at once, would not be sufficient; for 9 multiplied into 245 (the number of grains consumed by each candle in 150 minutes) amounts to no more than 2205 grains; whereas the quantity of wax necessary to be burnt, in order to produce the given quantity of Heat, was found to be 2303.8 grains.

From the result of these computations it appears, that the quantity of Heat produced equably, or in a continual stream (if I may use that expression), by the friction of the blunt steel borer against the bottom of the hollow metallic cylinder, in the experiment under consideration, was *greater* than that produced equably in the combustion of *nine wax candles*, each $\frac{3}{4}$ of an inch in diameter, all burning together, or at the same time, with clear bright flames.

As the machinery used in this experiment could easily be carried round by the force of one horse (though, to render the work lighter, two horses were actually employed in doing it), these computations shew further how large a quantity of Heat might be produced, by proper mechanical contrivance, merely by the strength of a horse, without either fire, light, combustion, or chemical decomposition; and, in a case of necessity, the Heat thus produced might be used in cooking victuals.

But no circumstances can be imagined in which this method of procuring Heat would not be disadvantageous; for more Heat might be obtained by using the fodder necessary for the support of a horse as fuel.

As soon as the last-mentioned experiment (No. 3) was finished, the water in the wooden box was let off, and the box removed; and the borer being taken out of the cylinder, the scaly metallic powder which had been produced by the friction of the borer against the bottom of the cylinder was collected, and, being carefully weighed, was found to weigh 4145 grains, or about $8\frac{2}{3}$ oz., Troy.

As this quantity was produced in $2\frac{1}{2}$ hours, this gives 824 grains for the quantity produced *in half an hour*.

In the first experiment, which lasted only *half an hour*, the quantity produced was 837 grains.

In the experiment No. 1, the quantity of Heat generated in *half an hour* was found to be equal to that which would be required to heat 5 lb., avoirdupois, of ice-cold water 180 degrees, or cause it to boil.

According to the result of the experiment No. 3, the Heat generated in *half an hour* would have caused 5.31 lb. of ice-cold water to boil. But, in this last-mentioned experiment, the Heat generated being more effectually confined, less of it was lost; which accounts for the difference of the results of the two experiments.

It remains for me to give an account of one experiment more, which was made with this apparatus. I found, by the experiment No. 1, how much Heat was generated when the air had free access to the metallic surfaces which were rubbed together. By the experiment No. 2, I found that the quantity of Heat generated was not sensibly diminished when the free access of the air was prevented; and by the result of No. 3, it appeared that the generation of the Heat was not prevented or retarded by keeping the apparatus immersed in water. But as, in this last-mentioned experiment, the water, though it surrounded the hollow metallic cylinder on every side, externally, was not suffered to enter the cavity of its bore (being prevented by the piston), and consequently did not come into contact with the metallic surfaces where the Heat was generated; to see what effects would be produced by giving the water free access to these surfaces, I now made the

*Experiment No.* 4.

The piston which closed the end of the bore of the cylinder being removed, the blunt borer and the cylinder were once more put together; and the box being fixed in its place, and filled with water, the machinery was again put in motion.

There was nothing in the result of this experiment that renders it necessary for me to be very particular in my account of it. Heat was generated as in the former experiments, and, to all appearance, quite as rapidly; and I have no doubt but the water in the box would have been brought to boil, had the experiment been continued as long as the last. The only circumstance that surprised me was, to find how little difference was occasioned in the noise made by the borer in rubbing against the bottom of the bore of the cylinder, by filling the bore with water. This noise, which was very grating to the ear, and sometimes almost insupportable, was, as nearly as I could judge of it, quite as loud and as disagreeable when the surfaces rubbed together were wet with water as when they were in contact with air.

By meditating on the results of all these experiments, we are naturally brought to that great question which has so often been the subject of speculation among philosophers; namely, —

What is Heat? Is there any such thing as an *igneous fluid?* Is there anything that can with propriety be called *caloric?*

We have seen that a very considerable quantity of Heat may be excited in the friction of two metallic surfaces, and given off in a constant stream or flux *in all directions* without interruption or intermission, and without any signs of diminution or exhaustion.

From whence came the Heat which was continually given off in this manner in the foregoing experiments? Was it furnished by the small particles of metal, detached from the larger solid masses, on their being rubbed together? This, as we have already seen, could not possibly have been the case.

Was it furnished by the air? This could not have been the case; for, in three of the experiments, the machinery being kept immersed in water, the access of the air of the atmosphere was completely prevented.

Was it furnished by the water which surrounded the machinery? That this could not have been the case is evident: *first*, because this water was continually *receiving Heat* from the machinery, and could not at the same time be *giving to*, and *receiving Heat from*, the same body; and, *secondly*, because there was no chemical decomposition of any part of this water. Had any such decomposition taken place (which, indeed, could not reasonably have been expected), one of its component elastic fluids (most probably inflammable air) must at the same time have been set at liberty, and, in making its escape into the atmosphere, would have been detected; but though I frequently examined the water to see if any air-bubbles rose up through it, and had even made preparations for catching them, in order to examine them, if any should appear, I could perceive none; nor was there any sign of decomposition of any kind whatever, or other chemical process, going on in the water.

Is it possible that the Heat could have been supplied by means of the iron bar to the end of which the blunt steel borer was fixed? or by the small neck of gun-metal by which the hollow cylinder was united to the cannon? These suppositions appear more improbable even than either of those before mentioned; for Heat was continually going off, or *out of the machinery*, by both these passages, during the whole time the experiment lasted.

And, in reasoning on this subject, we must not forget to consider that most remarkable circumstance, that the source of the Heat generated by friction, in these experiments, appeared evidently to be *inexhaustible.*

It is hardly necessary to add, that anything which any *insulated* body, or system of bodies, can continue to furnish *without limitation*, cannot possibly be *a material substance;* and it appears to me to be extremely difficult, if not quite impossible, to form any distinct idea of anything capable of being excited and communicated in the manner the Heat was excited and communicated in these experiments, except it be MOTION.

I am very far from pretending to know how, or by what means or mechanical contrivance, that particular kind of motion in bodies which has been supposed to constitute Heat is excited, continued, and propagated; and I shall not presume to trouble the Society with mere conjectures, particularly on a subject which, during so many thousand years, the most enlightened philosophers have endeavoured, but in vain, to comprehend.

But, although the mechanism of Heat should, in fact, be one of those mysteries of nature which are beyond the reach of human intelligence, this ought by no means to discourage us or even lessen our ardour, in our attempts to investigate the laws of its operations. How far can we advance in any of the paths which science has opened to us before we find ourselves enveloped in those thick mists which on every side bound the horizon of the human intellect? But how ample and how interesting is the field that is given us to explore!

Nobody, surely, in his sober senses, has ever pretended to understand the mechanism of gravitation; and yet what sublime discoveries was our immortal Newton enabled to make, merely by the investigation of the laws of its action!

The effects produced in the world by the agency of Heat are probably *just as extensive*, and quite as important, as those which are owing to the tendency of the particles of matter towards each other; and there is no doubt but its operations are, in all cases, determined by laws equally immutable.

Before I finish this Essay, I would beg leave to observe, that although, in treating the subject I have endeavoured to investigate, I have made no mention of the names of those who have gone over the same ground before me, nor of the success of their labours, this omission has not been owing to any want of respect for my predecessors, but was merely to avoid prolixity, and to be more at liberty to pursue, without interruption, the natural train of my own ideas.

## DESCRIPTION OF THE FIGURES.

FIG. 1 shews the cannon used in the foregoing experiments in the state it was in when it came from the foundry.

Fig. 2 shews the machinery used in the experiments No. 1 and No. 2. The cannon is seen fixed in the machine used for boring cannon. W is a strong iron bar (which, to save room in the drawing, is represented as broken off), which bar, being united with machinery (not expressed in the figure) that is carried round by horses, causes the cannon to turn round its axis.

*m* is a strong iron bar, to the end of which the blunt borer is fixed; which, by being forced against the bottom of the bore of the short hollow cylinder that remains connected by a small cylindrical neck to the end of the cannon, is used in generating Heat by friction.

Fig. 3 shews, on an enlarged scale, the same hollow cylinder that is represented on a smaller scale in the foregoing figure. It is here seen connected with the wooden box (*g*, *h*, *i*, *k*) used in the experiments No. 3 and No. 4, when this hollow cylinder was immersed in water.

*p*, which is marked by dotted lines, is the piston which closed the end of the bore of the cylinder.

*n* is the blunt borer seen sidewise.

*d*, *e*, is the small hole by which the thermometer was introduced that was used for ascertaining the Heat of the cylinder. To save room in the drawing, the cannon is represented broken off near its muzzle; and the iron

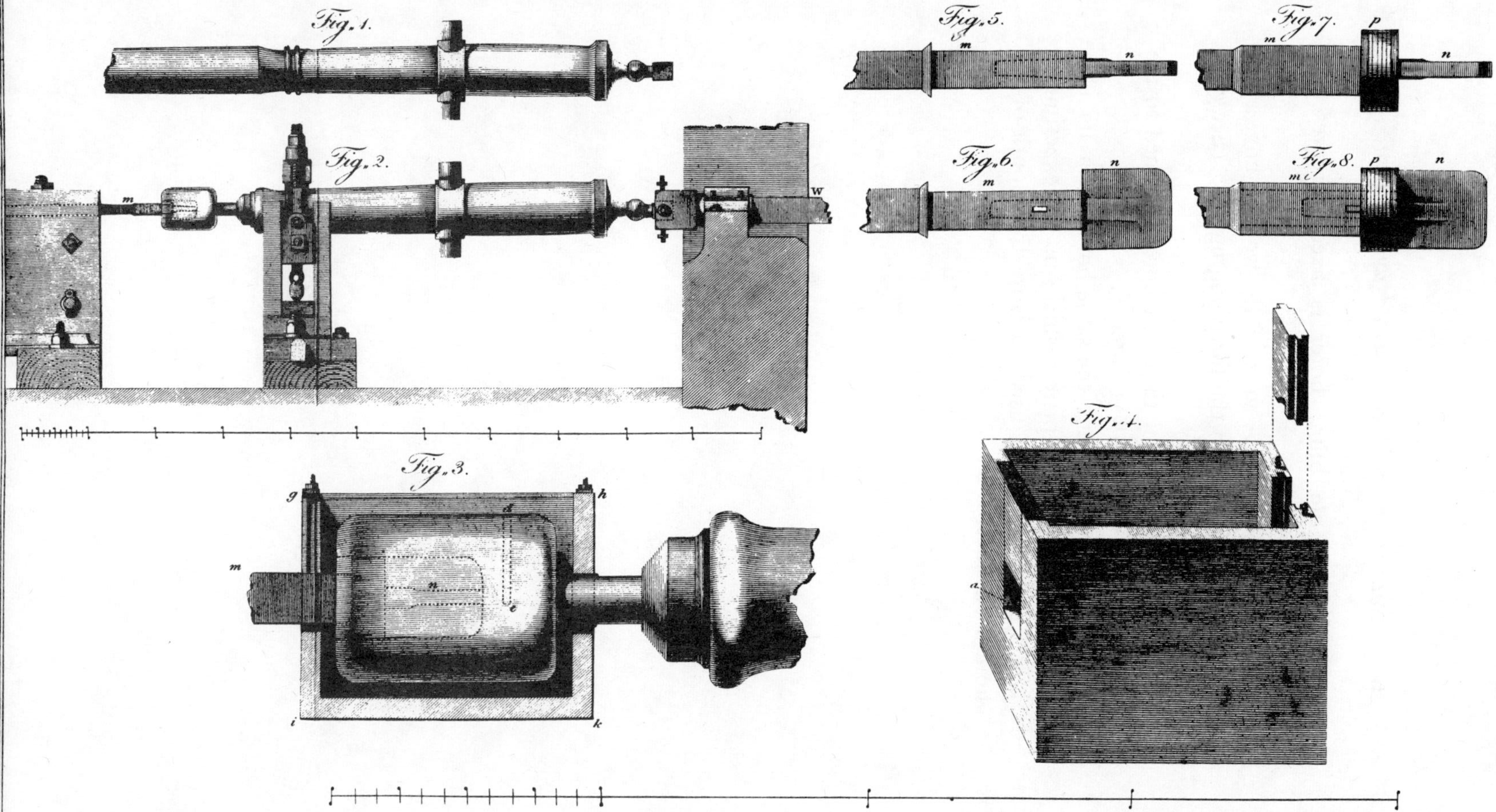

Fig. 1.
Fig. 2.
m
W
Fig. 3.
g
h
i
k
m
n
d
e
Fig. 4.
a
Fig. 5.
m
n
Fig. 6.
m
n
Fig. 7.
m
p
n
Fig. 8.
m
p
n

bar to which the blunt borer is fixed is represented broken off at *m*.

Fig. 4 is a perspective view of the wooden box, a section of which is seen in the foregoing figure. (See *g*, *h*, *i*, *k*, Fig. 3.)

Fig. 5 and 6 represent the blunt borer *n*, joined to the iron bar *m*, to which it was fastened.

Fig. 7 and 8 represent the same borer, with its iron bar, together with the piston which, in the experiments No. 2 and No. 3, was used to close the mouth of the hollow cylinder.

# AN INQUIRY

CONCERNING

## THE WEIGHT ASCRIBED TO HEAT.

THE various experiments which have hitherto been made with a view to determine the question, so long agitated, relative to the weight which has been supposed to be gained, or to be lost, by bodies upon their being heated, are of a nature so very delicate, and are liable to so many errors, not only on account of the imperfections of the instruments made use of, but also of those, much more difficult to appreciate, arising from the vertical currents in the atmosphere, caused by the hot or the cold body which is placed in the balance, that it is not at all surprising that opinions have been so much divided, relative to a fact so very difficult to ascertain.

It is a considerable time since I first began to meditate upon this subject, and I have made many experiments with a view to its investigation; and in these experiments I have taken all those precautions to avoid errors which a knowledge of the various sources of them, and an earnest desire to determine a fact which I conceived to be of importance to be known, could inspire; but though all my researches tended to convince me more and more that *a body acquires no additional weight upon being heated*, or, rather, that heat has no effect whatever upon the weights of bodies, I have been so sensible of the delicacy of the inquiry, that I was for a long time afraid to form a decided opinion upon the subject.

Being much struck with the experiments recorded in the Transactions of the Royal Society, Vol. LXXV., made by Dr. Fordyce, upon the weight said to be acquired by water upon being frozen; and being possessed of an excellent balance, belonging to his Most Serene Highness the Elector Palatine Duke of Bavaria; early in the beginning of the winter of the year 1787, — as soon as the cold was sufficiently intense for my purpose, — I set about to repeat those experiments, in order to convince myself whether the very extraordinary fact related might be depended on; and with a view to removing, as far as was in my power, every source of error and deception, I proceeded in the following manner.

Having provided a number of glass bottles, of the form and size of what in England is called a Florence flask, — blown as thin as possible, — and of the same shape and dimensions, I chose out from amongst them two, which, after using every method I could imagine of comparing them together, appeared to be so much alike as hardly to be distinguished from each other.

Into one of these bottles, which I shall call A, I put 4107.86 grains Troy of pure distilled water, which filled it about half full; and into the other, B, I put an equal weight of weak spirit of wine; and, sealing both the bottles hermetically, and washing them, and wiping them perfectly clean and dry on the outside, I suspended them to the arms of the balance, and placed the balance in a large room, which for some weeks had been regularly heated every day by a German stove, and in which the air was kept up to the temperature of 61° of Fahrenheit's thermometer, with very little variation. Having suffered the bottles, with their contents, to remain in this situation till I conceived they must have acquired the temperature of the circumambient air, I wiped them afresh, with a very clean, dry cambric handkerchief, and brought them into the most exact equilibrium possible, by attaching a small piece of very fine silver wire to the arm of the balance to which the bottle which was the lightest was suspended.

Having suffered the apparatus to remain in this situation about twelve hours longer, and finding no alteration in the relative weights of the bottles, — they continuing all this time to be in the most perfect equilibrium, — I now removed them into a large uninhabited room, fronting the north, in which the air, which was very quiet, was at the temperature of 29° F.; the air without doors being at the same time at 27°; and going out of the room, and locking the door after me, I suffered the bottles to remain forty-eight hours, undisturbed, in this cold situation, attached to the arms of the balance as before.

At the expiration of that time, I entered the room, — using the utmost caution not to disturb the balance, — when, to my great surprise, I found that the bottle A very sensibly preponderated.

The water which this bottle contained was completely frozen into one solid body of ice; but the spirit of wine, in the bottle B, showed no signs of freezing.

I now very cautiously restored the equilibrium by adding small pieces of the very fine wire of which gold lace is made, to the arm of the balance to which the bottle B was suspended, when I found that the bottle A had augmented its weight by $\frac{1}{35904}$ part of its whole weight at the beginning of the experiment; the weight of the bottle with its contents having been 4811.23 grains Troy (the bottle weighing 703.37 grains, and the water 4107.86 grains), and it requiring now $\frac{134}{1000}$ parts of a grain, added to the opposite arm of the balance, to counterbalance it.

Having had occasion, just at this time, to write to my friend, Sir Charles Blagden, upon another subject, I added a postscript to my letter, giving him a short account of this experiment, and telling him how "*very contrary to my expectation*" the result of it had turned out; but I soon after found that I had been too hasty in my communication. Sir Charles, in his answer to my letter, expressed doubts respecting the fact; but, before his letter had reached me, I had learned from my own experience how very dangerous it is in philosophical investigations to draw conclusions from single experiments.

Having removed the balance, with the two bottles attached to it, from the cold into the warm room (which still remained at the temperature of 61°), the ice in the bottle A gradually thawed; and, being at length totally reduced to water, and this water having acquired the temperature of the surrounding air, the two bottles, after being wiped perfectly clean and dry, were found to weigh as at the beginning of the experiment, before the water was frozen.

This experiment, being repeated, gave nearly the same result, — the water appearing when frozen to be heavier than in its fluid state; but some irregularity in the manner in which the water lost the additional weight which it had appeared to acquire upon being frozen when it was afterwards thawed, as also a sensible difference in the quantities of weight apparently acquired in the different experiments, led me to suspect that the experiment could not be depended on for deciding the fact in question. I therefore set about to repeat it, with some variations and improvements; but before I give an account of my further investigations relative to this subject, it may not be amiss to mention the method I pursued for discovering whether the appearances mentioned in the foregoing experiments might not arise from the imperfections of my balance; and it may likewise be proper to give an account, in this place, of an intermediate experiment which I made, with a view to discover, by a shorter route, and in a manner less exceptionable than that above mentioned, whether bodies actually lose or acquire any weight upon acquiring an additional quantity of latent heat.

My suspicions respecting the accuracy of the balance arose from a knowledge — which I acquired from the maker of it — of the manner in which it was constructed.

The three principal points of the balance having been determined, as nearly as possible, by measurement, the axes of motion were firmly fixed in their places, in a right line, and, the beam being afterwards finished, and its two arms brought to be in equilibrio, the balance was proved, by suspending weights, which before were known to be exactly equal, to the ends of its arms.

If with these weights the balance remained in equilibrio, it was considered as a proof that the beam was just; but if one arm was found to preponderate, the other was gradually lengthened, by beating it upon an anvil, until the difference of the lengths of the arms was reduced to nothing, or until equal weights, suspended to the two arms, remained in equilibrio; care being taken before each trial to bring the two ends of the beam to be in equilibrio, by reducing with a file the thickness of the arm which had been lengthened.

Though in this method of constructing balances the most perfect equality in the lengths of the arms may be obtained, and consequently the greatest possible accuracy, when used at a time when the temperature of the air is the same as when the balance was made, yet, as it may happen that, in order to bring the arms of the balance to be of the same length, one of them may be much more hammered than the other, I suspected it might be possible that the texture of the metal forming the two arms might be rendered so far different by this operation as to occasion a difference in their expansions with heat; and that this difference might occasion a sensible error in the balance, when, being charged with a great weight, it should be exposed to a considerable change of temperature.

To determine whether the apparent augmentation of weight, in the experiments above related, arose in any degree from this cause, I had only to repeat the experiment, causing the two bottles A and B to change places upon the arms of the balance; but, as I had already found a sensible difference in the results of different repetitions of the same experiment, made as nearly as possible under the same circumstances, and as it was above all things of importance to ascertain the accuracy of my balance, I preferred making a particular experiment for that purpose.

My first idea was, to suspend to the arms of the balance, by very fine wires, two equal globes of glass, filled with mercury, and, suffering them to remain in my room till they should have acquired the known temperature of the air in it, to have removed them afterward into the cold, and to have seen if they still remained in equilibrio under such difference of temperature; but, considering the obstinacy with which moisture adheres to the surface of glass, and being afraid that somehow or other, notwithstanding all my precautions, one of the globes might acquire or retain more of it than the other, and that by that means its apparent weight might be increased; and having found by a former experiment, of which an account is given in one of the preceding papers (that on the Moisture absorbed from the Atmosphere by various Substances[1]), that the gilt surfaces of metals do not attract moisture, instead of the glass globes filled with mercury, I made use of two equal solid globes of brass, well gilt and burnished, which I suspended to the arms of the balance by fine gold wires.

These globes, which weighed 4975 grains each, being wiped perfectly clean, and having acquired the temperature (61°) of my room, in which they were exposed more than twenty-four hours, were brought into the most scrupulous equilibrium, and were then removed, attached to the arms of the balance, into a room in which the air was at the temperature of 26°, where they were left all night.

The result of this trial furnished the most satisfactory proof of the accuracy of the balance; for, upon entering the room, I found the equilibrium as perfect as at the beginning of the experiment.

Having thus removed my doubts respecting the accuracy of my balance, I now resumed my investigations relative to the augmentation of weight which fluids have been said to acquire upon being congealed.

In the experiments which I had made, I had, as I then imagined, guarded as much as possible against every source of error and deception. The bottles being of the same size, neither any occasional alteration in the pressure of the atmosphere during the experiment, nor the necessary and unavoidable difference in the densities of the air in the hot and in the cold rooms in which they were weighed, could affect their apparent weights; and their shapes and their quantities of surface being the same, and as they remained for such a considerable length of time in the heat and cold to which they were exposed, I flattered myself that the quantities of moisture remaining attached to their surfaces could not be so different as sensibly to affect the results of the experiments. But, in regard to this last circumstance, I afterwards found reason to conclude that my opinion was erroneous.

Admitting the fact stated by Dr. Fordyce, — and which my experiments had hitherto rather tended to corroborate than to contradict, — I could not conceive any other cause for the augmentation of the apparent weight of water upon its being frozen than the loss of so great a proportion of its latent heat as that fluid is known to evolve when it congeals; and I concluded that, if the loss of latent heat added to the weight of one body, it must of necessity produce the same effect on another, and consequently, that the augmentation of the quantity of latent heat must in all bodies and in all cases diminish their apparent weights.

To determine whether this is actually the case or not, I made the following experiment.

Having provided two bottles, as nearly alike as possible, and in all respects similar to those made use of in the experiments above mentioned, into one of them I put 4012.46 grains of water, and into the other an equal weight of mercury; and, sealing them hermetically, and suspending them to the arms of the balance, I suffered them to acquire the temperature of my room, 61°; then, bringing them into a perfect equilibrium with each other, I removed them into a room in which the air was at the temperature of 34°, where they remained twenty-four hours. But there was not the least appearance of either of them acquiring or losing any weight.

Here it is very certain that the quantity of heat lost by the water must have been very considerably greater than that lost by the mercury, the specific quantities of latent heat in water and in mercury having been determined to be to each other as 1000 to 33; but this difference in the quantities of heat lost produced no sensible difference on the weights of the fluids in question.

Had any difference of weight really existed, had it been no more than *one millionth* part of the weight of either of the fluids, I should certainly have discovered it; and had it amounted to so much as $\frac{1}{700000}$ part of that weight, I should have been able to have measured it, so sensible and so very accurate is the balance which I used in these experiments.

I was now much confirmed in my suspicions that the apparent augmentation of the weight of the water upon its being frozen, in the experiments before related, arose from some accidental cause: but I was not able to conceive what that cause could possibly be, unless it were either a greater quantity of moisture attached to the external surface of the bottle which contained the water than to the surface of that containing the spirits of wine, or some vertical current or currents of air caused by the bottles, or one of them not being exactly of the temperature of the surrounding atmosphere.

Though I had foreseen, and, as I thought, guarded sufficiently against, these accidents, by making use of bottles of the same size and form, and which were blown of the same kind of glass and at the same time, and by suffering the bottles in the experiments to remain for so considerable a length of time exposed to the different degrees of heat and of cold which alternately they were made to acquire; yet, as I did not know the relative conducting powers of ice and of spirit of wine with respect to heat, or, in other words, the degrees of facility or difficulty with which they acquire the temperature of the medium in which they are exposed, or the time taken up in that operation, and, consequently, was not *absolutely certain* as to the equality of the temperatures of the contents of the bottles at the time when their weights were compared, I determined now to repeat the experiments, with such variations as should put the matter in question out of all doubt.

I was the more anxious to assure myself of the real temperatures of the bottles and their contents, as any difference in their temperatures might vitiate the experiment, not only by causing unequal currents in the air, but also by causing, at the same time, a greater or less quantity of moisture to remain attached to the glass.

To remedy these evils, and also to render the experiment more striking and satisfactory in other respects, I proceeded in the following manner: —

Having provided three bottles, A, B, and C, as nearly alike as possible, and resembling in all respects those already described, into the first, A, I put 4214.28 grains of water, and a small thermometer, made on purpose for the experiment, and suspended in the bottle in such a manner that its bulb remained in the middle of the mass of water; into the second bottle, B, I put a like weight of spirit of wine, with a like thermometer; and, into the bottle C, I put an equal weight of mercury.

These bottles, being all hermetically sealed, were placed in a large room, in a corner far removed from the doors and windows, and where the air appeared to be perfectly quiet; and, being suffered to remain in this situation more than twenty-four hours, the heat of the room (61°) being kept up all that time with as little variation as possible, and the contents of the bottles A and B appearing, by their inclosed thermometers, to be exactly at the same temperature, the bottles were all wiped with a very clean, dry, cambric handkerchief; and, being afterwards suffered to remain exposed to the free air of the room a couple of hours longer, in order that any inequalities in the quantities of heat, or of the moisture attached to their surfaces, which might have been occasioned by the wiping, might be corrected by the operation of the atmosphere by which they were surrounded, they were all weighed, and were brought into the most exact equilibrium with each other, by means of small pieces of very fine silver wire, attached to the necks of those of the bottles which were the lightest.

This being done, the bottles were all removed into a room in which the air was at 30°, where they were suffered to remain, perfectly at rest and undisturbed, forty-eight hours; the bottles A and B being suspended to the arms of the balance, and the bottle C suspended, at an equal height, to the arm of a stand constructed for that purpose, and placed as near the balance as possible, and a very sensible thermometer suspended by the side of it.

At the end of forty-eight hours, during which time the apparatus was left in this situation, I entered the room, opening the door very gently for fear of disturbing the balance; when I had the pleasure to find the three thermometers, *viz.* that in the bottle A, — which was now inclosed in a solid cake of ice, — that in the bottle B, and that suspended in the open air of the room, all standing at the same point, 29° F., and the bottles A and B *remaining in the most perfect equilibrium.*

To assure myself that the play of the balance was free, I now approached it very gently, and caused it to vibrate; and I had the satisfaction to find, not only that it moved with the utmost freedom, but also, when its vibration ceased, that it rested precisely at the point from which it had set out.

I now removed the bottle B from the balance, and put the bottle C in ts place; and I found that *that* likewise remained of the same apparent weight as at the beginning of the experiment, being in the same perfect equilibrium with the bottle A as at first.

I afterwards removed the whole apparatus into a warm room, and causing the ice in the bottle A to thaw, and suffering the three bottles to remain till they and their contents had acquired the exact temperature of the surrounding air, I wiped them very clean, and, comparing them together, I found their weights remained unaltered.

This experiment I afterwards repeated several times, and always with precisely the same result, — the water *in no instance* appearing to gain, or to lose, the least weight upon being frozen or upon being thawed; neither were the relative weights of the fluids in either of the other bottles in the least changed by the various degrees of heat and of cold to which they were exposed.

If the bottles were weighed at a time when their contents were not *precisely of the same temperature*, they would frequently appear to have gained, or to have lost, something of their weights; but this doubtless arose from the vertical currents which they caused in the atmosphere, upon being heated or cooled in it, or to unequal quantities of moisture attached to the surfaces of the bottles, or to both these causes operating together.

As I knew that the conducting power of mercury, with respect to heat, was considerably greater than either that of water or that of spirit of wine, while its capacity for receiving heat is much less than that of either of them, I did not think it necessary to inclose a thermometer in the bottle C, which contained the mercury; for it was evident that, when the contents of the other two bottles should appear, by their thermometers, to have arrived at the temperature of the medium in which they were exposed, the contents of the bottle C could not fail to have acquired it also, and even to have arrived at it before them; for the time taken up in the heating or in the cooling of any body, is, *cæteris paribus*, as the capacity of the body to receive and retain heat, *directly*, and as its conducting power, *inversely*.

The bottles were suspended to the balance by silver wires about two inches long, with hooks at the ends of them; and, in removing and changing the bottles, I took care not to touch the glass. I likewise avoided upon all occasions, and particularly in the cold room, coming near the balance with my breath, or touching it, or any part of the apparatus, with my naked hands.

Having determined that water does not acquire or lose any weight upon being changed from a state of *fluidity* to that of *ice*, and *vice versâ*, I shall now take my final leave of a subject which has long occupied me, and which has cost me much pains and trouble; being fully convinced, from the results of the above-mentioned experiments, that if heat be in fact a *substance*, or matter, — a fluid *sui generis*, as has been supposed, — which, passing from one body to another, and being accumulated, is the immediate cause of the phenomena we observe in heated bodies, — of which, however, I cannot help entertaining doubts, — it must be something so infinitely rare, even in its most condensed state, as to baffle all our attempts to discover its gravity. And if the opinion which has been adopted by many of our ablest philosophers, that heat is nothing more than an intestine vibratory motion of the constituent parts of heated bodies, should be well founded, it is clear that the weights of bodies can in no wise be affected by such motion.

It is, no doubt, upon the supposition that heat is a substance distinct from the heated body, and which is accumulated in it, that all the experiments which have been undertaken with a view to determine the weight which bodies have been supposed to gain or to lose upon being heated or cooled, have been made; and upon this supposition, — but without, however, adopting it entirely, as I do not conceive it to be sufficiently proved, — all my researches have been directed.

The experiments with *water* and with *ice* were made in a manner which I take to be perfectly unexceptionable, in which no foreign cause whatever could affect the results of them; and the quantity of heat which water is known to part with, upon being frozen, is so considerable, that if this loss has no effect upon its apparent weight, it may be presumed that we shall never be able to contrive an experiment by which we can render the weight of heat sensible.

Water, upon being frozen, has been found to lose a quantity of heat amounting to 140 degrees of Fahrenheit's thermometer; or — which is the same thing — the heat which a given quantity of water, previously cooled to the temperature of freezing, actually loses upon being changed to ice, if it were to be imbibed and retained by an equal quantity of water, at the given temperature (that of freezing), would heat it 140 degrees, or would raise it to the temperature of $(32^\circ + 140)$ $172^\circ$ of Fahrenheit's thermometer, which is only $40^\circ$ short of that of boiling water; consequently, any given quantity of water, at the temperature of freezing, upon being actually frozen, loses almost as much heat as, added to it, would be sufficient to make it boil.

It is clear, therefore, that the difference in the quantities of heat contained by the water in its fluid state and heated to the temperature of $61^\circ$ F., and by the ice, in the experiments before mentioned, was very nearly equal to that between water in a state of boiling, and the same at the temperature of freezing.

But this quantity of heat will appear much more considerable when we consider the great capacity of water to contain heat, and the great apparent effect which the heat that water loses upon being frozen would produce were it to be imbibed by, or communicated to, any body whose power of receiving and retaining heat is much less.

The capacity of water to receive and retain heat — or what has been called its specific quantity of latent heat — has been found to be to that of gold as 1000 to 50, or as 20 to 1; consequently, the heat which any given quantity of water loses upon being frozen, were it to be communicated to an equal weight of gold at the temperature of freezing, the gold, instead of being heated 162 degrees, would be heated $140 \times 20 = 2800$ degrees, or, would be raised to a *bright red heat.*

It appears, therefore, to be clearly proved by my experiments, that a quantity of heat equal to that which 4214 grains (or about $9\frac{3}{4}$ oz.) of gold would require to heat it from the temperature of freezing water to be *red hot*, has no sensible effect upon a balance capable of indicating so small a variation of weight as that of $\frac{1}{1000000}$ part of the body in question; and, if the weight of gold is neither augmented nor lessened by *one millionth part*, upon being heated from the point of *freezing water* to that of a *bright red heat*, I think we may very safely conclude, that ALL ATTEMPTS TO DISCOVER ANY EFFECT OF HEAT UPON THE APPARENT WEIGHTS OF BODIES WILL BE FRUITLESS.

## SUPPLEMENT.

THE foregoing paper having been originally drawn up for the purpose of being laid before the Royal Society, my respect for that learned body induced me to confine my observations to such points as I conceived to be new; and I took no notice whatever of a considerable number of experiments which I had made in the course of my investigations, because they were very similar to experiments that had before been made by other persons; and because their results did not appear to me to afford sufficient grounds to form any decisive opinion respecting the matter in question. There were, however, among my experiments, two or three of which I shall now give an account, which will probably be thought sufficiently interesting to deserve being mentioned.

Most of the experiments, from the results of which philosophers had been induced to form their opinions respecting the *ponderability of heat*, had been made by weighing the same given body at different temperatures. Thus, solid globes of metal — cannon-balls, for instance — had frequently been weighed when cold, and then, being heated red-hot, had been again weighed at that high temperature, and, from the apparent difference of the weight of the ball when cold and when red-hot, conclusions had been formed respecting the weight or levity of heat. But had the numerous causes of error in these most difficult experiments been less evident than they are, yet the results of the experiments of this kind which have hitherto been made by different persons have been so various. and contradictory that no reliance whatever can be placed on them.

When a hot body is suspended in the air to the arm of a balance in order to its being weighed, as it continually gives off heat to the fluid in contact with it, this communication of heat occasions a strong ascending current of air to be formed over and by the sides of the hot body, which current cannot fail to affect the result of the experiment, and render the conclusions drawn from it fallacious. To prevent, if possible, these causes of error, the following experiments were contrived.

The hot body to be weighed, which was a small metallic ball, heated red-hot, was placed in the scale of the balance in a small hemispherical porcelain cup, which had a slender foot, or stand, about one inch high; and this cup, with the hot ball in it, was covered over by a porcelain coffee-cup, turned upside down, which, without touching the hot ball, confined the heated air which surrounded it. This coffee-cup and the porcelain stand were very exactly balanced, by weights in the opposite side, before the ball was introduced.

The following experiment was made at Munich on the 20th of April, 1785. The weather being cloudy, with intervals of sunshine, the thermometer in my room stood at 52° F., and the barometer 26 inches 4 lines, French measure.

*At 30 minutes after noon*, a small bullet, or grape-shot, of cast-iron, very well formed, and apparently solid, having been well washed and cleaned by scouring with sand, and thoroughly dried, was exposed in a clean vessel of porcelain in the midst of a mixture of pounded ice and sea salt, till it had acquired the temperature of 25° F. (7 degrees below the point of freezing water), when it was carefully weighed, and found to weigh very exactly $773\frac{15}{64}$ grains Troy.

*At* 1 *h.* 30 *m. P. M.*, the same bullet having been exposed 30 minutes in a clean, dry vessel of porcelain, placed in a sand heat, at the temperature of 212° F., or that of boiling water, was again weighed, while yet hot, and found to weigh no more than $773\frac{5}{64}$ grains.

*At* 2 *h.* 0 *m. P. M.*, the bullet having now been exposed 15 minutes, in a clean new Hessian crucible, well covered, to the heat of a strong charcoal fire, and being thoroughly red-hot, was found to weigh $773\frac{31}{64}$ grains.

The bullet, being yet red-hot, was put again into the crucible, and being once more exposed to the fire, which now burned very bright, *at* 2 *h.* 20 *m.* it had acquired a white heat, and began to show signs of melting, some small bubbles appearing upon its surface. In this state it was taken from the fire, and very carefully weighed sixteen times successively, at different intervals, when it was found to weigh as follows: —

| Time when weighed. | | Was found to weigh. |
|---|---|---|
| At 2 h. | 20 m. | $773\frac{35}{64}$ grains. |
| 2 | 21 | $773\frac{23}{64}$ " |
| 2 | 23 | $773\frac{13}{64}$ " |
| 2 | 26 | $772\frac{61}{64}$ " |
| 2 | 29 | $773\frac{3}{64}$ " |
| 2 | 32 | $773\frac{15}{64}$ " |
| 2 | 43 | $773\frac{63}{64}$ " |
| 2 | 46 | $774\frac{7}{64}$ " |
| 2 | 49 | $774\frac{11}{64}$ " |
| 2 | 52 | $774\frac{13}{64}$ " |
| 2 | 56 | $774\frac{19}{64}$ " |
| 2 | 58 | $774\frac{23}{64}$ " |

| Time when weighed. | | Was found to weigh. |
| --- | --- | --- |
| At 3 h. 1 m. | . . . . . . . | $774\frac{25}{64}$ grains. |
| 3 18 | . . . . . . | $774\frac{39}{64}$ " |
| * 3 25 | . . . . . . . | $774\frac{34}{64}$ " |
| 6 15 | . . . . . . | $774\frac{60}{64}$ " |

Immediately after this last-mentioned weighing of the bullet, the whole of the apparatus appearing to have acquired the temperature of the air in the room (60° F.), the bullet was taken away, and the porcelain stand and cup were again balanced in the scale, when it appeared that they had lost $\frac{1}{4}$ of a grain in weight during the preceding experiment. This apparent loss of weight I could ascribe to nothing but to the thorough drying of the cups in the experiment with the red-hot bullet, and to the drying of the silk cords by which the scale containing the cup and stand was suspended to the arm of the balance.

This weight $= \frac{1}{4}$ of a grain, which was required to balance the scales at the end of the experiment, being added to the apparent weight of the bullet at 6 h. 15 m. $= 774\frac{61}{64}$, its true weight at that time appears to have been $775\frac{13}{64}$.

The weight of the bullet at 1 h. 30 m. having been no more than $773\frac{5}{46}$ grains, and at 6 h. 15 m. it being found to weigh $775\frac{13}{64}$ grains, it appears that it had gained in weight by being heated red-hot $775\frac{13}{64} - 773\frac{5}{64} = 2\frac{1}{8}$ grains.

This augmentation of weight doubtless arose from the partial oxidation of the iron. It certainly did not arise *from the heat*, for it remained after the bullet had become cold.

* Just before this weighing, the coffee-cup which covered the bullet in the scale had been removed for a moment to look at the bullet, and then immediately replaced. What it was that escaped on this occasion I will not undertake to say, but certain it is that its weight amounted to $\frac{5}{64}$ of a grain, at the least.

*An Account of an Experiment made with a Bullet of Fine Gold.*

*Munich*, *23d April*, 1785. — Weather cloudy, with intervals of sunshine; thermometer in my room at 65° F; barometer at 26 inches 4 lines.

A small bullet of fine gold, equal in value to 10 German ducats, which I procured from the master of the mint, being weighed in the open air of my room, was found to weigh $477\frac{157}{256}$ grains.

The small open china cup, in which the bullet was weighed, was exactly counterbalanced by a weight = $240\frac{248}{256}$ grains.

At 10 h. 5 m. A. M. the bullet, heated to a clear red heat approaching to whiteness, and weighed in the cup, open to the air, the bullet and the cup together were found to have lost of their weight $\frac{123}{256}$ of a grain.

Removing the bullet immediately, I found that the cup, or rather the cup and the scale in which it was placed, together had lost in weight $\frac{120}{256}$ parts of a grain.

Consequently, the bullet must have lost of its weight by being heated red-hot; or it appeared to be lighter when red-hot than when cold by $\frac{3}{256}$ parts of a grain, or $\frac{1}{40749}$ part of its whole weight.

Upon repeating the experiment I had nearly the same result; but upon varying it, by covering the heated bullet in the scale, in different ways, I found such variations in the results as convinced me that the apparent diminution of weight above mentioned might easily have arisen from currents in the atmosphere, and consequently that no dependence can be placed in experiments of that kind for deciding the fact relative to the weights of heated bodies, or the ponderability of heat.

I afterwards contrived an apparatus for making the experiment in a different and more unexceptionable manner. I provided three hollow globes of brass, very thin, and one larger than the other, and which being made to open in the middle, like a tobacco-box, could be placed one within another. In the centre globe I intended to place a solid bullet of pure gold, red-hot. Between the centre globe and that next it, I proposed to leave a space equal to the diameter of the heated bullet, filled with air; and the space between the second globe and the third I meant to have filled with pounded ice; and I proposed to have made the experiment at a time when the heat of the atmosphere should be just equal to that of freezing water; and in this manner I conceived that I should be able to avoid the currents in the air, whose effects I had found so distressing in my former experiments. But when I considered that the whole of the heat contained by the red-hot bullet would not be sufficient to thaw one half of the ice which surrounded it, and that, when the bullet should be cooled to the temperature of the ice, the whole mass of metal, of ice, and of water, would still remain *at the point of freezing;* and, moreover, that weighing the water produced by the ice would, in fact, be weighing the heat which before existed in the red-hot bullet, it first occurred to me that the point in question might much more readily be determined by simply weighing a quantity of ice at the temperature of freezing, and weighing the same again when changed into water.

I therefore left my apparatus unfinished, and turned my whole attention to the experiments of which an account has been given in the former part of this paper.

OF THE

# PROPAGATION OF HEAT

IN

# VARIOUS SUBSTANCES:

BEING

*An Account of a Number of* NEW EXPERIMENTS *made with a View to the Investigation of the* CAUSES *of the* WARMTH *of* NATURAL *and* ARTIFICIAL CLOTHING.

## INTRODUCTION.

THIS Essay contains nothing that will be new to philosophical readers; for it is little more than the substance of two Papers which have already appeared in the Philosophical Transactions of the Royal Society of London; one in the year 1786, and the other (for which the Author had the honour to receive from the Society the Copleian Annual Medal) in the year 1792.[2]

As reference has frequently been made to these Papers in several of the preceding Essays; and as many of the experiments of which an account is given in them are not only interesting in themselves, but are necessary to be known in all their details in order to judge of several important conclusions that have been founded on their results, the Author has thought that it would not be improper to republish them under the present form. He was also desirous of adding the substance of those Papers to his Sixth and Seventh Essays, in order that all that he has written on the *Science* of *Heat* might be brought together in one volume.

The Essays which are destined to compose the next volume (many of which are already in great forwardness) are all on practical subjects of a popular nature, and of general utility; and on that account it was judged best to keep them separate from those contained in this volume, which partake more of the nature of abstruse philosophical investigations.

Various unforeseen events have contributed to retard the publication of the promised Essays on Kitchen Fire-places — on Cottage Fire-places — and on Clothing; but the Author has well-founded hopes of being able to bring them forward in the course of a few months.

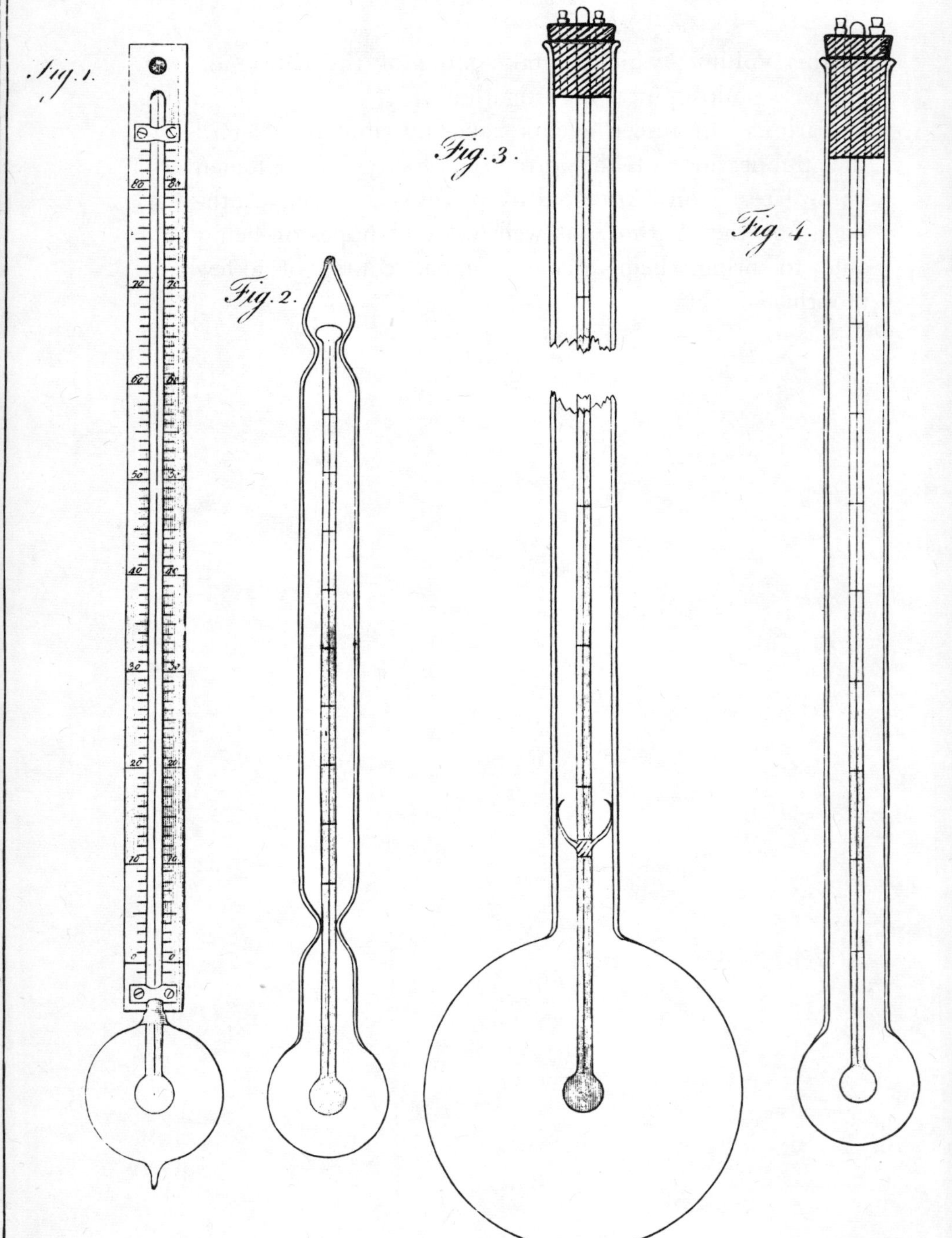


Scale ½ an Inch to the Inch

# OF THE PROPAGATION OF HEAT IN VARIOUS SUBSTANCES.

## CHAPTER I.

*An Account of the Instruments that were prepared for making the proposed Experiments. — A Thermometer is constructed whose Bulb is surrounded by a* TORRICELLIAN VACUUM. — *Heat is found to pass in a Torricellian Vacuum with greater Difficulty than in Air. — Relative conducting Powers of a Torricellian Vacuum and of Air with regard to Heat determined by Experiment. — Relative conducting Powers of dry Air and of moist Air. — Relative conducting Powers of Air of different Degrees of Density. — Relative conducting Powers of* MERCURY ; WATER ; AIR ; *and a* TORRICELLIAN VACUUM.

March 9, 1786

*Dear Sir,

I have at length begun the course of experiments upon heat which I have so long had in contemplation, and I have already made a discovery, which, if not new to you, is perfectly so to me, and which I think may. lead to a further knowledge respecting the nature of heat.

EXAMINING the conducting power of air, and of various other fluid and solid bodies, with regard to Heat, I was led to examine the conducting power of the *Torricellian vacuum.* From the striking analogy between the electric fluid and Heat respecting their conductors and non-conductors (having found that bodies, in general, which are conductors of the electric fluid, are likewise good conductors of Heat, and, on the contrary, that electric bodies, or such as are bad conductors of the electric fluid, are likewise bad conductors of Heat), I

*A letter from Count Rumford to Sir Joseph Banks, president of the Royal Society.

was led to imagine that the Torricellian vacuum, which is known to afford so ready a passage to the electric fluid, would also have afforded a ready passage to Heat.

The common experiments of heating and cooling bodies under the receiver of an air-pump I conceive to be inadequate to determining this question; not only on account of the impossibility of making a perfect void of air by means of the pump, but also on account of the moist vapour, which, exhaling from the wet leather and the oil used in the machine, expands under the receiver, and fills it with a watery fluid, which, though extremely rare, is yet capable of conducting a great deal of Heat: I had recourse, therefore, to other contrivances.

I took a thermometer, unfilled, the diameter of whose bulb (which was globular) was just half an inch, Paris measure, and fixed it in the center of a hollow glass ball of the diameter of $1\frac{3}{4}$ Paris inch, in such a manner that, the short neck or opening of the ball being soldered fast to the tube of the thermometer $7\frac{1}{2}$ lines above its bulb, the bulb of the thermometer remained fixed in the center of the ball, and consequently was cut off from all communication with the external air. In the bottom of the glass ball was fixed a small hollow tube or point, which projecting outwards was soldered to the end of a common barometer tube about 32 inches in length, and by means of this opening the space between the internal surface of the glass ball and the bulb of the thermometer was filled with hot mercury, which had been previously freed of air and moisture by boiling. The ball, and also the barometrical tube attached to it, being filled with mercury, the tube was carefully inverted, and its open end placed in a bowl in which there was a quantity of mercury. The instrument now became a barometer, and the mercury descending from the ball (which was now uppermost) left the space surrounding the bulb of the thermometer free of air. The mercury having totally quitted the glass ball, and having sunk in the tube to the height of 28 inches (being the height of the mercury in the common barometer at that time), with a lamp and a blow-pipe I melted the tube together, or sealed it hermetically, about three quarters of an inch below the ball, and, cutting it at this place with a fine file, I separated the ball from the long barometrical tube. The thermometer being afterwards filled with mercury in the common way, I now possessed a thermometer whose bulb was confined in the center of a *Torricellian vacuum*, and which served at the same time as the body to be heated, and as the instrument for measuring the Heat communicated.

### *Experiment No.* 1.

With this instrument (see Fig. 1) I made the following experiment. Having plunged it into a vessel filled with water, warm to the 18th degree of Reaumur's scale, and suffered it to remain there till it had acquired the temperature of the water, that is to say, till the mercury in the inclosed thermometer stood at 18°, I took it out of this vessel and plunged it suddenly into a vessel of boiling water, and holding it in the water (which was kept constantly boiling) by the end of the tube, in such a manner that the glass ball, in the center of which was the bulb of the thermometer, was just submerged, I observed the number of degrees to which the mercury in the thermometer had arisen at different periods of time, counted from the moment of its immersion. Thus, after it had remained in the boiling water 1 min. 30 sec. I found the mercury had risen from

18° to 27°. After 4 minutes had elapsed, it had risen to $44^\circ\frac{9}{10}$; and at the end of 5 minutes it had risen to $48^\circ\frac{2}{10}$.

## *Experiment No.* 2.

Taking it now out of the boiling water I suffered it to cool gradually in the air, and after it had acquired the temperature of the atmosphere, which was that of 15° R. (the weather being perfectly fine), I broke off a little piece from the point of the small tube which remained at the bottom of the glass ball, where it had been hermetically sealed, and of course the atmospheric air rushed immediately into the ball. The ball surrounding the bulb of the thermometer being now filled with air (instead of being emptied of air, as it was in the before-mentioned experiment), I resealed the end of the small tube at the bottom of the glass ball hermetically, and by that means cut off all communication between the air confined in the ball and the external air; and with the instrument so prepared I repeated the experiment before mentioned, that is to say, I put it into water warmed to 18°, and when it had acquired the temperature of the water, I plunged it into boiling water, and observed the times of the ascent of the mercury in the thermometer. They were as follows: —

| | Time elapsed. | | Heat acquired. |
|---|---|---|---|
| Heat at the moment of being plunged into the boiling water | | | 18° R. |
| | M | S. | ° |
| After having remained in the boiling water | 0 | 45 | 27 |
| | 1 | 0 | $34\frac{4}{10}$ |
| | 2 | 10 | $44\frac{9}{10}$ |
| | 2 | 40 | $48\frac{2}{10}$ |
| | 4 | 0 | $56\frac{2}{10}$ |
| | 5 | 0 | $60\frac{9}{10}$ |

From the result of these experiments it appears, evidently, that the Torricellian vacuum, which affords so ready a passage to the electric fluid, so far from being a good conductor of Heat, is a much worse conductor of it than common air, which of itself is reckoned among the worst; for in the last experiment, when the bulb of the thermometer was surrounded with air, and the instrument was plunged into boiling water, the mercury rose from 18° to 27° in 45 seconds; but in the former experiment, when it was surrounded by a Torricellian vacuum, it required to remain in the boiling water 1 minute 30 seconds = 90 seconds, to acquire that degree of heat. In the vacuum it required 5 minutes to rise to $48°\frac{2}{10}$; but in air it rose to that height in 2 minutes 40 seconds; and the proportion of the times in the other observations is nearly the same, as will appear by the following table.

| | The bulb of the thermometer placed in the center of the glass ball, and | | | |
|---|---|---|---|---|
| | surrounded by a Torricellian vacuum. (Exp. No. 1.) | | surrounded by air. (Exp. No. 2.) | |
| | Time elapsed. | Heat acquired. | Time elapsed. | Heat acquired. |
| Upon being plunged into boiling water | | 18° | | 18° |
| | M. S. | ° | M. S. | ° |
| After remaining in it | 1 30 | 27 | 0 45 | 27 |
| | —— | —— | 1 0 | $30\frac{4}{10}$ |
| | 4 0 | $44\frac{9}{10}$ | 2 10 | $44\frac{9}{10}$ |
| | 5 0 | $48\frac{2}{10}$ | 2 40 | $48\frac{2}{10}$ |
| | —— | —— | 4 0 | $56\frac{2}{10}$ |
| | —— | —— | 5 0 | $60\frac{9}{10}$ |

These experiments were made at Manheim, upon the first day of July, 1785, in the presence of Professor Hemmer, of the Electoral Academy of Sciences of Manheim, and Charles Artaria, meteorological instrument maker to the Academy, by whom I was assisted in making them.

Finding the construction of the instrument made use of in these experiments attended with much trouble and risk, on account of the difficulty of soldering the glass ball to the tube of the thermometer without at the same time either closing up, or otherwise injuring, the bore of the tube, I had recourse to another contrivance much more commodious, and much easier in the execution.

At the end of a glass tube or cylinder about eleven inches in length, and near three quarters of an inch in diameter internally, I caused a hollow globe to be blown $1\frac{1}{2}$ inch in diameter, with an opening in the bottom of it corresponding with the bore of the tube, and equal to it in diameter, leaving to the opening a neck or short tube, about an inch in length. Having a thermometer prepared, whose bulb was just half an inch in diameter, and whose freezing point fell at about $2\frac{3}{4}$ inches above its bulb, I graduated its tube according to Reaumur's scale, beginning at 0°, and marking that point, and also every tenth degree above it to 80°, with threads of fine silk bound round it, which being moistened with lac varnish adhered firmly to the tube. This thermometer I introduced into the glass cylinder and globe just described, by the opening in the bottom of the globe, having first choaked the cylinder at about 2 inches from its junction with the globe by heating it, and crowding its sides inwards towards its axis, leaving only an opening sufficient to admit the tube of the thermometer. The thermometer being introduced into the cylinder in such a manner that the center of its bulb coincided with the center of the globe, I marked a place in the cylinder, about three quarters of an inch above the 80th degree or boiling point upon the tube of the inclosed thermometer, and taking out the thermometer, I choaked the cylinder again in this place. Introducing now the thermometer for the last time, I closed the opening at the bottom of the globe at the lamp, taking care before I brought it to the fire, to turn the cylinder upside down, and to let the bulb of the thermometer fall into the cylinder till it rested upon the lower choak in the cylinder. By this means the bulb of the thermometer was removed more than 3 inches from the flame of the lamp. The opening at the bottom of the globe being now closed, and the bulb of the thermometer being suffered to return into the globe, the end of the cylinder was cut off to within about half an inch of the upper choak. This being done, it is plain that the tube of the thermometer projected beyond the end of the cylinder. Taking hold of the end of the tube, I placed the bulb of the thermometer as nearly as possible in the center of the globe, and observing and marking a point in the tube immediately above the upper choak of the cylinder, I turned the cylinder upside down, and suffering the bulb of the thermometer to enter the cylinder, and rest upon the first or lower choak, (by which means the end of the tube of the thermometer came further out of the cylinder,) the end of the tube was cut off at the mark just mentioned, (care having first been taken to melt the internal cavity or bore of the tube together at that place,) and a small solid ball of glass, a little larger than the internal diameter or opening of the choak, was soldered to the end of the tube, forming a little button or knob, which resting upon the upper choak of the cylinder served to suspend the thermometer in such a manner that the center of its bulb coincided with the center of the globe in which it was shut up. The end of the cylinder above the upper choak being now heated and drawn out to a point, or rather being formed into the figure of the frustum of a hollow cone, the end of it was soldered to the end of a barometrical tube, by the help of which the cavity of the cylinder and globe containing the thermometer was completely voided of air with mercury; when, the end of the cylinder being hermetically sealed, the barometrical tube was detached from it with a file, and the thermometer was left completely shut up in a Torricellian vacuum, the centre of the bulb of the thermometer being confined in the centre of the glass globe, without touching it in any part, by means of the two choaks in the cylinder, and the button upon the end of the tube. (See Fig. 2.)

Of these instruments I provided myself with two, as nearly as possible of the same dimensions; the one, which I shall call No. 1, being voided of air, in the manner above described; the other, No. 2, being filled with air, and hermetically sealed.

With these two instruments (see Fig. 2) I made the following experiments upon the 11th of July last at Manheim, between the hours of ten and twelve, the weather being very fine and clear, the mercury in the barometer standing at 27 inches 11 lines, Reaumur's thermometer at 15°, and the quill hygrometer of the Academy of Manheim at 47°.

*Experiments No. 3, 4, 5, and 6.*

Putting both the instruments into a mixture of pounded ice and water, I let them remain there till the mercury in the inclosed thermometers rested at the point 0°, that is to say, till they had acquired exactly the temperature of the cold mixture; and then taking them out of it I plunged them suddenly into a large vessel of boiling water, and observed the time required for the mercury to rise in the thermometers from ten degrees to ten degrees, from 0° to 80°, taking care to keep the water constantly boiling during the whole of this time, and taking care also to keep the instruments immersed to the same depth, that is to say, just so deep that the point 0° of the inclosed thermometer was even with the surface of the water.

These experiments I repeated twice with the utmost care; and the following table gives the result of them.

*Thermometer No. 1.*

Its bulb half an inch in diameter, shut up in the center of a hollow glass globe, 1½ inch in diameter, *void of air*, and hermetically sealed.

*Taken out of freezing water, and plunged into boiling water.*

| Time elapsed. | | Heat acquired. |
|---|---|---|
| Exp. No. 3. | Exp. No 4. | |
| | | 0° |
| M. S. | M. S. | 0 |
| 0 51 | 0 51 | 10 |
| 0 59 | 0 59 | 20 |
| 1 1 | 1 2 | 30 |
| 1 18 | 1 22 | 40 |
| 1 24 | 1 23 | 50 |
| 2 0 | 1 51 | 60 |
| 3 30 | 3 6 | 70 |
| 11 41 | 10 27 | 80 |
| 22 44 | 21 1 | = total time of heating from 0° to 80°. |

Total time from 0° to 70°:

| | M. S. |
|---|---|
| In Exp. No. 3 = | 11 3 |
| In Exp. No. 4 = | 10 34 |
| Medium = | 10 48½ |

*Thermometer No. 2.*

Its bulb half an inch in diameter, shut up in the center of a hollow glass globe, 1½ inch in diameter, *filled with air*, and hermetically sealed.

*Taken out of freezing water, and plunged into boiling water.*

| Time elapsed. | | Heat acquired. |
|---|---|---|
| Exp. No. 5. | Exp. No. 6. | |
| | | 0° |
| M. S. | M. S. | 0 |
| 0 30 | 0 30 | 10 |
| 0 35 | 0 37 | 20 |
| 0 41 | 0 41 | 30 |
| 0 49 | 0 53 | 40 |
| 1 1 | 0 59 | 50 |
| 1 24 | 1 20 | 60 |
| 2 45 | 2 25 | 70 |
| 9 10 | 9 38 | 80 |
| 16 55 | 17 3 | = total time of heating from 0° to 80°. |

Total time from 0° to 70°:

| | M. S |
|---|---|
| In Exp. No. 5 = | 7 45 |
| In Exp. No. 6 = | 7 25 |
| Medium = | 7 35 |

It appears from these experiments that the conducting power of air to that of the Torricellian vacuum, under the circumstances described, is as $7\frac{35}{60}$ to $10\frac{48\frac{1}{2}}{60}$ inversely, or as 1000 to 702 nearly; for, the quantities of Heat communicated being equal, the intensity of the communication is as the times inversely.

In these experiments the Heat passed through the surrounding medium *into* the bulb of the thermometer: in order to reverse the experiment, and make the Heat pass *out of* the thermometer, I put the instruments into boiling water, and let them remain therein till they had acquired the temperature of the water, that is to say, till the mercury in the inclosed thermometers stood at 80°; and then, taking them out of the boiling water, I plunged them suddenly into a mixture of water and pounded ice, and moving them about continually in this mixture, I observed the times employed in cooling as follows: —

*Thermometer No. 1.*
Surrounded by a Torricellian vacuum.
*Taken out of boiling water, and plunged into freezing water.*

| Time elapsed. Exp. No. 7. | Time elapsed. Exp. No. 8. | Heat lost. |
|---|---|---|
| | | 80° |
| M. S. | M. S. | 0 |
| 1 2 | 0 54 | 70 |
| 0 58 | 1 2 | 60 |
| 1 17 | 1 18 | 50 |
| 1 46 | 1 37 | 40 |
| 2 5 | 2 16 | 30 |
| 3 14 | 3 10 | 20 |
| 5 42 | 5 59 | 10 |
| Not observed. | Not observed. | 0 |

Total time of cooling from 80° to 10°.

| | M. S. |
|---|---|
| In Exp. No. 7 = | 16 4 |
| In Exp. No. 8 = | 16 16 |
| Medium = | 16 10 |

*Thermometer No. 2.*
Surrounded by air.
*Taken out of boiling water, and plunged into freezing water.*

| Time elapsed. Exp. No. 9. | Time elapsed. Exp. No. 10. | Heat lost. |
|---|---|---|
| | | 80° |
| M. S. | M. S. | 0 |
| 0 33 | 0 33 | 70 |
| 0 39 | 0 34 | 60 |
| 0 44 | 0 44 | 50 |
| 0 55 | 0 55 | 40 |
| 1 17 | 1 18 | 30 |
| 1 57 | 1 57 | 20 |
| 3 44 | 3 40 | 10 |
| 40 10 | Not observed. | 0 |

Total time of cooling from 80° to 10°.

| | M. S. |
|---|---|
| In Exp. No. 9 = | 9 49 |
| In Exp. No. 10 = | 9 41 |
| Medium = | 9 45 |

By these experiments it appears that the conducting power of air is to that of the Torricellian vacuum as $9\frac{45}{60}$ to $16\frac{10}{60}$ inversely, or as 1000 to 603.

To determine whether the same law would hold good when the heated thermometers, instead of being plunged into freezing water, were suffered to cool in the open air, I made the following experiments. The thermometers No. 1 and No. 2 being again heated in boiling water, as in the last experiments, I took them out of the water, and suspended them in the middle of a large room, where the air (which appeared to be perfectly at rest, the windows and doors being all shut) was warm to the 16th degree of Reaumur's thermometer, and the times of cooling were observed as follows: —

| (Exp. No. 11.) *Thermometer No.* 1. Surrounded by a Torricellian vacuum. *Heated to* 80°, *and suspended in the open air warm to* 16°. | | (Exp. No. 12.) *Thermometer No.* 2. Surrounded by air. *Heated to* 80°, *and suspended in the open air warm to* 16°. | |
|---|---|---|---|
| Time elapsed. | Heat lost. | Time elapsed. | Heat lost. |
| M. S. | 80° | M. S. | 80° |
| | 0 | | 0 |
| Not observed. | 70 | Not observed. | 70 |
| 1 24 | 60 | 0 51 | 60 |
| 1 44 | 50 | 1 5 | 50 |
| 2 28 | 40 | 1 34 | 40 |
| 4 16 | 30 | 2 41 | 30 |
| 10 12 = total time employed in cooling from 70° to 30°. | | 6 11 = total time employed in cooling from 70° to 30°. | |

Here the difference in the conducting powers of air and of the Torricellian vacuum appears to be nearly the same as in the foregoing experiments, being as $6\frac{11}{60}$ to $10\frac{12}{60}$ inversely, or as 1000 to 605. I could not observe the time of cooling from 80° to 70°, being at that time busied in suspending the instruments.

As it might possibly be objected to the conclusions drawn from these experiments that, notwithstanding all the care that was taken in the construction of the two instruments made use of that they should be perfectly alike, yet they might in reality be so far different, either in shape or size, as to occasion a very sensible error in the result of the experiments; to remove these doubts I made the following experiments:—

In the morning towards eleven o'clock, the weather being remarkably fine, the mercury in the barometer standing at 27 inches 11 lines, Reaumur's thermometer at 15°, and the hygrometer at 47°, I repeated the experiment No. 3 (of heating the thermometer No. 1 in boiling water, &c.), and immediately afterwards opened the cylinder containing the thermometer at its upper end, where it had been sealed, and letting the air into it, I resealed it hermetically, and repeated the experiment again with the same instrument, the thermometer being now surrounded with air, like the thermometer No. 2.

The result of these experiments, which may be seen in the following table, shews evidently that the error arising from the difference of the shapes or dimensions of the two instruments in question was inconsiderable, if not totally imperceptible.

(Exp. No. 13.)

*Thermometer No.* 1.

Its bulb half an inch in diameter, shut up in the center of a glass globe, $1\frac{1}{2}$ inch in diameter, *void of air*, and hermetically sealed.

*Taken out of freezing water and plunged into boiling water.*

| Time elapsed. M. S. | Heat acquired. 0° |
|---|---|
| 0 55 | 10 |
| 0 55 | 20 |
| 1 7 | 30 |
| 1 15 | 40 |
| 1 29 | 50 |
| 2 2 | 60 |
| 3 21 | 70 |
| 13 44 | 80 |

24 48 = total time of heating from 0° to 80°.

Total time from 0° to 70° = 11′ 4″.

(Exp. No. 14.)

*The same Thermometer (No.* 1).

The glass globe, containing the bulb of the thermometer, being now *filled with air*, and hermetically sealed.

*Taken out of freezing water and plunged into boiling water.*

| Time elapsed. M. S. | Heat acquired. 0° |
|---|---|
| 0 32 | 10 |
| 0 32 | 20 |
| 0 43 | 30 |
| 0 50 | 40 |
| 1 1 | 50 |
| 1 24 | 60 |
| 2 38 | 70 |
| 10 25 | 80 |

18 5 = total time of heating from 0° to 80°.

Total time from 0° to 70° = 7′ 40″.

It appears, therefore, from these experiments, that the conducting power of common atmospheric air is to that of the Torricellian vacuum as $7\frac{40}{60}$ to $11\frac{4}{60}$ inversely, or as 1000 to 602; which differs but very little from the result of all the foregoing experiments.

Notwithstanding that it appeared, from the result of these last experiments, that any difference there might possibly have been in the forms or dimensions of the instruments No. 1 and No. 2 could hardly have produced any sensible error in the result of the experiments in question; I was willing, however, to see how far any considerable alterations of size in the instrument would affect the experiment: I therefore provided myself with another instrument, which I shall call *Thermometer No.* 3, different from those already described in size, and a little different in its construction.

The bulb of the thermometer was of the same form and size as in the instruments No. 1 and No. 2, that is to say, it was globular, and half an inch in diameter; but the glass globe, in the center of which it was confined, was much larger, being 3 inches 7½ lines in diameter; and the bore of the tube of the thermometer was much finer, and consequently its length and the divisions of its scale were greater. The divisions were marked upon the tube with threads of silk of different colours at every tenth degree, from 0° to 80°, as in the before-mentioned instruments. The tube or cylinder belonging to the glass globe was 8 lines in diameter, a little longer than the tube of the thermometer, and perfectly cylindrical from its upper end to its junction with the globe, being without any choak; the thermometer being confined in the center of the globe by a different contrivance, which was as follows. To the opening of the cylinder was fitted a stopple of dry wood, covered with a coating of hard varnish, through the centre or axis of which passed the end of the tube of the thermometer; this stopple confined the tube in the axis of the cylinder at its upper end. To confine it at its lower end, there was fitted to it a small steel spring, a little below the point 0°; which, being fastened to the tube of the thermometer, had three elastic points projecting outwards, which, pressing against the inside of the cylinder, confined the thermometer in its place. The total length of this instrument, from the bottom of the globe to the upper end of the cylinder, was 18 inches, and the freezing point upon the thermometer fell about 3 inches above the bulb; consequently this point lay about 1½ inch above the junction of the cylinder with the globe, when the thermometer was confined in its place, the center of its bulb coinciding with the center of the globe. Through the stopple which closed the end of the cylinder passed two small glass tubes, about a line in diameter, which being about a line longer than the stopple were closed occasionally with small stopples fitted to their bores. These tubes (which were fitted exactly in the holes bored in the great stopple of the cylinder to receive them, and fixed in their places with cement) served to convey air, or any other fluid, into the glass ball, without its being necessary to remove the stopple closing the end of the cylinder; which stopple, in order to prevent the position of the thermometer from being easily deranged, was cemented in its place.

I have been the more particular in the description of these instruments, as I conceive it to be absolutely necessary to have a perfect idea of them in order to judge of the experiments made with them, and of their results.

With the instrument last described (which I have called *Thermometer No.* 3) I made the following experiment. It was upon the 18th of July, 1785, in the afternoon, the weather variable, alternate clouds and sunshine; wind strong at S. E. with now and then a sprinkling of rain; barometer at 27 inches $10\frac{1}{2}$ lines, thermometer at $18^{\circ}\frac{1}{4}$, and hygrometer variable from $44^{\circ}$ to extreme moisture.

In order to compare the result of the experiment made with this instrument with those made with the thermometer No. 2, I have placed together in the same table the different experiments made with them.

(Exp. No. 15.)

*Thermometer No.* 3.

Its bulb half an inch in diameter, shut up in the center of a glass tube, 3 inches $7\frac{1}{2}$ lines in diameter, and surrounded by *air*.

*Taken out of freezing water and plunged into boiling water.*

| Time elapsed. | Heat acquired. |
|---|---|
| | 0° |
| M. S. | 0 |
| 0 33 | 10 |
| 0 38 | 20 |
| 0 54 | 30 |
| 0 51 | 40 |
| 1 7 | 50 |
| 1 28 | 60 |
| 2 28 | 70 |
| 9 0 | 80 |
| 16 59 | |

16 59 = total time of heating from 0° to 80°.
Time from 0° to 70° = 7′ 59″.

(Exp. No. 5 and No. 6.)

*Thermometer No.* 2.

Its bulb half an inch in diameter, shut up in the center of a glass globe, $1\frac{1}{2}$ inch in diameter, and surrounded by *air*.

*Taken out of freezing water and plunged into boiling water.*

| Time elapsed. | | | Heat acquired. |
|---|---|---|---|
| Exp. No. 5. | Exp. No. 6. | Medium. | |
| | | | 0° |
| M. S. | M. S. | M. S. | 0 |
| 0 30 | 0 30 | 0 30 | 10 |
| 0 35 | 0 37 | 0 36 | 20 |
| 0 41 | 0 41 | 0 41 | 30 |
| 0 49 | 0 53 | 0 51 | 40 |
| 1 1 | 0 59 | 1 0 | 50 |
| 1 24 | 1 20 | 1 22 | 60 |
| 2 45 | 2 25 | 2 35 | 70 |
| 9 10 | 9 38 | 9 24 | 80 |
| 16 55 | 17 3 | 16 59 | |

16 59 = total time of heating from 0° to 80°.
Time from 0° to 70° = 7′ 35″.

If the agreement of these experiments with the thermometers No. 2 and No. 3 surprised me, I was not less surprised with their disagreement in the experiment which follows: —

## *Experiment No.* 16.

Taking the thermometer No. 3 out of the boiling water, I immediately suspended it in the middle of a large room, where the air, which was quiet, was at the temperature of $18°\frac{1}{4}$ R. and observed the times of cooling as follows: —

| Time elapsed. | Heat lost. |
|---|---|
| | 80° |
| M. S. | 0 |
| 1 55 | 70 |
| 0 12 | 60 |
| 0 33 | 50 |
| 2 15 | 40 |
| 4 0 | 30 |
| 9 55 | |

9 55 = total time of cooling from 80° to 30°.

Time from 70° to 30° = 8′ 0″; but in the experiment No. 12, with the thermometer No. 2, the time employed in cooling from 70° to 30° was only 6′ 11″. In this experiment, with the thermometer No. 3, the time employed in cooling from 60° to 30° was 7′ 48″; but in the above-mentioned experiment, with the thermometer No. 2 it was only 5′ 20″. It is true, the air of the room was somewhat cooler when the former experiment was made, than when this latter was made, with the thermometer No. 3; but this difference of temperature, which was only $2^\circ\frac{1}{4}$ (in the former case the thermometer in the room standing at 16°, and in the latter at $18^\circ\frac{1}{4}$), certainly could not have occasioned the whole of the apparent difference in the results of the experiments.

Does air receive Heat more readily than it parts with it? This is a question highly deserving of further investigation, and I hope to be able to give it a full examination in the course of my projected inquiries; but leaving it for the present, I shall proceed to give an account of the experiments which I have already made. Conceiving it to be a step of considerable importance towards coming at a further knowledge of the nature of Heat, to ascertain, by indisputable evidence, its passage through the Torricellian vacuum, and to determine, with as much precision as possible, the law of its motions in that medium; and being apprehensive that doubts might arise with respect to the experiments before described, on account of the contact of the tubes of the inclosed thermometers in the instruments made use of with the containing glass globes, or rather with their cylinders: by means of which (it might be suspected) that a certain quantity, if not all the Heat acquired, might possibly be communicated; to put this matter beyond all doubt, I made the following experiment.

In the middle of a glass body, of a pear-like form, about 8 inches long, and $2\frac{1}{2}$ inches in its greatest diameter, I suspended a small mercurial thermometer, $5\frac{1}{2}$ inches long, by a fine thread of silk, in such a manner that neither the bulb of the thermometer, nor its tube, touched the containing glass body in any part. The tube of the thermometer was graduated, and marked with fine threads of silk of different colours, bound round it, as in the thermometers belonging to the other instruments already described, and the thermometer was suspended in its place by means of a small steel spring, to which the end of the thread of silk which held the thermometer being attached, it (the spring) was forced into a small globular protuberance or cavity, blown in the upper extremity of the glass body, about half an inch in diameter, where, the spring remaining, the thermometer necessarily remained suspended in the axis of the glass body. There was an opening at the bottom of the glass body, through which the thermometer was introduced; and a barometrical tube being soldered to this opening, the inside of the glass body was voided of air by means of mercury; and this opening being afterwards sealed hermetically, and the barometrical tube being taken away, the thermometer was left suspended in a Torricellian vacuum.

In this instrument, as the inclosed thermometer did not touch the containing glass body in any part, on the contrary, being distant from its internal surface an inch or more in every part, it is clear that whatever Heat passed *into* or *out of* the thermometer must have passed *through* the surrounding Torricellian vacuum; for it cannot be supposed that the fine thread of silk, by which the thermometer was suspended, was capable of conducting any Heat at all, or at least any sensible quantity. I therefore flattered myself with hopes of being able, with the assistance of this instrument, to determine positively with regard to the passage of Heat in the Torricellian vacuum: and this I think I have done, notwithstanding an unfortunate accident that put it out of my power to pursue the experiment so far as I intended.

This instrument being fitted to a small stand or foot of wood, in such a manner that the glass body remained in a perpendicular situation, I placed it in my room, by the side of another inclosed thermometer (No. 2) which was surrounded by air, and observed the effects produced on it by the variation of Heat in the atmosphere. I soon discovered, by the motion of the mercury in the inclosed thermometer, that the Heat passed through the Torricellian vacuum; but it appeared plainly, from the sluggishness or great insensibility of the thermometer, that the Heat passed with much greater difficulty in this medium than in common air. I now plunged both the thermometers into a bucket of cold water; and I observed that the mercury in the thermometer surrounded by air descended much faster than that in the thermometer surrounded by the Torricellian vacuum. I took them out of the cold water, and plunged them into a vessel of hot water (having no conveniences at hand to repeat the experiment in due form with the freezing and with the boiling water); and the thermometer surrounded by the Torricellian vacuum appeared still to be much more insensible or sluggish than that surrounded by air.

These trials were quite sufficient to convince me of the passage of Heat in the Torricellian vacuum, and also of the greater difficulty of its passage in that medium than in common air; but not satisfied to rest my inquiries here, I took the first opportunity that offered, and set myself to repeat the experiments which I had before made with the instruments No. 1 and No. 2. I plunged this instrument into a mixture of pounded ice and water, where I let it remain till the mercury in the inclosed thermometer had descended to 0°; when, taking it out of this cold mixture, I plunged it suddenly into a vessel of boiling water, and prepared myself to observe the ascent of the mercury in the inclosed thermometer, as in the foregoing experiments; but unfortunately the moment the end of the glass body touched the boiling water, it cracked with the Heat at the point where it had been hermetically sealed, and the water rushing into the body spoiled the experiment: and I have not since had an opportunity of providing myself with another instrument to repeat it.

It having been my intention from the beginning to examine the conducting powers of the artificial airs or gases, the thermometer No. 3 was constructed with a view to those experiments; and having now provided myself with a stock of those different kinds of airs, I began with *fixed air*, with which, by means of water, I filled the globe and cylinder containing the thermometer; and stopping up the two holes in the great stopple closing the end of the cylinder, I exposed the instrument in freezing water till the mercury in the inclosed thermometer had descended to 0°; when, taking it out of the freezing water, I plunged it into a large vessel of boiling water, and prepared myself to observe the times of heating, as in the former cases; but an accident happened, which suddenly put a stop to the experiment. Immediately upon plunging the instrument into the boiling water, the mercury began to rise in the thermometer with such uncommon celerity that it had passed the first division upon the tube (which marked the 10th degree, according to Reaumur's scale) before I was aware of its being yet in motion; and having thus missed the opportunity of observing the time elapsed when the mercury arrived at that point, I was preparing to observe its passage of the next, when all of a sudden the stopple closing the end of the cylinder was blown up the chimney with a great explosion, and the thermometer, which, being cemented to it by its tube, was taken along with it, was broken to pieces, and destroyed in its fall.

This unfortunate experiment, though it put a stop for the time to the inquiries proposed, opened the way to other researches not less interesting. Suspecting that the explosion was occasioned by the rarefaction of the water which remained attached to the inside of the globe and cylinder after the operation of filling them with fixed air, and thinking it more than probable that the uncommon celerity with which the mercury rose in the thermometer was principally owing to the same cause, I was led to examine the conducting power of *moist air*, or air saturated with water.

For this experiment I provided myself with a new thermometer No. 4, the bulb of which, being of the same form as those already described (*viz.* globular), was also of the same size, or half an inch in diameter. To receive this thermometer a glass cylinder was provided, 8 lines in diameter, and about 14 inches long, and terminated at one end by a globe $1\frac{1}{2}$ inch in diameter. In the center of this globe the bulb of the thermometer was confined, by means of the stopple which closed the end of the cylinder; which stopple, being near 2 inches long, received the end of the tube of the thermometer into a hole bored through its center or axis, and confined the thermometer in its place, without the assistance of any other apparatus. Through this stopple two other small holes were bored, and lined with thin glass tubes, as in the thermometer No. 3, opening a passage into the cylinder, which holes were occasionally stopped up with stopples of cork; but to prevent accidents, such as I have before experienced from an explosion, great care was taken not to press these stopples into their places with any considerable force, that they might the more easily be blown out by any considerable effort of the confined air, or vapour.

Though in this instrument the thermometer was not altogether so steady in its place as in the thermometers No. 1, No. 2, and No. 3, the elasticity of the tube, and the weight of the mercury in the bulb of the thermometer, occasioning a small vibration or trembling of the thermometer upon any sudden motion or jar; yet I preferred this method to the others, on account of the lower part of this thermometer being entirely free, or suspended in such a manner as not to touch, or have any communication with, the lower part of the cylinder or the globe; for though the quantity of Heat received by the tube of the thermometer at its contact with the cylinder at its choaks, in the instruments No. 1 and No. 2, or with the branches of the steel spring in No. 3, and from thence communicated to the bulb, must have been exceedingly small; yet I was desirous to prevent even that, and every other possible cause of error or inaccuracy.

Does humidity augment the conducting power of air?

To determine this question I made the following experiments, the weather being clear and fine, the mercury in the barometer standing at 27 inches 8 lines, the thermometer at 19°, and the hygrometer at 44°.

| (Exp. No. 17.) *Thermometer No. 4.* Surrounded by air dry to the 44th degree of the quill hygrometer of the Manheim Academy. *Taken out of freezing water and plunged into boiling water.* | | (Exp. No. 18.) *The same Thermometer (No. 4).* Surrounded by air rendered as moist as possible by wetting the inside of the cylinder and globe with water. *Taken out of freezing water and plunged into boiling water.* | |
|---|---|---|---|
| Time elapsed. | Heat acquired. 80° | Time elapsed. | Heat acquired. 80° |
| M. S. | 0 | M. S. | 0 |
| 0 34 | 10 | 0 6 | 10 |
| 0 39 | 20 | 0 4 | 20 |
| 0 44 | 30 | 0 5 | 30 |
| 0 51 | 40 | 0 9 | 40 |
| 1 6 | 50 | 0 18 | 50 |
| 1 35 | 60 | 0 26 | 60 |
| 2 40 | 70 | 0 43 | 70 |
| Not observed. | 80 | 7 45 | 80 |
| 8 9 = total time of heating from 0° to 70° | | 1 51 = total time of heating from 0° to 70°. | |

From these experiments it appears that the conducting power of air is very much increased by humidity. To see if the same result would obtain when the experiment was reversed, I now took the thermometer with the *moist air* out of the boiling water, and plunged it into freezing water; and moving it about continually from place to place in the freezing water, I observed the times of cooling, as set down in the following table. N. B. To compare the result of this experiment with those made with *dry air*, I have placed on one side in the following table the experiment in question, and on the other side the experiment No. 10, made with the thermometer No. 2.

| (Exp. No. 19.)<br>*Thermometer No.* 4.<br>Surrounded by moist air.<br>*Taken out of boiling water and plunged into freezing water.* | | (Exp. No. 10.)<br>*Thermometer No.* 2.<br>Surrounded by dry air.<br>*Taken out of boiling water and plunged into freezing water.* | |
|---|---|---|---|
| Time elapsed. | Heat lost. | Time elapsed. | Heat lost. |
| | 80° | | 80° |
| M. S. | 0 | M. S. | 0 |
| 0 4 | 70 | 0 33 | 70 |
| 0 14 | 60 | 0 34 | 60 |
| 0 31 | 50 | 0 44 | 50 |
| 0 52 | 40 | 0 55 | 40 |
| 1 22 | 30 | 1 18 | 30 |
| 2 3 | 20 | 1 57 | 20 |
| 4 2 | 10 | 3 40 | 10 |
| 9 8 = total time of cooling from 80° to 10°. | | 9 41 = total time of cooling from 80° to 10°. | |

Though the difference of the whole times of cooling from 80° to 10° in these two experiments appears to have been very small, yet the difference of the times taken up by the first twenty or thirty degrees from the boiling point is very remarkable, and shows with how much greater facility Heat passes in moist air than in dry air. Even the slowness with which the mercury in the thermometer No. 4 descended in this experiment from the 30th to the 20th, and from the 20th to the 10th degree, I attribute in some measure to the great conducting power of the moist air with which it was surrounded; for the cylinder containing the thermometer and the moist air being not wholly submerged in the freezing water, that part of it which remained out of the water was necessarily surrounded by the air of the atmosphere; which, being much warmer than the water, communicated of its Heat to the glass; which, passing from thence into the contained moist air as soon as that air became colder than the external air, was, through that medium, communicated to the bulb of the inclosed thermometer, which prevented its cooling so fast as it would otherwise have done. But when the weather becomes cold, I propose to repeat this experiment with variations, in such a manner as to put the matter beyond all doubt. In the mean time I cannot help observing, with what infinite wisdom and goodness Divine Providence appears to have guarded us against the evil effects of excessive Heat and Cold in the atmosphere; for if it were possible for the air to be equally damp during the severe cold of the winter months as it sometimes is in summer, its conducting power, and consequently its apparent coldness, when applied to our bodies, would be so much increased, by such an additional degree of moisture, that it would become quite intolerable; but, happily for us, its power to hold water in solution is diminished, and with it its power to rob us of our animal heat, in proportion as its coldness is increased. Everybody knows how very disagreeable a very moderate degree of cold is when the air is very damp; and from hence it appears, why the thermometer is not always a just measure of the apparent or sensible Heat of the atmosphere. If colds or catarrhs are occasioned by our bodies being robbed of our animal heat, the reason is plain why those disorders prevail most during the cold autumnal rains, and upon the breaking up of the frost in the spring. It is likewise plain from whence it is that sleeping in damp beds, and inhabiting damp houses, is so very dangerous; and why the evening air is so pernicious in summer and in autumn, and why it is not so during the hard frosts of winter. It has puzzled many very able philosophers and physicians to account for the manner in which the extraordinary degree or rather *quantity* of Heat is generated which an animal body is supposed to lose, when exposed to the cold of winter, above what it communicates to the surrounding atmosphere in warm summer weather; but is it not more than probable that the difference of the quantities of Heat, actually lost or communicated, is infinitely less than what they have imagined? These inquiries are certainly very interesting; and they are undoubtedly within the reach of well-contrived and well-conducted experiments. But taking my leave for the present of this curious subject of investigation, I hasten to the sequel of my experiments.

Finding so great a difference in the conducting powers of common air and of the Torricellian vacuum, I was led to examine the conducting powers of common air of different degrees of density. For this experiment I prepared the thermometer No. 4, by stopping up one of the small glass tubes passing through the stopple, and opening a passage into the cylinder, and by fitting a valve to the external overture of the other. The instrument, thus prepared, being put under the receiver of an air-pump, the air passed freely out of the globe and cylinder upon working the machine, but the valve above described prevented its return upon letting air into the receiver. The gage of the air-pump shewed the degree of rarity of the air under the receiver, and consequently of that filling the globe and cylinder, and immediately surrounding the thermometer.

With this instrument, the weather being clear and fine, the mercury in the barometer standing at 27 inches 9 lines, the thermometer at 15°, and the hygrometer at 47°, I made the following experiments.

| (Exp. No. 20.) *Thermometer No. 4.* Surrounded by common air, barometer standing at 27 inches 9 lines. *Taken out of freezing water, and plunged into boiling water.* | | (Exp. No. 21.) *Thermometer No. 4.* Surrounded by air rarefied by pumping till the barometer-gage stood at 6 inches $11\frac{1}{2}$ lines. *Taken out of freezing water, and plunged into boiling water.* | | (Exp. No. 22.) *Thermometer No. 4.* Surrounded by air rarefied by pumping till the barometer-gage stood at 1 inch 2 lines. *Taken out of freezing water, and plunged into boiling water.* | |
|---|---|---|---|---|---|
| Time elapsed. | Heat acquired. 0° | Time elapsed | Heat acquired. 0° | Time elapsed. | Heat acquired. 0° |
| M. S. | 0 | M. S. | 0 | M. S. | 0 |
| 0 31 | 10 | 0 31 | 10 | 0 29 | 10 |
| 0 40 | 20 | 0 38 | 20 | 0 36 | 20 |
| 0 41 | 30 | 0 44 | 30 | 0 49 | 30 |
| 0 47 | 40 | 0 51 | 40 | 1 1 | 40 |
| 1 4 | 50 | 1 7 | 50 | 1 1 | 50 |
| 1 25 | 60 | 1 19 | 60 | 1 24 | 60 |
| 2 28 | 70 | 2 27 | 70 | 2 31 | 70 |
| 10 17 | 80 | 10 21 | 80 | Not observed. | 80 |
| 7 36 = total time of heating from 0° to 70°. | | 7 37 = total time of heating from 0° to 70°. | | 7 51 = total time of heating from 0° to 70°. | |

The result of these experiments, I confess, surprised me not a little; but the discovery of truth being the sole object of my inquiries (having no favourite theory to defend) it brings no disappointment along with it, under whatever unexpected shape it may appear. I hope that further experiments may lead to the discovery of the cause why there is so little difference in the conducting powers of air of such very different degrees of rarity, while there is so great a difference in the conducting powers of air, and of the Torricellian vacuum. At present I shall not venture any conjectures upon the subject; but in the mean time I dare to assert that the experiments I have made may be depended on.

The time of my stay at Manheim being expired (having had the honour to attend thither his most Serene Highness the Elector Palatine, reigning Duke of Bavaria, in his late journey) I was prevented from pursuing these inquiries further at that time; but I shall not fail to recommence them the first leisure moment I can find, which I fancy will be about the beginning of the month of November. In the mean time, to enable myself to pursue them with effect, I am sparing neither labor nor expence to provide a complete apparatus necessary for my purpose; and his Electoral Highness has been graciously pleased to order M. Artaria (who is in his service) to come to Munich to assist me. With such a patron as his most Serene Highness, and with such an assistant as Artaria, I shall go on in my pursuits with cheerfulness. Would to God that my labours might be as useful to others as they will be pleasant to me!

I shall conclude this chapter with a short account of some experiments I have made to determine the conducting powers of water and of mercury; and with a table, showing at one view the conducting powers of all the different mediums which I have examined.

Having filled the glass globe inclosing the bulb of the thermometer No. 4, first with water, and then with mercury, I made the following experiments, to ascertain the conducting powers of those two fluids.

(Exp. No. 23.)
*Thermometer No.* 4.
Surrounded by water.
*Taken out of freezing water, and plunged into boiling water.*

| Time elapsed. | Heat acquired. |
|---|---|
| | 0° |
| M. S. | 0 |
| 0 19 | 10 |
| 0 8 | 20 |
| 0 9 | 30 |
| 0 11 | 40 |
| 0 15 | 50 |
| 0 21 | 60 |
| 0 34 | 70 |
| 2 13 | 80 |
| 1 57 = total time of heating from 0° to 70°. | |

(Exp. No. 24, 25, and 26.)
*Thermometer No.* 4.
Surrounded by mercury.
*Taken out of freezing water, and plunged into boiling water.*

| Time elapsed. | | | Heat acquired. |
|---|---|---|---|
| Exp. No. 24. | Exp. No. 25. | Exp. No. 26. | |
| | | | 0° |
| M. S. | M. S. | M. S. | 0 |
| 0 5 | 0 5 | 0 5 | 10 |
| 0 4 | 0 2 | 0 5 | 20 |
| 0 2 | 0 2 | 0 4 | 30 |
| 0 4 | 0 5 | 0 5 | 40 |
| 0 4 | 0 4 | 0 7 | 50 |
| 0 7 | 0 4 | 0 8 | 60 |
| 0 15 | 0 9 | 0 14 | 70 |
| Not observed. | 0 58 | Not observed. | 80 |
| 0 41 | 0 31 | 0 48 = | total times of heating from 0° to 70°. |

The total times of heating from 0° to 70° in the three experiments with mercury being 41 seconds, 31 seconds, and 48 seconds, the mean of these times is 40 seconds; and as in the experiment with water the time employed in acquiring the same degree of Heat was 1′ 57″ = 117 seconds, it appears from these experiments that the conducting power of mercury to that of water, under the circumstances described, is as $36\frac{2}{3}$ to 117 inversely, or as 1000 to 342. And hence it is plain, why mercury *appears* so much hotter, and so much colder, to the touch than water, when in fact it is of the same temperature: for the force or violence of the sensation of what appears *hot* or *cold* depends not entirely upon the temperature of the body exciting in us those sensations, or upon the degree of Heat it actually possesses, but upon the *quantity* of Heat it is capable of communicating to us, or receiving from us, in any given short period of time, or as the intensity of the communication; and this depends in a great measure upon the conducting powers of the bodies in question.

The sensation excited in us when we touch anything that appears to us to be *hot* is the entrance of heat into our bodies; that of *cold* is its exit; and whatever contributes to facilitate or accelerate this communication adds to the violence of the sensation. And this is another proof that the thermometer cannot be a just measure of the intensity of the *sensible* Heat, or Cold, existing in bodies; or rather, that the touch does not afford us a just indication of their *real* temperatures.

*A* TABLE *of the* CONDUCTING POWERS *of the undermentioned* MEDIUMS *as determined by the foregoing Experiments.*

| *Therm. No. 1.* | *Thermometer No. 4.* | | | | | | |
|---|---|---|---|---|---|---|---|
| *Taken out of freezing water and plunged into boiling water.* | | | | | | | |
| Time elapsed. | | | | | | | |
| Torricellian Vacuum (Exp. No. 3, 4, and 13). | Common air, density = 1 (Exp. No. 20). | Rarefied air, density = ¼ (Exp. No. 21). | Rarefied air, density = 1/24 (Exp. No. 22). | Moist air (Exp. No. 18). | Water. (Exp. No. 23). | Mercury (Exp. No. 24, 25, and 26). | Heat acquired. |
| | | | | | | | 0° |
| M. S. | M. S. | M. S. | M. S. | M. S. | M. S. | M. S. | 0 |
| 0 52 | 0 31 | 0 31 | 0 29 | 0 6 | 0 19 | 0 5 | 10 |
| 0 58 | 0 40 | 0 38 | 0 36 | 0 4 | 0 8 | 0 3⅔ | 20 |
| 1 3 | 0 41 | 0 44 | 0 49 | 0 5 | 0 9 | 0 2⅔ | 30 |
| 1 18 | 0 47 | 0 51 | 1 1 | 0 9 | 0 11 | 0 4⅔ | 40 |
| 1 25 | 1 4 | 1 7 | 1 1 | 0 18 | 0 15 | 0 5 | 50 |
| 1 58 | 1 25 | 1 19 | 1 24 | 0 26 | 0 21 | 0 6⅓ | 60 |
| 3 19 | 2 28 | 2 27 | 2 31 | 0 43 | 0 34 | 0 12⅔ | 70 |
| 11 57 | 10 17 | 10 21 | —— | 7 45 | 2 13 | 0 58 | 80 |
| 10 53 | 7 36 | 7 37 | 7 51 | 1 51 | 1 57 | 0 40 | = total times of heating from 0° to 70°. |

In determining the relative conducting powers of these mediums, I have compared the times of the heating of the thermometer from 0° to 70° instead of taking the whole times from 0° to 80°, and this I have done on account of the small variation in the Heat of the boiling water arising from the variation of the weight of the atmosphere, and also on account of the very slow motion of the mercury between the 70th and the 80th degrees, and the difficulty of determining the precise moment when the mercury arrives at the 80th degree.

Taking now the conducting power of mercury = 1000, the conducting powers of the other mediums, as determined by these experiments, will be as follows, *viz.*: —

| | |
|---|---|
| Mercury . . . . . . . | 1000 |
| Moist air . . . . . . | 330 |
| Water . . . . . . . | 313 |
| Common air, density = 1 . . | $80\frac{41}{100}$ |
| Rarefied air, density = $\frac{1}{4}$ . . . | $80\frac{23}{100}$ |
| Rarefied air, density = $\frac{1}{24}$ . . | 78 |
| The Torricellian vacuum . . . | 55 |

And in these proportions are the quantities of Heat which these different mediums are capable of transmitting in any given time; and consequently these numbers express the relative *sensible* temperatures of the mediums, as well as their conducting powers. How far these decisions will hold good under a variation of circumstances, experiment only can determine. This is certainly a subject of investigation not less curious in itself than it is interesting to mankind; and I wish that what I have done may induce others to turn their attention to this long neglected field of experimental inquiry. For my own part I am determined not to quit it.

In the further prosecution of these inquiries, I do not mean to confine myself solely to the determining of the conducting powers of Fluids; on the contrary, solids, and particularly such bodies as are made use of for cloathing, will be principal subjects of my future experiments. I have indeed already begun these researches, and have made some progress in them; but I forbear to anticipate a matter which will be the subject of a future communication.

---

## CHAPTER II.

*The relative Warmth of various Substances used in making artificial Cloathing, determined by Experiment. — Relative Warmth of Coverings of the same Thickness, and formed of the same Substance, but of different Densities. — Relative Warmth of Coverings formed of equal Quantities of the same Substance, disposed in different Ways. — Experiments made with a View to determining how far the Power which certain Bodies possess of confining Heat depends on their chymical Properties. — Experiments with* Charcoal *— with* Lampblack *— with* Wood-ashes *— Striking Experiments with* Semen Lycopodii. *— All these Experiments indicate that the Air which occupies the Interstices of Substances used in forming Coverings for confining Heat, acts a very important Part in that Operation. — Those Substances appear to prevent the Air from conducting the Heat. — An Inquiry concerning the Manner in which this is effected. — This Inquiry leads to a decisive Experiment from the Result of which it appears that* Air is a perfect Non-conductor of Heat. *— This Discovery affords the* *means of explaining a variety of interesting Phenomena in the Economy of Nature.*

Munich, June, 1787

Dear Sir,

Since my last communication upon the subject of HEAT, which the Royal Society have done me the honour to publish in their Transactions, I have made some further progress in the investigation of that most interesting subject of which I propose to give you an account in this letter.

THE confining and directing of Heat are objects of such vast importance in the economy of human life, that I have been induced to confine my researches chiefly to those points, conceiving that very great advantages to mankind could not fail to be derived from the discovery of any new facts relative to these operations.

If the laws of the communication of Heat from one body to another were known, measures might be taken with certainty, in all cases, for confining it, and directing its operations, and this would not only be productive of great economy in the articles of fuel and cloathing, but would likewise greatly increase the comforts and conveniences of life, — objects of which the philosopher should never lose sight.

The route which I have followed in this inquiry is that which I thought bid fairest to lead to useful discoveries. Without embarrassing myself with any particular theory, I have formed to myself a plan of experimental investigation, which I conceived would conduct me to the knowledge of *certain facts*, of which we are now ignorant, or very imperfectly informed, and with which it is of consequence that we should be made acquainted.

The first great object which I had in view in this inquiry was to ascertain, if possible, the cause of the warmth of certain bodies, or the circumstances upon which their power of confining Heat depends. This, in other words, is no other than to determine the cause of the conducting and non-conducting power of bodies, with regard to Heat.

To this end, I began by determining by actual experiment the relative conducting powers of various bodies of very different natures, both fluids and solids; of some of which experiments I have already given an account in the paper above mentioned, which is published in the Transactions of the Royal Society for the year 1786:[3] I shall now, taking up the matter where I left it, give the continuation of the history of my researches.

Having discovered that the Torricellian vacuum is a much worse conductor of Heat than common air, and having ascertained the relative conducting powers of air, of water, and of mercury, under different circumstances, I proceeded to examine the conducting powers of various *solid bodies*, and particularly of such substances as are commonly made use of for cloathing.

The method of making these experiments was as follows: a mercurial thermometer (see Fig. 4), whose bulb was about $\frac{55}{100}$ of an inch in diameter, and its tube about 10 inches in length, was suspended in the axis of a cylindrical glass tube, about $\frac{3}{4}$ of an inch in diameter, ending with a globe $1\frac{6}{10}$ inch in diameter, in such a manner that the center of the bulb of the thermometer occupied the center of the globe; and the space between the internal surface of the globe and the surface of the bulb of the thermometer being filled with the substance whose conducting power was to be determined, the instrument was heated in boiling water, and afterwards, being plunged into a freezing mixture of pounded ice and water, the times of cooling were observed, and noted down.

The tube of the thermometer was divided at every tenth degree from 0°, or the point of freezing, to 80°, that of boiling water; and these divisions being marked upon the tube with the point of a diamond, and the cylindrical tube being left empty, the height of the mercury in the tube of the thermometer was seen through it.

The thermometer was confined in its place by means of a stopple of cork, about 1½ inch long, fitted to the mouth of the cylindrical tube, through the center of which stopple the end of the tube of the thermometer passed, and in which it was cemented.

The operation of introducing into the globe the substances whose conducting powers are to be determined, is performed in the following manner: the thermometer being taken out of the cylindrical tube, about two thirds of the substance which is to be the subject of the experiment are introduced into the globe; after which, the bulb of the thermometer is introduced a few inches into the cylinder; and, after it, the remainder of the substance being placed round about the tube of the thermometer; and, lastly, the thermometer being introduced farther into the tube, and being brought into its proper place, that part of the substance which, being introduced last, remains in the cylindrical tube above the bulb of the thermometer, is pushed down into the globe, and placed equally round the bulb of the thermometer by means of a brass wire which is passed through holes made for that purpose in the stopple closing the end of the cylindrical tube.

As this instrument is calculated merely for measuring the passage of Heat in the substance whose conducting power is examined, I shall give it the name of *passage-thermometer*, and I shall apply the same appellation to all other instruments constructed upon the same principles, and for the same use, which I may in future have occasion to mention; and as this instrument has been so particularly described, both here, and in my former paper upon the subject of Heat, in speaking of any others of the same kind in future it will not be necessary to enter into such minute details. I shall, therefore, only mention their *sizes*, or the diameters of their bulbs, the diameters of their globes, the diameters of their cylinders, and the lengths and divisions of their tubes, taking it for granted that this will be quite sufficient to give a clear idea of the instrument.

In most of my former experiments, in order to ascertain the conducting power of any body, the body being introduced into the globe of the passage-thermometer, the instrument was cooled to the temperature of freezing water, after which, being taken out of the ice-water, it was plunged suddenly into boiling water, and the times of heating from ten to ten degrees were observed and noted; and I said that these times were as the conducting power of the body inversely; but in the experiments of which I am now about to give an account, I have in general reversed the operation; that is to say, instead of observing the times of heating, I have first heated the body in boiling water, and then plunging it into a mixture of pounded ice and ice-cold water, I have noted the times taken up in cooling.

I have preferred this last method to the former, not only on account of the greater ease and convenience with which a thermometer, plunged into ice and water, may be observed, than when placed in a vessel of boiling water, and surrounded by hot steam, but also on account of the greater accuracy of the experiment, the heat of boiling water varying with the variations of the pressure of the atmosphere; consequently, the experiments made upon different days will have different results, and of course, strictly speaking, cannot be compared together; but the temperature of pounded ice and water is ever the same, and of course the results of the experiments are uniform.

In heating the thermometer, I did not in general bring it to the temperature of the boiling water, as this temperature, as I have just observed, is variable; but when the mercury had attained the 75° of its scale, I immediately took it out of the boiling water, and plunged it into the ice and water; or, which I take to be still more accurate, suffering the mercury to rise a degree or two above 75°, and then taking it out of the boiling water, I held it over the vessel containing the pounded ice and water, ready to plunge it into that mixture the moment the mercury, descending, passes the 75°.

Having a watch at my ear which beat half seconds (which I counted), I noted the time of the passage of the mercury over the divisions of the thermometer, marking 70° and every tenth degree from it, descending to 10° of the scale. I continued the cooling to 0°, or the temperature of the ice and water, in very few instances, as this took up much time, and was attended with no particular advantage, the determination of the times taken up in cooling 60 degrees of Reaumur's scale — that is to say, from 70° to 10° — being quite sufficient to ascertain the conducting power of any body whatever.

During the time of cooling in ice and water, the thermometer was constantly moved about in this mixture from one place to another; and there was always so much pounded ice mixed with the water that the ice appeared above the surface of the water, — the vessel, which was a large earthen jar, being first quite filled with pounded ice, and the water being afterwards poured upon it, and fresh quantities of pounded ice being added as the occasion required.

Having described the apparatus made use of in these experiments, and the manner of performing the different operations, I shall now proceed to give an account of the experiments themselves.

My first attempt was to discover the relative conducting powers of such substances as are commonly made use of for cloathing; accordingly, having procured a quantity of *raw silk*, as spun by the worm, *sheep's-wool*, *cotton-wool*, *linen* in the form of the finest lint, being the scrapings of very fine Irish linen, the finest part of the *fur of the beaver* separated from the skin, and from the long hair, the finest part of the *fur of a white Russian hare*, and *eider-down*, — I introduced successively 16 grains in weight of each of these substances into the globe of the passage-thermometer, and placing it carefully and equally round the bulb of the thermometer, I heated the thermometer in boiling water, as before described, and taking it out of the boiling water, plunged it into pounded ice and water, and observed the times of cooling.

But as the interstices of these bodies thus placed in the globe were filled with air, I first made the experiment with air alone, and took the result of that experiment as a standard by which to compare all the others; the results of three experiments with air were as follows: —

| The Bulb of the Thermometer surrounded by Air. | | | | |
|---|---|---|---|---|
| Heat lost. | Exp. No. 1. Time elapsed. | Exp. No. 2. Time elapsed. | Heat acquired. | Exp. No. 3. Time elapsed. |
| 70° | — | — | 10° | — |
| 60 | 38″ | 38″ | 20 | 39″ |
| 50 | 46 | 46 | 30 | 43 |
| 40 | 59 | 59 | 40 | 53 |
| 30 | 80 | 79 | 50 | 67 |
| 20 | 122 | 122 | 60 | 96 |
| 10 | 231 | 230 | 70 | 175 |
| Total times | 576 | 574 | — | 473 |

The following table shews the results of the experiments with the various substances therein mentioned: —

| Heat lost. | Air. | Raw silk, 16 grs. | Sheep's-wool, 16 grs. | Cotton-wool, 16 grs. | Fine lint, 16 grs. | Beavers' fur, 16 grs. | Hares' fur, 16 grs. | Eider-down, 16 grs. |
|---|---|---|---|---|---|---|---|---|
| | Exp. 1. | Exp. 4. | Exp. 5. | Exp. 6. | Exp. 7. | Exp. 8. | Exp. 9. | Exp. 10. |
| 70° | — | — | — | — | — | — | — | — |
| 60 | 38″ | 94″ | 79″ | 83″ | 80″ | 99″ | 97″ | 98″ |
| 50 | 46 | 110 | 95 | 95 | 93 | 116 | 117 | 116 |
| 40 | 59 | 133 | 118 | 117 | 115 | 153 | 144 | 146 |
| 30 | 80 | 185 | 162 | 152 | 150 | 185 | 193 | 192 |
| 20 | 122 | 273 | 238 | 221 | 218 | 265 | 270 | 268 |
| 10 | 231 | 489 | 426 | 378 | 376 | 478 | 494 | 485 |
| Total times | 576 | 1284 | 1118 | 1046 | 1032 | 1296 | 1315 | 1305 |

Now the *warmth* of a body, or its power to confine Heat, being as its power of resisting the passage of Heat through it (which I shall call its *non-conducting power*); and the time taken up by any body in cooling, which is surrounded by any medium through which the Heat is obliged to pass, being, *cæteris paribus*, as the resistance which the medium opposes to the passage of the Heat, it appears that the *warmth* of the bodies mentioned in the foregoing table are as the times of cooling, — the *conducting powers* being inversely as those times, as I have formerly shown.

From the results of the foregoing experiments it appears that, of the seven different substances made use of, hares' fur and eider-down were the warmest; after these came beavers' fur, raw silk, sheep's-wool, cotton-wool, and, lastly, lint, or the scrapings of fine linen; but I acknowledge that the differences in the warmth of these substances were much less than I expected to have found them.

Suspecting that this might arise from the volumes or solid contents of the substances being different (though their weights were the same), arising from the difference of their specific gravities; and as it was not easy to determine the specific gravities of these substances with accuracy, in order to see how far any known difference in the volume or quantity of the same substance, confined always in the same space, would add to or diminish the time of cooling, or the apparent warmth of the covering, I made the three following experiments.

In the first, the bulb of the thermometer was surrounded by 16 grains of eider-down; in the second by 32 grains; and in the third by 64 grains; and in all these experiments the substance was made to occupy exactly the same space, viz. the whole internal capacity of the glass globe, in the center of which the bulb of the thermometer was placed; consequently, the thickness of the covering of the thermometer remained the same, while its density was varied in proportion to the numbers 1, 2, and 4.

The results of these experiments were as follows: —

| The Bulb of the Thermometer being surrounded by Eider-down. | | | |
|---|---|---|---|
| Heat lost. | 16 grains. | 32 grains. | 64 grains. |
| | (Exp. No. 11.) | (Exp. No. 12.) | (Exp. No. 13) |
| 70° | — | — | — |
| 60 | 97″ | 111″ | 112″ |
| 50 | 117 | 128 | 130 |
| 40 | 145 | 157 | 165 |
| 30 | 192 | 207 | 224 |
| 20 | 267 | 304 | 326 |
| 10 | 486 | 565 | 658 |
| Total times | 1304 | 1472 | 1615 |

Without stopping at present to draw any particular conclusions from the results of these experiments, I shall proceed to give an account of some others, which will afford us a little further insight into the nature of some of the circumstances upon which the warmth of covering depends.

Finding, by the last experiments, that the density of the covering added so considerably to the warmth of it, its thickness remaining the same, I was now desirous of discovering how far the internal structure of it contributed to render it more or less pervious to Heat, its thickness and quantity of matter remaining the same. By internal structure, I mean the disposition of the parts of the substance which forms the covering; thus they may be extremely divided, or very fine, as raw silk as spun by the worms, and they may be equally distributed through the whole space they occupy; or they may be coarser, or in larger masses, with larger interstices, as the ravelings of cloth, or cuttings of thread.

If Heat passed *through* the substances made use of for covering, and if the warmth of the covering depended solely upon the difficulty which the Heat meets with in its passage through the substances, *or solid parts*, of which they are composed, — in that case, the warmth of covering would be always, *cæteris paribus*, as the quantity of materials of which it is composed; but that this is not the case, the following, as well as the foregoing, experiments clearly evince.

Having, in the experiment No. 4, ascertained the warmth of 16 grains of raw silk, I now repeated the experiment with the same quantity, or weight, of the ravelings of white taffety, and afterwards with a like quantity of common sewing-silk, cut into lengths of about two inches.

The following table shows the results of these three experiments: —

| Heat lost. | Raw silk, 16 grs. | Ravelings of taffety, 16 grs. | Sewing-silk cut into lengths, 16 grs. |
|---|---|---|---|
| | Exp. 4. | Exp. 14. | Exp. 15. |
| 70° | — | — | — |
| 60 | 94″ | 90″ | 67″ |
| 50 | 110 | 106 | 79 |
| 40 | 133 | 128 | 99 |
| 30 | 185 | 172 | 135 |
| 20 | 273 | 246 | 195 |
| 10 | 489 | 427 | 342 |
| Total times | 1284 | 1169 | 917 |

Here, notwithstanding that the quantities of the silk were the same in the three experiments, and though in each of them it was made to occupy the same space, yet the warmth of the coverings which were formed were very different, owing to the different disposition of the material.

The raw silk was very fine, and was very equally distributed through the space it occupied, and it formed a warm covering.

The ravelings of taffety were also fine, but not so fine as the raw silk, and of course the interstices between its threads were greater, and it was less warm; but the cuttings of sewing-silk were very coarse, and consequently it was very unequally distributed in the space in which it was confined; and it made a very bad covering for confining Heat.

It is clear from the results of the five last experiments, that the air which occupies the interstices of bodies, made use of for covering, acts a very important part in the operation of confining Heat; yet I shall postpone the examination of that circumstance till I shall have given an account of several other experiments, which, I think, will throw still more light upon that subject.

But, before I go any further, I will give an account of three experiments, which I made, or, rather, the same experiment which I repeated three times the same day, in order to see how far they may be depended on, as being regular in their results.

The glass globe of the passage-thermometer being filled with 16 grains of cotton-wool, the instrument was heated and cooled three times successively, when the times of cooling were observed as follows: —

| Heat lost. | Exp. 16. | Exp. 17. | Exp. 18. |
|---|---|---|---|
| 70° | — | — | — |
| 60 | 82″ | 84″ | 83″ |
| 50 | 96 | 95 | 95 |
| 40 | 118 | 117 | 116 |
| 30 | 152 | 153 | 151 |
| 20 | 221 | 221 | 220 |
| 10 | 380 | 377 | 377 |
| Total times | 1049 | 1047 | 1042 |

The differences of the times of cooling in these three experiments were extremely small ; but regular as these experiments appear to have been in their results, they were not more so than the other experiments made in the same way, many of which were repeated two or three times, though, for the sake of brevity, I have put them down as single experiments.

But to proceed in the account of my investigations relative to the causes of the warmth of warm cloathing. Having found that the fineness and equal distribution of a body or substance made use of to form a covering to confine Heat contributes so much to the warmth of the covering, I was desirous, in the next place, to see the effect of condensing the covering, its quantity of matter remaining the same, but its thickness being diminished in proportion to the increase of its density.

The experiment I made for this purpose was as follows : I took 16 grains of common sewing-silk, neither very fine nor very coarse, and winding it about the bulb of the thermometer in such a manner that it entirely covered it, and was as nearly as possible of the same thickness in every part, I replaced the thermometer in its cylinder and globe, and heating it in boiling water, cooled it in ice and water, as in the foregoing experiments. The results of the experiment were as may be seen in the following table; and in order that it may be compared with those made with the same quantity of silk differently disposed of, I have placed those experiments by the side of it: —

| Heat lost. | Raw silk, 16 grs. | Fine ravelings of taffety, 16 grs. | Sewing-silk cut into lengths, 16 grs. | Sewing-silk, 16 grs. wound round the bulb of the thermometer. |
|---|---|---|---|---|
| | Exp. No. 4. | Exp. No. 14. | Exp. No. 15. | Exp. No. 19. |
| 70° | — | — | — | — |
| 60 | 94″ | 90″ | 67″ | 46″ |
| 50 | 110 | 106 | 79 | 62 |
| 40 | 133 | 128 | 99 | 85 |
| 30 | 185 | 172 | 135 | 121 |
| 20 | 273 | 246 | 195 | 191 |
| 10 | 489 | 427 | 342 | 399 |
| Total times. | 1284 | 1169 | 917 | 904 |

It is not a little remarkable, that, though the covering formed of sewing-silk wound round the bulb of the thermometer in the 19th experiment appeared to have so little power of confining the Heat when the instrument was very hot, or when it was first plunged into the ice and water, yet afterwards, when the Heat of the thermometer approached much nearer to that of the surrounding medium, its power of confining the Heat which remained in the bulb of the thermometer appeared to be even greater than that of the silk in the experiment No. 15, the time of cooling from 20° to 10° being in the one 399″, and in the other 342″. The same appearance was observed in the following experiments, in which the bulb of the thermometer was surrounded by threads of *wool*, of *cotton*, and of *linen*, or *flax*, wound round it, in the like manner as the sewing-silk was wound round it in the last experiment.

The following table shows the results of these experiments, with the threads of various kinds; and that they may the more easily be compared with those made with the same quantity of the same substances in a different form, I have placed the accounts of these experiments by the side of each other. I have also added the account of an experiment, in which 16 grains of fine linen cloth were wrapped round the bulb of the thermometer, going round it nine times, and being bound together at the top and bottom of it, so as completely to cover it.

| Heat lost. | *Sheep's-wool*, 16 grains, surrounding the bulb of the thermometer. | *Woollen thread*, 16 grains, wound round the bulb of the thermometer. | *Cotton-wool*, 16 grains, surrounding the bulb of the thermometer. | *Cotton thread*, 16 grains, wound round the bulb of the thermometer. | *Lint*, 16 grains, surrounding the bulb of the thermometer. | *Linen thread*, 16 grains, wound round the bulb of the thermometer. | *Linen cloth*, 16 grains, wrapped round the bulb of the thermometer. |
|---|---|---|---|---|---|---|---|
| | Exp. 5. | Exp. 20. | Exp. 6. | Exp. 21. | Exp. 7. | Exp. 22. | Exp. 23. |
| 70° | — | — | — | — | — | — | — |
| 60 | 79″ | 46″ | 83″ | 45″ | 80″ | 46″ | 42″ |
| 50 | 95 | 63 | 95 | 60 | 93 | 62 | 56 |
| 40 | 118 | 89 | 117 | 83 | 115 | 83 | 74 |
| 30 | 162 | 126 | 152 | 115 | 150 | 117 | 108 |
| 20 | 238 | 200 | 221 | 179 | 218 | 180 | 168 |
| 10 | 426 | 410 | 378 | 370 | 376 | 385 | 338 |
| Total times | 1118 | 934 | 1046 | 852 | 1032 | 873 | 783 |

That thread wound tight round the bulb of the thermometer should form a covering less warm than the same quantity of wool, or other raw materials of which the thread is made, surrounding the bulb of the thermometer in a more loose manner, and consequently occupying a greater space, is no more than what I expected, from the idea I had formed of the causes of the warmth of covering; but I confess I was much surprised to find that there is so great a difference in the relative warmth of these two coverings, when they are employed to confine great degrees of Heat, and when the Heat they confine is much less in proportion to the temperature of the surrounding medium. This difference was very remarkable; in the experiments with sheep's-wool, and with woollen thread, the warmth of the covering formed of 16 grains of the former was to that formed of 16 grains of the latter, when the bulb of the thermometer was heated to 70° and cooled to 60°, as 79 to 46 (the surrounding medium being at 0°); but afterwards, when the thermometer had only fallen from 20° to 10° of Heat, the warmth of the wool was to that of the woollen thread only as 426 to 410; and in the experiments with lint, and with linen thread, when the Heat was much abated, the covering of the thread appeared to be even warmer than that of the lint, though in the beginning of the experiments, when the Heat was much greater, the lint was warmer than the thread, in the proportion of 80 to 46.

From hence it should seem that a covering may, under certain circumstances, be very good for confining small degrees of warmth, which would be but very indifferent when made use of for confining a more intense Heat, and *vice versa.* This, I believe, is a new fact; and I think the knowledge of it may lead to further discoveries relative to the causes of the warmth of coverings, or the manner in which Heat makes its passage through them. But I forbear to enlarge upon this subject, till I shall have given an account of several other experiments, which I think throw more light upon it, and which will consequently render the investigation easier and more satisfactory.

With a view to determine how far the power which certain bodies appear to possess of confining Heat, when made use of as covering, depends upon the natures of those bodies, considered as chymical substances, or upon the chymical principles of which they are composed, I made the following experiments.

As charcoal is supposed to be composed almost entirely of phlogiston, I thought that, if that principle was the cause either of the conducting power or the non-conducting power of the bodies which contain it, I should discover it by making the experiment with charcoal, as I had done with various other bodies. Accordingly, having filled the globe of the passage-thermometer with 176 grains of that substance in very fine powder (it having been pounded in a mortar, and sifted through a fine sieve), the bulb of the thermometer being surrounded by this powder, the instrument was heated in boiling water, and being afterwards plunged into a mixture of pounded ice and water, the times of cooling were observed as mentioned in the following table. I afterwards repeated the experiment with lampblack, and with very pure and very dry wood-ashes; the results of which experiments were as under mentioned: —

| Heat lost. | The Bulb of the Thermometer surrounded by | | | |
|---|---|---|---|---|
| | 176 grains of fine powder of charcoal. | 176 grains of fine powder of charcoal. | 195 grains of lampblack. | 307 grains of pure dry wood-ashes. |
| | Exp. No. 24. | Exp. No. 25. | Exp. No. 26. | Exp. No. 27. |
| 70° | — | — | — | — |
| 60 | 79″ | 91″ | 124″ | 96″ |
| 50 | 95 | 91 | 118 | 92 |
| 40 | 100 | 109 | 134 | 107 |
| 30 | 139 | 133 | 164 | 136 |
| 20 | 196 | 192 | 237 | 185 |
| 10 | 331 | 321 | 394 | 311 |
| Total times | 940 | 937 | 1171 | 927 |

The experiment No. 25 was simply a repetition of that numbered 24, and was made immediately after it; but, in moving the thermometer about in the former experiment, the powder of charcoal which filled the globe was shaken a little together, and to this circumstance I attribute the difference in the results of the two experiments.

In the experiments with lampblack and with wood-ashes, the times taken up in cooling from 70° to 60° were greater than those employed in cooling from 60° to 50°; this most probably arose from the considerable quantity of Heat contained by these substances, which was first to be disposed of, before they could receive and communicate to the surrounding medium that which was contained by the bulb of the thermometer.

The next experiment I made was with *semen lycopòdii*, commonly called witch-meal, a substance which possesses very extraordinary properties. It is almost impossible to wet it; a quantity of it strewed upon the surface of a basin of water, not only swims upon the water without being wet, but it prevents other bodies from being wet which are plunged into the water through it; so that a piece of money, or other solid body, may be taken from the bottom of the basin by the naked hand without wetting the hand; which is one of the tricks commonly shown by the jugglers in the country: this meal covers the hand, and, descending along with it to the bottom of the basin, defends it from the water. This substance has the appearance of an exceeding fine, light, and very moveable yellow powder, and it is very inflammable; so much so, that, being blown out of a quill into the flame of a candle, it flashes like gunpowder, and it is made use of in this manner in our theatres for imitating lightning.

Conceiving that there must have been a strong attraction between this substance and air, and suspecting, from some circumstances attending some of the foregoing experiments, that the warmth of a covering depends not merely upon the fineness of the substance of which the covering is formed, and the disposition of its parts, but that it arises in some measure from a certain attraction between the substance and the air which fills its interstices, I thought that an experiment with *semen lycopodii* might possibly throw some light upon this matter; and in this opinion I was not altogether mistaken, as will appear by the results of the three following experiments.

| The Bulb of the Thermometer surrounded by 256 Grs. of *Semen Lycopodii.* | | | | |
|---|---|---|---|---|
| Heat lost. | Cooled. Exp. No. 28. | Cooled. Exp. No. 29. | Heat acquired. | Heated. Exp. No. 30. |
| 70° | — | — | 0° | — |
| 60 | 146″ | 157″ | 10 | 230″ |
| 50 | 162 | 160 | 20 | 68 |
| 40 | 175 | 170 | 30 | 63 |
| 30 | 209 | 203 | 40 | 76 |
| 20 | 284 | 288 | 50 | 121 |
| 10 | 502 | 513 | 60 | 316 |
| — | — | — | 70 | 1585 |
| Total times | 1478 | 1491 | — | 2459 |

In the last experiment (No. 30), the result of which was so very extraordinary, the instrument was cooled to 0° in thawing ice, after which it was plunged suddenly into boiling water, where it remained till the inclosed thermometer had acquired the Heat of 70°, which took up no less than 2459 seconds, or above 40 minutes; and it had remained in the boiling water full a minute and a half before the mercury in the thermometer showed the least sign of rising. Having at length been put into motion, it rose very rapidly 40 or 50 degrees, after which its motion gradually abating became so slow, that it took up 1585 seconds, or something more than 26 minutes, in rising from 60° to 70°, though the temperature of the medium in which it was placed during the whole of this time was very nearly 80°; the mercury in the barometer standing but little short of 27 Paris inches.

All the different substances which I had yet made use of in these experiments for surrounding or covering the bulb of the thermometer, fluids excepted, had, in a greater or in a less degree confined the Heat, or prevented its passing into or out of the thermometer so rapidly as it would have done, had there been nothing but air in the glass globe, in the center of which the bulb of the thermometer was suspended. But the great question is, how, or in what manner, they produced this effect?

And first, it was not in consequence of their own non-conducting powers, simply considered; for if, instead of being only bad conductors of Heat, we suppose them to have been totally impervious to Heat, their volumes or solid contents were so exceedingly small in proportion to the capacity of the globe in which they were placed, that, had they had no effect whatever upon the air filling their interstices, that air would have been sufficient to have conducted all the Heat communicated in less time than was actually taken up in the experiment.

The diameter of the globe being 1.6 inches, its contents amounted to 2.14466 cubic inches; and the contents of the bulb of the thermometer being only 0.08711 of a cubic inch (its diameter being 0.55 of an inch), the space between the bulb of the thermometer and the internal surface of the globe amounted to 2.14466 — 0.08711 = 2.05755 cubic inches; the whole of which space was occupied by the substances by which the bulb of the thermometer was surrounded in the experiments in question.

But though these substances occupied this space, they were far from *filling it;* by much the greater part of it being filled by the air which occupied the interstices of the substances in question. In the experiment No. 4, this space was occupied by 16 grains of raw silk; and as the specific gravity of raw silk is to that of water as 1734 to 1000, the volume of this silk was equal to the volume of 9.4422 grains of water; and as 1 cubic inch of water weighs 253.185 grains, its volume was equal to $\frac{9.4422}{253.1850} = 0.037294$ of a cubic inch; and, as the space it occupied amounted to 2.05755 cubic inches, it appears that the silk filled no more than about $\frac{1}{55}$ part of the space in which it was confined, the rest of that space being filled with air.

In the experiment No. 1, when the space between the bulb of the thermometer and the glass globe, in the center of which it was confined, was filled with nothing but air, the time taken up by the thermometer in cooling from 70° to 10° was 576 seconds; but in the experiment No. 4, when this same space was filled with 54 parts air, and 1 part raw silk, the time of cooling was 1284 seconds.

Now, supposing that the silk had been totally incapable of conducting any Heat at all, if we suppose, at the same time, that it had no power to prevent the air remaining in the globe from conducting it, in that case its presence in the globe could only have prolonged the time of cooling in proportion to the quantity of the air it had displaced to the quantity remaining, that is to say, as 1 is to 54, or a little more than 10 seconds. But the time of cooling was actually prolonged 708 seconds (for in the experiment No. 1 it was 576 seconds, and in the experiment No. 4 it was 1284 seconds, as has just been observed); and this shows that the silk not only did not conduct the Heat itself, but that it prevented the air by which its interstices were filled from conducting it; or, at least, it greatly weakened its power of conducting it.

The next question which arises is, how air can be prevented from conducting Heat? and this necessarily involves another, which is, How does air conduct Heat?

If air conducted Heat, as it is probable that the metals and water, and all other solid bodies and unelastic fluids, conduct it, — that is to say, if, its particles remaining in their places, the Heat passed from one particle to another, through the whole mass, as there is no reason to suppose that the propagation of Heat is necessarily in right lines, I cannot conceive how the interposition of so small a quantity of any solid body as $\frac{1}{55}$ part of the volume of the air could have effected so remarkable a diminution of the conducting power of the air, as appeared in the experiment (No. 4) with raw silk, above-mentioned.

If air and water conducted Heat in the same *manner*, it is more than probable that their conducting powers might be impaired by the same means; but when I made the experiment with water, by filling the glass globe, in the center of which the bulb of the thermometer was suspended, with that fluid, and afterwards varied the experiment by adding 16 grains of raw silk to the water, I did not find that the conducting power of the water was sensibly impaired by the presence of the silk.*

But we have just seen that the same silk, mixed with an equal volume of air, diminished its conducting power in a very remarkable degree; consequently, there is great reason to conclude that water and air conduct Heat in a *different manner*.

But the following experiment, I think, puts the matter beyond all doubt.

It is well known that the power which air possesses of holding water in solution is augmented by Heat, and diminished by cold, and that, if hot air is saturated with water, and if this air is afterwards cooled, a part of its water is necessarily deposed.

I took a cylindrical bottle of very clear transparent glass, about 8 inches in diameter, and 12 inches high, with a short and narrow neck, and, suspending a small piece of linen rag, moderately wet, in the middle of it, I plunged it into a large vessel of water, warmed to about 100° of Fahrenheit's thermometer, where I suf-

* The experiment here mentioned was made in the year 1787; but the result of a more careful investigation of the subject has since shown that Heat is not propagated in water in the manner here supposed. (See Essay VII, Edition of 1798.)[4]

fered it to remain till the contained air was not only warm, but thoroughly saturated with the moisture which it attracted from the linen rag, the mouth of the bottle being well stopped up during this time with a good cork; this being done, I removed the cork for a moment, to take away the linen rag, and, stopping up the bottle again immediately, I took it out of the warm water, and plunged it into a large cylindrical jar, about 12 inches in diameter, and 16 inches high, containing just so much ice-cold water, that, when the bottle was plunged into it, and quite covered by it, the jar was quite full.

As the jar was of very fine transparent glass, as well as the bottle, and as the cold water contained in the jar was perfectly clear, I could see what passed in the bottle most distinctly; and having taken care to place the jar upon a table near the window, in a very favourable light, I set myself to observe the appearances which should take place, with all that anxious expectation which a conviction that the result of the experiment must be decisive naturally inspired.

I was certain that the air contained in the bottle could not part with its Heat, without at the same time—that is to say, *at the same moment*, and *in the same place*—parting with a portion of its water; if, therefore, the Heat penetrated the mass of air from the center to the surface, or *passed through it* from particle to particle, in the same manner as it is probable that it passes through water, and all other unelastic fluids,* by far the greater part of the air contained in the bottle would part with its Heat,

* This opinion respecting the manner in which Heat is propagated in water, and other unelastic fluids, was afterwards found to be erroneous, as has been shown in the preceding Essay.[5]

when *not actually in contact with the glass*, and a proportional part of its water being let fall at the same time, and in the *same place*, would necessarily descend in the form of rain; and, though this rain might be too fine to be visible in its descent, yet I was sure I should find it at the bottom of the bottle, if not in visible drops of water, yet in that kind of cloudy covering which cold glass acquires from a contact with hot steam or watery vapour.

But if the particles of air, instead of communicating their Heat from one to another, from the center to the surface of the bottle, each in its turn, and for itself, came to the surface of the bottle, and there deposited its Heat and its water, I concluded that the cloudiness occasioned by this deposit of water would appear all over the bottle, or, at least, not more of it at the bottom than at the sides, but rather less; and this I found to be the case in fact.

The cloudiness first made its appearance upon the sides of the bottle, near the top of it; and from thence it gradually spread itself downwards, till, growing fainter as it descended lower, it was hardly visible at the distance of half an inch from the bottom of the bottle; and upon the bottom itself, which was nearly flat, there was scarcely the smallest appearance of cloudiness.

These appearances, I think, are easy to be accounted for. The air immediately in contact with the glass being cooled, and having deposited a part of its water upon the surface of the glass, at the same time that it communicates to it its Heat, slides downwards by the sides of the bottle in consequence of its increased specific gravity, and, taking its place at the bottom of the bottle, forces the whole mass of hot air upwards; which, in its turn, coming to the sides of the bottle, *there* deposits its Heat and its water, and afterwards bending its course downwards, this circulation is continued till all the air in the bottle has acquired the exact temperature of the water in the jar.

From hence it is clear why the first appearance of condensed vapour is near the top of the bottle, as also why the greatest collection of vapour is in that part, and that so very small a quantity of it is found nearer the bottom of the bottle.

This experiment confirmed me in an opinion which I had for some time entertained, that, though the particles of air individually, or each for itself, are capable of receiving and *transporting* Heat, yet air in a quiescent state, or as a fluid whose parts are at rest with respect to each other, is not capable of conducting it, or giving it a passage; in short, that Heat is incapable of *passing through a mass of air*, penetrating from one particle of it to another, and that it is to this circumstance that its non-conducting power is principally owing.

It is also to this circumstance, in a great measure, that it is owing that its non-conducting power, or its apparent warmth when employed as a covering for confining Heat, is so remarkably increased upon its being mixed with a small quantity of any very fine, light, solid substance, such as the raw silk, fur, eider-down, &c., in the foregoing experiments; for as I have already observed, though these substances, in the very small quantities in which they were made use of, could hardly have prevented, in any considerable degree, the air from conducting or giving a *passage* to the Heat, had it been capable of passing through it, yet they might very much impede it in the operation of transporting it.

But there is another circumstance which it is necessary to take into the account, and that is the attraction which subsists between air and the bodies above mentioned, and other like substances, constituting natural and artificial cloathing. For, though the incapacity of air to give a passage to Heat in the manner solid bodies permit it to pass through them may enable us to account for its warmth under certain circumstances, yet the bare admission of this principle does not seem to be sufficient to account for the very extraordinary degrees of warmth which we find in furs and in feathers, and in various other kinds of natural and artificial cloathing; nor even that which we find in snow; for if we suppose the particles of air to be at liberty to *carry off* the Heat which these bodies are meant to confine, without any other obstruction or hindrance than that arising from their *vis inertiæ*, or the force necessary to put them in motion, it seems probable that the succession of fresh particles of cold air, and the consequent loss of Heat, would be much more rapid than we find it to be in fact.

That an attraction, and a very strong one, actually subsists between the particles of air and the fine hair or furs of beasts, the feathers of birds, wool, &c., appears by the obstinacy with which these substances retain the air which adheres to them, even when immersed in water, and put under the receiver of an air-pump; and that this attraction is essential to the warmth of these bodies, I think is very easy to be demonstrated.

In furs, for instance, the attraction between the particles of air and the fine hairs in which it is concealed being greater than the increased elasticity or repulsion of those particles with regard to each other, arising from the Heat communicated to them by the animal body, the air in the fur, though heated, is not easily displaced; and this coat of confined air is the real barrier which defends the animal body from the external cold. This air cannot *carry off* the Heat of the animal, because it is itself confined, by its attraction to the hair or fur; and it transmits it with great difficulty, if it transmits at all, as has been abundantly shown by the foregoing experiments.

Hence it appears why those furs which are the finest, longest, and thickest, are likewise the warmest; and how the furs of the beaver, of the otter, and of other like quadrupeds which live much in water, and the feathers of water-fowls, are able to confine the Heat of those animals in winter, notwithstanding the extreme coldness and great conducting power of the water in which they swim. The attraction between these substances and the air which occupies their interstices is so great that this air is not dislodged even by the contact of water, but, remaining in its place, it defends the body of the animal at the same time from being wet, and from being robbed of its Heat by the surrounding cold fluid, and it is possible that the pressure of this fluid upon the covering of air confined in the interstices of the fur, or feathers, may at the same time increase its warmth, or non-conducting power, in such a manner that the animal may not, in fact, lose more heat when in water than when in air: for we have seen, by the foregoing experiments, that, under certain circumstances, the warmth of a covering is increased by bringing its component parts nearer together, or by increasing its density even at the expense of its thickness. But this point will be further investigated hereafter.

Bears, wolves, foxes, hares, and other like quadrupeds, inhabitants of cold countries, which do not often take the water, have their fur much thicker upon their backs than upon their bellies. The heated air occupying the interstices of the hairs of the animal tending naturally to rise upwards, in consequence of its increased elasticity, would escape with much greater ease from the backs of quadrupeds than from their bellies, had not Providence wisely guarded against this evil by increasing the obstructions in those parts, which entangle it and confine it to the body of the animal. And this, I think, amounts almost to a proof of the principles assumed relative to the manner in which Heat is carried off by air, and the causes of the non-conducting power of air, or its apparent warmth, when, being combined with other bodies, it acts as a covering for confining Heat.

The snows which cover the surface of the earth in winter, in high latitudes, are doubtless designed by an all-provident Creator as a garment to defend it against the piercing winds from the polar regions, which prevail during the cold season.

These winds, notwithstanding the vast tracts of continent over which they blow, retain their sharpness as long as the ground they pass over is covered with snow; and it is not till, meeting with the ocean, they acquire, from a contact with its waters, the Heat which the snows prevent their acquiring from the earth, that the edge of their coldness is taken off, and they gradually die away and are lost.

The winds are always found to be much colder when the ground is covered with snow than when it is bare, and this extraordinary coldness is vulgarly supposed to be communicated to the air by the snow; but this is an erroneous opinion, for these winds are in general much colder than the snow itself.

They retain their coldness because the snow prevents them from being warmed at the expence of the earth; and this is a striking proof of the use of the snows in preserving the Heat of the earth during the winter in cold latitudes.

It is remarkable that these winds seldom blow from the poles directly towards the equator, but from the land towards the sea. Upon the eastern coast of North America the cold winds come from the northwest; but upon the western coast of Europe they blow from the northeast.

That they should blow towards those parts where they can most easily acquire the Heat they are in search of, is not extraordinary; and that they should gradually cease and die away, upon being warmed by a contact with the waters of the ocean, is likewise agreeable to the nature and causes of their motion; and if I might be allowed a conjecture respecting the principal use of the seas, or the reason why the proportion of water upon the surface of our globe is so great, compared to that of the land, it is to maintain a more equal temperature in the different climates, by heating or cooling the winds which at certain periods blow from the great continents.

That cold winds actually grow much milder upon passing over the sea, and that hot winds are refreshed by a contact with its waters, is very certain; and it is equally certain that the winds from the ocean are, in all climates, much more temperate than those which blow from the land.

In the islands of Great Britain and Ireland, there is not the least doubt but the great mildness of the climate is entirely owing to their separation from the neighbouring continent by so large a tract of sea; and in all similar situations, in every part of the globe, similar causes are found to produce similar effects.

The cold northwest winds which prevail upon the coast of North America during the winter seldom extend above 100 leagues from the shore, and they are always found to be less violent, and less piercing, as they are further from the land.

These periodical winds from the continents of Europe and North America prevail most towards the end of the month of February, and in the month of March; and I conceive that they contribute very essentially towards bringing on an early spring, and a fruitful summer, particularly when they are very violent in the month of March, and if at that time the ground is well covered with snow. The whole atmosphere of the polar regions being, as it were, transported into the ocean by these winds, is there warmed and saturated with water: and, a great accumulation of air upon the sea being the necessary consequence of the long continuance of these cold winds from the shore, upon their ceasing the warm breezes from the sea necessarily commence, and, spreading themselves upon the land far and wide, assist the returning sun in dismantling the earth of the remains of her winter garment, and in bringing forward into life all the manifold beauties of the new-born year.

This warmed air which comes in from the sea, having acquired its Heat from a contact with the ocean, is, of course, saturated with water; and hence the warm showers of April and May, so necessary to a fruitful season.

The ocean may be considered as the great reservoir

and equalizer of Heat; and its benign influences in preserving a proper temperature in the atmosphere operate in all seasons and in all climates.

The parching winds from the land under the torrid zone are cooled by a contact with its waters, and, in return, the breezes from the sea, which at certain hours of the day come in to the shores in almost all hot countries, bring with them refreshment, and, as it were, new life and vigor both to the animal and vegetable creation, fainting and melting under the excessive Heats of a burning sun. What a vast tract of country, now the most fertile upon the face of the globe, would be absolutely barren and uninhabitable on account of the excessive Heat, were it not for these refreshing sea-breezes! And is it not more than probable, that the extremes of heat and of cold in the different seasons in the temperate and frigid zones would be quite intolerable, were it not for the influence of the ocean in preserving an equability of temperature?

And to these purposes the ocean is wonderfully well adapted, not only on account of the great power of water to absorb Heat, and the vast depth and extent of the different seas (which are such that one summer or one winter could hardly be supposed to have any sensible effect in heating or cooling this enormous mass); but also on account of the continual circulation which is carried on in the ocean itself by means of the currents which prevail in it. The waters under the torrid zone being carried by these currents towards the polar regions, are there cooled by a contact with the cold winds, and, having thus communicated their Heat to these inhospitable regions, return towards the equator, carrying with them refreshment for those parching climates.

The wisdom and goodness of Providence have often been called in question with regard to the distribution of land and water upon the surface of our globe, the vast extent of the ocean having been considered as a proof of the little regard that has been paid to man in this distribution. But the more light we acquire respecting the real constitution of things, and the various uses of the different parts of the visible creation, the less we shall be disposed to indulge ourselves in such frivolous criticisms.

# THE PROPAGATION OF HEAT
## IN FLUIDS.

# PART I.

## OF A REMARKABLE LAW

WHICH HAS BEEN FOUND TO OBTAIN IN THE CONDENSATION OF WATER WITH COLD, WHEN IT IS NEAR THE TEMPERATURE AT WHICH IT FREEZES; AND OF THE WONDERFUL EFFECTS WHICH ARE PRODUCED BY THE OPERATION OF THAT LAW IN THE ECONOMY OF NATURE.

TOGETHER WITH

## CONJECTURES

RESPECTING THE FINAL CAUSE OF THE SALTNESS OF THE SEA.

---

## CHAPTER I.

*Danger of admitting received Opinions in Philosophical Investigations without Examination. — The free Passage of* HEAT, *in all Bodies, in all Directions, never yet called in Question. — Heat does not, however, pass in this Manner in all Bodies without Exception. —* AIR *and* WATER, *and probably all other* FLUIDS, *are, in fact,* NON-CONDUCTORS OF HEAT. — *Accidental Discoveries which led to an experimental Investigation of this curious Subject. — The internal Motions among the Particles of Fluids rendered visible. — The Propagation of Heat in Fluids obstructed and retarded by everything which obstructs the internal Motions of their Particles; hence there is Reason to conclude that Heat is propagated in them only in Consequence of those Motions, — that it is transported by them, not suffered to pass through them. —* FURS *and* FEATHERS, *and all other like Substances, which, in Air, form warm Covering for confining Heat, found, by Experiment, to produce the same Effects in Water. — These Effects are probably produced in both Fluids in the same Manner, namely, by obstructing the Motions of their Particles in the Operation of transporting the Heat. — The conducting Power of Water remarkably impaired by mixing with it such Substances as render it viscous and diminish its Fluidity. — These Discoveries respecting the Manner in which Heat is propagated in Water throw much Light on several of the most interesting Operations in the Economy of Nature. — They enable us to account in a satisfactory Manner for the Preservation of Trees and other Vegetables, and of Fruits, during the Winter, in cold Climates.*

IT is certain that there is nothing more dangerous in philosophical investigations, than to take anything for granted, however unquestionable it may appear, till it has been proved by direct and decisive experiment.

I have very often, in the course of my philosophical researches, had occasion to lament the consequences of my inattention to this most necessary precaution.

There is not, perhaps, any phænomenon that more frequently falls under our observation than the Propagation of Heat. The changes of the temperature of sensible bodies, of solids, liquids, and elastic fluids, are going on perpetually under our eyes, and there is no fact which one would not as soon think of calling in question as to doubt of the free passage of Heat, in all directions, through all kinds of bodies. But, however obviously this conclusion appears to flow from all that we observe and experience in the common course of life, yet it is certainly not true; and to the erroneous opinion respecting this matter, which has been universally entertained by the *learned* and by the *unlearned*, and which has, I believe, never even been called in question, may be attributed the little progress that has been made in the investigation of the science of Heat, — a science, assuredly, of the utmost importance to mankind!

Under the influence of this opinion I, many years ago, began my experiments on Heat; and had not an accidental discovery drawn my attention with irresistible force and fixed it on the subject, I probably never should have entertained a doubt of the free passage of Heat *through air*; and even after I had found reason to conclude, from the results of experiments which to me appeared to be perfectly decisive, that air is a *non-conductor* of Heat, or that Heat cannot pass through it without being transported by its particles, which, in this process, act individually or independently of each other; yet, so far from pursuing the subject and contriving experiments to ascertain the manner in which Heat is communicated in other bodies, I was not sufficiently awakened to suspect it to be even possible that this quality could extend farther than to elastic Fluids.

With regard to liquids, so entirely persuaded was I that Heat could pass freely *in them* in all directions, that I was perfectly blinded by this prepossession, and rendered incapable of seeing the most striking and most evident proofs of the fallacy of this opinion.

I have already given an account, in one of my late publications (Essay on the Management of Fire and the Economy of Fuel),[6] of the manner in which I was led to discover that *steam* and *flame* are *non-conductors* of Heat. I shall now lay before the public an account of a number of experiments I have lately made, which seem to show that *water*, and probably all other liquids, and Fluids of every kind, possess the same property. That is to say, that, although the particles of any Fluid, *individually*, can receive Heat from other bodies or communicate it to them, yet among these particles themselves all *interchange* and *communication* of Heat is absolutely impossible.

It may, perhaps, be thought not altogether uninteresting to be acquainted with the various steps by which I was led to an experimental investigation of this curious subject of enquiry.

When dining, I had often observed that some particular dishes retained their Heat much longer than others, and that apple-pies, and apples and almonds mixed (a dish in great repute in England), remained hot a surprising length of time.

Much struck with this extraordinary quality of retaining Heat which apples appeared to possess, it frequently occurred to my recollection; and I never burnt my mouth with them, or saw others meet with the same misfortune, without endeavouring, but in vain, to find out some way of accounting in a satisfactory manner for this surprising phænomenon.

About four years ago, a similar accident awakened my attention, and excited my curiosity still more: being engaged in an experiment which I could not leave, in a room heated by an iron stove, my dinner, which consisted of a bowl of thick rice-soup, was brought into the room, and as I happened to be too much engaged at the time to eat it, in order that it might not grow cold, I ordered it to be set down on the top of the stove; about an hour afterwards, as near as I can remember, beginning to grow hungry, and seeing my dinner standing on the stove, I went up to it and took a spoonful of the soup, which I found almost cold and quite thick. Going, by accident, deeper with the spoon the second time, this second spoonful burnt my mouth.* This accident recalled very forcibly to my mind the recollection of the hot apples and almonds with which I had so often burned my mouth a dozen years before in England; but even this, though it surprised me very much, was not sufficient to open my eyes, and to remove my prejudices respecting the conducting power of water.

Being at Naples in the beginning of the year 1794, among the many natural curiosities which attracted my attention, I was much struck with several very interesting phænomena which the hot baths of Baiæ presented to my observation, and among them there was one which quite astonished me: standing on the sea-shore near the baths, where the hot steam was issuing out of every crevice of the rocks, and even rising up out of the ground, I had the curiosity to put my hand into the water. As the waves which came in from the sea followed each other without intermission, and broke over the even surface of the beach, I was not surprised to find the water cold; but I was more than surprised, when, on running the ends of my fingers through the cold water into the sand, I found the heat so intolerable that I was obliged instantly to remove my hand. The sand was perfectly wet, and yet the temperature was so very different at the small distance of two or three inches! I could not reconcile this with the supposed great conducting power of water. I even found that the top of the sand was, to all appearance, quite as cold as the water which flowed over it, and this increased my astonishment still more. I then, for the first time, began to doubt of the conduct-

* It is probable that the stove happened to be nearly cold when the bowl was set down upon it, and that the soup had grown almost cold; when a fresh quantity of fuel being put into the stove, the Heat had been suddenly increased.

ing power of water, and resolved to set about making experiments to ascertain the fact. I did not, however, put this resolution into execution till about a month ago, and should perhaps never have done it, had not another unexpected appearance again called my attention to it, and excited afresh all my curiosity.

In the course of a set of experiments on the communication of Heat, in which I had occasion to use thermometers of an uncommon size (their globular bulbs being above four inches in diameter) filled with various kinds of liquids, having exposed one of them, which was filled with spirits of wine, in as great a heat as it was capable of supporting, I placed it in a window, where the sun happened to be shining, to cool; when, casting my eye on its tube, which was quite naked (the divisions of its scale being marked in the glass with a diamond), I observed an appearance which surprised me, and at the same time interested me very much indeed. I saw the whole mass of the liquid in the tube in a most rapid motion, running swiftly in two opposite directions, *up* and *down* at the same time. The bulb of the thermometer, which is of copper, had been made two years before I found leisure to begin my experiments, and having been left unfilled, without being closed with a stopple, some fine particles of dust had found their way into it, and these particles, which were intimately mixed with the spirits of wine, on their being illuminated by the sun's beams, became perfectly visible (as the dust in the air of a darkened room is illuminated and rendered visible by the sunbeams which come in through a hole in the window-shutter), and by their motion discovered the violent motions by which the spirits of wine in the tube of the thermometer was agitated.

This tube, which is $\frac{43}{400}$ of an inch in diameter internally, and very thin, is composed of very transparent, colourless glass, which rendered the appearance clear and distinct and exceedingly beautiful. On examining the motion of the spirits of wine with a lens, I found that the ascending current occupied the *axis of the tube*, and that it descended by the *sides of the tube.*

On inclining the tube a little, the *rising* current moved out of the axis and occupied that side of the tube which was uppermost, while the *descending* current occupied the whole of the lower side of it.

When the cooling of the spirits of wine in the tube was hastened by wetting the tube with ice-cold water, the velocities of both the ascending and the descending currents were sensibly accelerated.

The velocity of these currents was gradually lessened as the thermometer was cooled, and when it had acquired nearly the temperature of the air of the room, the motion ceased entirely.

By wrapping up the bulb of the thermometer in furs, or any other warm covering, the motion might be greatly prolonged.

I repeated the experiment with a similar thermometer of equal dimensions, filled with linseed-oil, and the appearances, on setting it in the window to cool, were just the same. The directions of the currents, and the parts they occupied in the tube, were the same, and their motions were to all appearance quite as rapid as those in the thermometer which was filled with spirits of wine.

Having now no longer any doubt with respect to the cause of these appearances, being persuaded that the motion in these liquids was occasioned by their particles *going individually*, and *in succession*, to give off their Heat to the cold sides of the tube in the same manner as I have shown in another place that the particles of air give off *their* Heat to other bodies, I was led to conclude that these, and probably all other liquids, are in fact *non-conductors* of Heat, and I went to work immediately to contrive experiments to put the matter out of all doubt.

On considering the subject attentively, it appeared to me that if liquids were in fact *non-conductors* of Heat, or if it be propagated in them *only* in consequence of the internal motions of their particles, in that case everything which tends to obstruct those motions ought certainly to retard the operation, and render the propagation of the Heat slower and more difficult. I had found that this is actually the case in respect to air, and though (under the influence of a strong and deep-rooted prejudice) I had, from the result of one imperfect experiment, too hastily concluded that it did not take place in regard to water, yet I now found strong reasons to call in question the result of that experiment, and to give the subject a careful and thorough investigation.

Thinking that the best mode of proceeding in this enquiry would be to adopt a method similar to that I had pursued in my experiments on the conducting power of Air, I prepared an apparatus suitable to that purpose. The first object I had in view being to discover whether the propagation of Heat through water was obstructed or not, by rendering the internal motion among the particles of the water, occasioned by their change of temperature, embarrassed and difficult, I contrived to make a certain quantity of Heat pass through a certain quantity of pure water confined in a certain space ; and, noting the time employed in this operation, I repeated the experiment again with the same apparatus, with this difference only, that in this second trial the water through which the Heat was made to pass, instead of being pure, was mixed with a small quantity of some fine substance (such as eider-down, for instance), which, without altering any of its chemical properties, or impairing its fluidity, served merely to obstruct and embarrass the motions of the particles of the water in transporting the Heat, in case Heat should be actually *transported* or *carried* in this manner, and not suffered to pass freely through liquids.

The body which received the Heat, and which served at the same time to measure the quantity of it communicated, was a very large cylindrical thermometer. (See Plate I.) The bulb of this thermometer, which is constructed of thin sheet-copper, is cylindrical, its two ends being hemispheres.

Its dimensions are as follows: —

| | | |
|---|---|---|
| Dimensions of the bulb of the thermometer. | Diameter . . . . . . | 1.84 inches. |
| | Length . . . . . . . | 4.99 " |
| | Capacity or contents . . | 13.2099 cubic inches. |
| | External superficies . . | 28.834 superficial inches. |

The thickness of the sheet-copper of which it is constructed is 0.03 of an inch. It weighs, empty, 1846 grains, and is capable of containing 3344 grains of water at the temperature of 55°. This copper bulb has a glass tube, 24 inches long, and $\frac{4}{10}$ of an inch in diameter, which is fitted by means of a good cork into a cylindrical tube or neck of copper, one inch long, and $\frac{65}{100}$ of an inch in diameter, belonging to the metallic bulb.

This thermometer, being filled with linseed-oil and its scale graduated, was fixed in the axis of a hollow cylinder constructed of thin sheet-copper, 11½ inches long, and 2.3435 inches in diameter internally. This cylinder, which is open at one end, is closed at the other with a hemispherical bottom, with its convex surface outwards. The cylinder weighs 2261 grains, and the sheet-brass, of which it is constructed, is 0.0128 of an inch in thickness.

The bulb of the thermometer was placed in the lower part of this brass cylindrical tube, and was confined in the middle or axis of it by means of three pins of wood, about $\frac{1}{10}$ of an inch in diameter, and $\frac{1}{4}$ of an inch long, which pins are fixed in tubes of thin sheet-brass $\frac{1}{10}$ of an inch in diameter, and $\frac{3}{20}$ of an inch in length. These short tubes, which are placed at proper distances on the inside of the large brass tube at its lower end, and firmly attached to it by solder, serve as sockets into which the ends of the wooden pins are fixed, which, pointing inwards or towards the axis of the large cylindrical tube, serve to confine the lower end of the bulb of the thermometer in its proper place. Its upper end is kept in its place, or the axis of the thermometer is made to coincide with the axis of the brass cylinder, by causing the tube of the thermometer to pass through a hole in the middle of a cork stopper which closes the end of the cylinder.

The bottom of the bulb of the thermometer does not repose on the hemispherical bottom of the brass cylinder, but is supported at the distance of $\frac{1}{4}$ of an inch above it, on the end of a wooden pin, like those just described, which pin is fixed in a socket in the middle of the bottom of the cylindrical tube and projects upwards. The ends of all these wooden pins which project beyond the sockets in which they are fixed are reduced to a blunt point. This was done to reduce as much as possible the points of contact between the ends of these pins and the bulb of the thermometer.

The thermometer being in its place, there is on every side a void space left between the bulb of the thermometer and the internal surface of the brass cylinder in which it is confined, the distance between the external surface of the bulb of the thermometer and the internal surface of the containing cylinder being 0.25175 of an inch. This space is designed to contain the water and other substance through which the Heat is made to pass *into*, or *out of*, the bulb of the thermometer, and the quantity of Heat which has passed is shewn by the height of the fluid in the tube of the thermometer. The quantity of water required to fill this space and to cover the upper end of the bulb of the thermometer to the height of about $\frac{1}{4}$ of an inch was found to weigh 2468 grains. As the thermometer was plunged into this water, it was, of course, in contact with it by its whole surface, which, as we have seen, is equal to 28.834 square inches.

The bulb of the thermometer being surrounded by water, or by any other liquid or mixture, the conducting power of which was to be ascertained, a cylinder of cork something less in diameter than the brass cylinder, about half an inch long, with a hole in its center, in which the tube of the thermometer passed freely, was thrust down into the brass cylinder, but not quite so low as to touch the surface of the water or other substance it contained. This cylinder, or disk, was supported in its proper place by three projecting brass points or pins which were fixed with solder to the outside of the metallic neck of the bulb of the thermometer.

As soon as this disk of cork is put into its place, the upper part of the hollow brass cylindrical tube is filled with eider-down, and it is closed above with its cork stopper, the tube of the thermometer, which passes through a fit hole in the middle of this stopper, projecting upwards. As the whole scale of the thermometer, from the point of freezing to that of boiling water, is above the upper surface of this stopper, all the changes of Heat to which the instrument is exposed can be observed at all times without deranging any part of the apparatus.

The thermometer is divided according to the scale of Fahrenheit, and its divisions are made to correspond with a very accurate mercurial thermometer made by Troughton.

The experiments with this instrument, which, for the sake of distinction, I shall call my *cylindrical passage thermometer*, were made in the following manner: The thermometer being fixed in its cylindrical brass tube in the manner above described, and surrounded by the substance the conducting power of which was to be ascertained, the instrument was placed in thawing ice, where it was suffered to remain till the thermometer fell to 32°. It was then taken out of the melting ice and immediately plunged into a large vessel of boiling water, and the conducting power of the substance which was the subject of the experiment was estimated by the time employed by the Heat in passing through it into the thermometer; the time being carefully noted when the liquid in the thermometer arrived at the 40th degree of its scale, and also when it came to every 20th degree above it.

As the slower Heat moves, or is transported, in any medium, the longer must of course be the time required for any given quantity of it to pass through it; and as the thermometer shows the changes which take place in the temperature of the body which is heated or cooled (namely, the liquid with which the thermometer is filled), in consequence of the passage of the Heat through the medium by which the thermometer is surrounded, the conducting power of that medium is shewn by the quickness of the ascent or descent of the thermometer, when, having been previously brought to a certain temperature, the instrument is suddenly removed and plunged into another medium at any other constant given temperature.

Having still fresh in my memory the accidents I had so often met with in eating hot apple-pies, I was very impatient, when I had completed this instrument, to see if apples, which, as I well knew, are composed almost entirely of water, really possess a greater power of retaining Heat than that liquid when it is pure or unmixed with other bodies. But before I made the experiment, in order that its result might be the more satisfactory, I determined in the following manner how much water there really is in apples, and what proportion their fibrous parts bear to their whole volume.

960 grains of stewed apples (the apples having been carefully pared and freed from their stems and seeds before they were stewed) were well washed in a large quantity of cold spring water, and the fibrous parts of the apples being suffered to subside to the bottom of the vessel, the clear part of the liquor was poured off, and the fibrous remainder being thoroughly dried was carefully weighed, and was found to weigh just 25 grains.

This fibrous remainder of the 960 grains of stewed apples being again washed in a fresh quantity of cold spring water, and afterwards very thoroughly dried by being exposed several days on a china plate placed on the top of a German stove, which was kept constantly hot, was again weighed, and was found to weigh no more than $18\frac{9}{16}$ grains.

From this experiment it appears that the fibrous parts of stewed apples amount to less than $\frac{1}{50}$ part of the whole mass, and there is abundant reason to conclude that the remainder, amounting to $\frac{49}{50}$ of the whole, is little else than pure water.

Having surrounded the bulb of my cylindrical passage thermometer with a quantity of these stewed apples (the consistence of the mass being such that it shewed no signs of fluidity), the instrument was placed in pounded ice which was melting, and when the thermometer indicated that the whole was cooled down to the temperature of 32°, the instrument was taken out of the melting ice and plunged into a large vessel of boiling water, and the water being kept boiling with the utmost violence during the whole time the experiment lasted, the times taken up in heating the thermometer from 20 to 20 degrees were observed and noted down in a table which had been previously prepared for that purpose.

This experiment having been repeated twice, and varied as often by first heating the instrument to the temperature of boiling water and then plunging it into melting ice, and observing the time taken up in the passage of the Heat *out* of the thermometer, I removed the stewed apples which surrounded the bulb of the thermometer, and, filling the space they had occupied with *pure water*, I now repeated the experiments again with that liquid. The following tables shew the results of these experiments.

| | Time the Heat was passing INTO the Thermometer. | | | |
|---|---|---|---|---|
| | Through Stewed Apples. | | Through Water. | |
| | Exp. No. 1. | Exp. No. 3. | Exp. No. 5. | Exp. No. 7. |
| | Seconds. | Seconds. | Seconds. | Seconds. |
| In heating the Thermometer from the temperature of 32° to that of . . . . 40 | 95 | 89 | 45 | 45 |
| from 40° to 60 | 75 | 67 | 36 | 35 |
| 60 to 80 | 61 | 56 | 34 | 31 |
| 100 | 65 | 60 | 30 | 30 |
| 120 | 73 | 66 | 37 | 36 |
| 140 | 90 | 82 | 44 | 44 |
| 160 | 121 | 113 | 63 | 60 |
| 180 | 188 | 170 | 93 | 90 |
| 200 | 360 | 364 | 226 | 215 |
| Total times in heating from 32° to 200° . . . . . | 1128 | 1057 | 608 | 586 |
| Times employed in heating the instrument 80 degrees, viz. from 80° to 160° . . . | 349″ | 321″ | 174″ | 170″ |
| Mean times in heating it from 80° to 160° . . . . . | In Stewed Apples 335″ | | In Water 172″ | |

The results of these experiments shew that Heat passes with much greater difficulty, or much slower, in *stewed apples* than in *pure water;* and as stewed apples are little else than water mixed with a very small proportion of fibrous and mucilaginous matter, this shews that the conducting power of water with regard to Heat *may be impaired.*

The results of the following experiments will serve to confirm this conclusion.

| | Time the Heat was passing OUT OF the Thermometer. | | | |
|---|---|---|---|---|
| | Through Stewed Apples. | | Through Water. | |
| | Exp. No. 2. | Exp. No. 4. | Exp. No. 6. | Exp. No. 8. |
| | Seconds. | Seconds. | Seconds. | Seconds. |
| In cooling the Thermometer from the temperature of 200° to that of . . . . 180 | 80 | 74 | 46 | 37 |
| from 180° to 160 | 75 | 72 | 42 | 37 |
| 160 to 140 | 84 | 83 | 43 | 43 |
| 120 | 107 | 101 | 54 | 51 |
| 100 | 141 | 136 | 73 | 73 |
| 80 | 198 | 190 | 112 | 105 |
| 60 | 321 | 307 | 200 | 204 |
| 40 | 775 | 733 | 483 | 461 |
| Total time in cooling from 200° to 40° . . . . . | 1781 | 1696 | 1053 | 1011 |
| Times employed in cooling the instrument 80 degrees, viz. from 160° to 80° . . . | 530″ | 510″ | 282″ | 272″ |
| Mean time in cooling it from 160° to 80° . . . . . | In Stewed Apples 520″ | | In Water 277″ | |

As the heating or cooling of the instrument goes on very slowly when it approaches to the temperature of the medium in which it is placed, while, on the other hand, this process is very rapid when, the temperature of the instrument being very different from that of the medium, it is first plunged into it, both these circumstances conspire to render the observations made at the extremities of the scale of the thermometer more subject to error, and consequently less satisfactory, than those made nearer the middle of it. In order that the general conclusions drawn from the result of the experiments might not be vitiated by the effects produced by these unavoidable inaccuracies, instead of estimating the celerity of the passage of the Heat by the times elapsed in heating and cooling the thermometer *through the whole length of its scale*, or between the point of freezing to that of boiling water, I have taken the times elapsed in heating and cooling it 80 *degrees in the middle of the scale*, viz. between 80° and 160°, as the measure of the conducting powers of the substances through which the Heat was made to pass.

I have, however, noted the times which elapsed in heating and cooling the instrument through a much larger interval, namely, through an interval of 168 degrees in *heating*, or from 32° to 200°, and in *cooling* through 160 degrees, or from 200° to 40°.

In respect to the *cooling* of the instrument, it is necessary that I should inform my reader, that, though I have not in the tables of the experiments mentioned any higher temperature than that of 200°, yet the instrument was always heated to the point of boiling water, which, under the pressure of the atmosphere at Munich, where the experiments were made, was commonly about 209½ deg. of Fahrenheit's scale. The instrument, being kept in boiling water till its thermometer appeared to be quite stationary, was then taken out of the water and instantly plunged into melting ice, and the time was observed and carefully noted down when the liquid in its thermometer passed the division of its scale which indicated 200°, as also when it arrived at the other divisions indicated in the tables.

With regard to the four last-mentioned experiments (No. 2, 4, 6, and 8), it will be found, on examination, that their results correspond very exactly with those before described; and they certainly prove in a very decisive manner this important fact, — *that a small proportion of certain substances, on being mixed with water, tend very powerfully to impair the conducting power of that Fluid in regard to Heat.*

In the experiments No. 1 and No. 2, which were both made on the same day, and in the order in which they are numbered, the Heat was considerably more obstructed in its passage through the mass of stewed apples which surrounded the thermometer than in the experiments No. 3 and No. 4, which were made on the following day. It is probable that this was occasioned by some change in the consistency of this soft mass of the stewed apples which had taken place while the instrument was left to repose in the interval between the experiments; but instead of stopping to show how this might be explained, I shall proceed to give an account of some experiments from the results of which we shall derive information that will be much more satisfactory than any speculations I could offer on that subject.

Supposing Heat to be propagated in water in the same manner as it is propagated in air and other elastic Fluids, namely, that it is *transported* by its particles, these particles being put in motion by the change which is produced in their specific gravity by the change of temperature, and that there is no communication whatever, or *interchange of their Heat*, among the particles of *the same Fluid;* in that case it is evident that the propagation of Heat in a Fluid may be obstructed in two ways, namely, by diminishing its *Fluidity* (which may be done by dissolving in it any mucilaginous substance); or, more simply, by merely embarrassing and obstructing the motion of its particles in the operation of transporting the Heat, which may be effected by mixing with the Fluid any solid substance (it must be a non-conductor of Heat, however) in small masses, or which has a very large surface in proportion to its solidity.

In the foregoing experiments with *stewed apples*, the passage of the Heat in the water (which constituted by far the greatest part of the mass) was doubtless obstructed in both these ways. The mucilaginous parts of the apples diminished very much the fluidity of the water, at the same time that the fibrous parts served to embarrass its internal motions.

In order to discover the *comparative effects* of these two causes, it was necessary to separate them, or to contrive experiments in which only one of them should be permitted to act at the same time. This I endeavoured to do in the following manner.

To ascertain the effects produced by diminishing the *fluidity* of water, I mixed with it a small quantity of starch, namely, 192 grains in weight to 2276 grains of water; and to determine the effects produced by merely *embarrassing* the water in its motions, I mixed with it an equal proportion (by weight) of *eider-down.* The starch was boiled with the water with which it was mixed, as was also the eider-down. This last-mentioned substance was boiled in the water in order to free it from air, which, as is well known, adheres to it with great obstinacy.

In order that these experiments may with greater facility be compared with those which were made with *stewed apples* and with *pure water*, I shall place their results all together in the following tables.

| | Time the Heat was in passing INTO the Thermometer. | | | |
|---|---|---|---|---|
| | Through a Mixture of 2276 Grains of Water and 192 Grains of STARCH. | Through a Mixture of 2276 Grains of Water and 192 Grains of EIDER-DOWN. | Through STEWED APPLES. | Through Pure WATER. |
| | Experiment No. 9. | Experiment No. 11. | Mean of Two Experiments, No. 1 and No. 3. | Mean of Two Experiments, No. 5 and No. 7. |
| | Seconds. | Seconds. | Seconds. | Seconds. |
| In heating the Thermometer from 32° to 40° | 101 | 83 | 92 | 45 |
| from 40 to 60 | 72 | 55 | 71 | 35½ |
| 60 to 80 | 64 | 49 | 58½ | 32½ |
| 100 | 63 | 52 | 62½ | 30 |
| 120 | 74 | 57 | 69½ | 36½ |
| 140 | 89 | 67 | 86 | 44 |
| 160 | 115 | 93 | 117 | 61½ |
| 180 | 178 | 133 | 179 | 91½ |
| 200 | 453 | 360 | 362 | 220½ |
| Total times in heating the instrument from 32° to 200° | 1109 | 949 | 1096½ | 597 |
| Times employed in heating the Thermometer 80 degrees, viz. from 80° to 160° | 341″ | 269″ | 335″ | 172″ |

| | Time the Heat was in passing OUT OF the Thermometer. | | | |
|---|---|---|---|---|
| | Through a Mixture of 2276 Grains of Water and 192 Grains of STARCH. | Through a Mixture of 2276 Grains of Water and 192 Grains of EIDER-DOWN. | Through STEWED APPLES. | Through Pure WATER. |
| | Experiment No. 10. | Experiment No. 12. | Mean of Two Experiments, No. 2 and No. 4. | Mean of Two Experiments, No. 6 and No 8. |
| | Seconds. | Seconds. | Seconds. | Seconds. |
| In cooling the Thermometer from 200° to 180° . . . | 69 | 68 | 77 | 41½ |
| from 180° to 160° | 66 | 61 | 73½ | 39½ |
| 160 to 140 | 74 | 72 | 83½ | 43 |
| 120 | 92 | 91 | 104 | 52½ |
| 100 | 119 | 120 | 138½ | 73 |
| 80 | 173 | 177 | 194 | 108½ |
| 60 | 283 | 279 | 314 | 202 |
| 40 | 672 | 673 | 754 | 472 |
| Total times in cooling from 200° to 40° | 1548 | 1541 | 1749½ | 1032 |
| Times employed in cooling the instrument 80 degrees, viz. from 160° to 80° | 468″ | 460″ | 520″ | 277″ |

As the results of these experiments prove, in the most decisive manner, that the propagation of Heat in water is retarded, not only by those things which diminish its fluidity, but also by those which, by mechanical means, and without forming any combination with it whatever, merely obstruct its internal motions, it appears to me that this proves, almost to a demonstration, that Heat is propagated in water *in consequence* of its internal motions, or that it is transported or *carried* by the particles of that liquid, and that it does not spread and expand in it, as has generally been imagined.

I have shewn in another place, and I believe I may venture to say I have proved,[7] that Heat is actually propagated in *air* in the same manner I here suppose it to be propagated in water, and if the conducting powers of both these fluids are found to be impaired by the *same means*, it affords very strong grounds to conclude that they both conduct Heat in the *same manner;* but this has been found to be actually the case.

*Eider-down*, which cannot affect the specific qualities of either of those fluids, and which certainly does no more when mixed with them than merely to obstruct and embarrass their internal motions, has been found to retard very much the propagation of Heat in both of them: on comparing these experiments with those I formerly made on the conducting power of air, it will even be found that the conducting power of water is nearly, if not quite, as much impaired by a mixture of *eider-down* as that of air.

In the course of my experiments on the various substances used in forming artificial clothing for confining Heat, I found that the thickness of a stratum of air, which served as a barrier to Heat, remaining the same, the passage of Heat through it was sometimes rendered more difficult by increasing the quantity of the light substance which was mixed with it to obstruct its internal motion.

To see if similar effects would be produced by the same means when Heat is made to pass through water, I repeated the experiments with *eider-down*, reducing the quantity of it mixed with the water to 48 grains, or *one quarter* of the quantity used in the experiments No. 11 and No. 12.

The results of these experiments, and a comparison of them with those before mentioned, may be seen in the following tables: —

| | Time the Heat was in passing INTO the Thermometer. | | |
|---|---|---|---|
| | Through Water with 48 Grains, or $\frac{1}{50}$ of its Bulk of EIDER-DOWN. | Through Water with 192 Grains, or $\frac{4}{50}$ of its Bulk of EIDER-DOWN. | Through Pure WATER. |
| | Experiment No. 13. | Experiment No. 11. | Mean of Two Experiments, No. 5 and No. 7. |
| | Seconds. | Seconds. | Seconds. |
| In heating the Thermometer from . . . 32° to 40° | 51 | 83 | 45 |
| from 40 to 60 | 47 | 55 | $35\frac{1}{2}$ |
| 60 to 80 | 39 | 49 | $32\frac{1}{2}$ |
| 100 | 40 | 52 | 30 |
| 120 | 45 | 57 | $36\frac{1}{2}$ |
| 140 | 56 | 67 | 44 |
| 160 | 74 | 93 | $61\frac{1}{2}$ |
| 180 | 118 | 133 | $91\frac{1}{2}$ |
| 200 | 293 | 360 | $220\frac{1}{2}$ |
| Total times in heating from 32° to 200° . . . . | 763 | 949 | 597 |
| Times employed in heating the instrument 80 degrees, or from 80° to 160° . . | 215″ | 269″ | 172″ |

| | Time the Heat was passing OUT OF the Thermometer. | | |
|---|---|---|---|
| | Through Water with 48 Grains, or $\frac{1}{50}$ of its Bulk of EIDER-DOWN. | Through Water with 192 Grains, or $\frac{4}{50}$ of its Bulk of EIDER-DOWN. | Through Pure WATER. |
| | Experiment No. 14. | Experiment No. 12. | Mean of Two Experiments, No. 6 and No. 8. |
| | Seconds. | Seconds. | Seconds. |
| In cooling the Thermometer from . . . 200° to 180° | 49 | 68 | 41½ |
| from 180 to 160 | 50 | 61 | 39½ |
| 160 to 140 | 56 | 72 | 43 |
| 120 | 70 | 91 | 52½ |
| 100 | 96 | 120 | 73 |
| 80 | 151 | 177 | 108½ |
| 60 | 262 | 279 | 202 |
| 40 | 661 | 673 | 472 |
| Total times in cooling from 200° to 40° . . . . | 1395 | 1541 | 1032 |
| Times employed in cooling the instrument 80 degrees, viz. from 160° to 80° . | 373″ | 460″ | 277″ |

The results of these experiments are extremely interesting. They not only make us acquainted with a new and very curious fact, namely, that feathers and other like substances, which, in air, are known to form very warm covering for confining Heat, not only serve the same purpose in water, but that their effect in preventing the passage of Heat is even greater in water than in air.

This discovery, if I do not deceive myself, throws a very broad light over some of the most interesting parts of the economy of Nature, and gives us much satisfactory information respecting the final causes of many phænomena which have hitherto been little understood.

As *liquid water* is the vehicle of Heat and nourishment, and consequently of life, in every living thing; and as water, left to itself, freezes, with a degree of cold much less than that which frequently prevails in cold climates, it is agreeable to the ideas we have of the wisdom of the Creator of the world to expect that effectual measures would be taken to preserve a sufficient quantity of that liquid in its fluid state to maintain life during the cold season: and this we find has actually been done; for both plants and animals are found to survive the longest and most severe winters; but the means which have been employed to produce this miraculous effect have not been investigated, — at least not in as far as they relate to vegetables.

But as animal and vegetable bodies are essentially different in many respects, it is very natural to suppose that the means would be different which are employed to preserve them against the fatal effects which would be produced in each by the congelation of their fluids.

Among organized bodies which live on the surface of the earth, and which, of course, are exposed to the vicissitudes of the seasons, we find that as the proportion of fluids to solids is greater, the greater is the Heat which is required for the support of life and health, and the less are they able to endure any considerable change of their temperature.

The proportion of Fluids to Solids is much greater in *animals* than in vegetables; and in order to preserve in them the great quantity of Heat which is necessary to the preservation of life, they are furnished with lungs, and are warmed by a process similar to that by which Heat is generated in the combustion of inflammable bodies.

Among *vegetables*, those which are the most succulent are *annual*. Not being furnished with lungs to keep the great mass of liquids warm, which fill their large and slender vessels, they live only while the genial influence of the sun warms them and animates their feeble powers, and they droop and die as soon as they are deprived of his support.

There are many tender plants to be found in cold countries, which die in the autumn, the roots of which remain alive during the winter, and send off fresh shoots in the ensuing spring. In these we shall constantly find the roots more compact and dense than the stalk, or with smaller vessels and a smaller proportion of Fluids.

Among the trees of the forest we shall constantly find that those which contain a great proportion of *thin watery liquids* not only shed their leaves every autumn, but are sometimes frozen, and actually killed, in severe frosts. Many thousands of the largest walnut-trees were killed by the frost in the Palatinate during the very cold winter in the year 1788; and it is well known that few, if any, of the deciduous plants of our temperate climate would be able to support the excessive cold of the frigid zone.

The trees which grow in those inhospitable climates, and which brave the colds of the severest winters, contain very little watery liquids. The sap which circulates in their vessels is thick and viscous, and can hardly be said to be fluid. Is there not the strongest reason to think that this was so contrived for the express purpose of preventing their being deprived of all their Heat, and killed by the cold during the winter?

We have seen by the foregoing experiments how much the propagation of Heat in a liquid is retarded by diminishing its fluidity; and who knows but this may continue to be the case as long as any degree of fluidity remains?

As the bodies and branches of trees are not covered in winter by the snow which protects their roots from the cold atmosphere, it is evident that extraordinary measures were necessary to prevent their being frozen. The bark of all such trees as are designed by nature to support great degrees of cold forms a very warm covering; but this precaution alone would certainly not have been sufficient for their protection. The sap in all trees which are capable of supporting a long continuance of frost grows thick and viscous on the approach of winter. What more important purpose could this change answer than that here indicated? And it would be more than folly to pretend that it answers no useful purpose at all.

We have seen by the results of the foregoing experiments how much the simple embarrassment of liquids in their internal motions tends to retard the propagation of Heat in them, and consequently its passage out of them; — and when we consider the extreme smallness of the vessels in which the sap moves in vegetables and particularly in large trees; when we recollect that the substance of which these small tubes are formed is one of the best non-conductors of Heat known; * and when we ad-

* I lately, by accident, had occasion to observe a very striking proof of the extreme difficulty with which Heat passes in wood. Being present at the foundery at Munich when cannons were casting, I observed that the founder used a wooden instrument for stirring the melted metal. It was a piece of oak plank, green or unseasoned, about ten inches square and two inches thick, with a long wooden handle which was fitted into a hole in the middle of it. As this instrument was frequently used, and sometimes remained a considerable time in the furnace, in which the Heat was most intense, I was surprised to find that it was not consumed; but I was still more surprised, on examining the part of the plank which had been immersed in the melted metal, to find that the Heat had penetrated it to so inconsiderable a depth, that, at the distance of one twentieth of an inch below its surface, the wood did not seem to have been in the least affected by it. The colour of the wood remained unchanged, and it did not appear to have lost even its moisture.

vert to the additional embarrassments to the passage of the Heat which arise from the increased viscosity of the sap in winter, and to the almost impenetrable covering for confining Heat which is formed by the bark, we shall no longer be at a loss to account for the preservation of trees during the winter, notwithstanding the long continuation of the hard frosts to which they are annually exposed.

On the same principles we may, I think, account in a satisfactory manner for the preservation of several kinds of fruit — such as apples and pears, for instance — which are known to support, without freezing, a degree of cold which would soon reduce an equal volume of *pure water* to a solid mass of ice.

At the same time that the compact skin of the fruit effectually prevents the evaporation of its fluid parts, which, as is well known, could not take place without occasioning a very great loss of Heat, the internal motions of those fluids are so much obstructed by the thin partitions of the innumerable small cells in which they are confined, that the communication of their Heat to the air ought, according to our hypothesis, to be extremely slow and difficult. These fruits do, however, freeze at last, when the cold is very intense; but it must be remembered that they are composed almost entirely of liquids, and of such liquids as do not grow viscous with cold, and, moreover, that they were evidently not designed to support for a long time very severe frosts.

Parsnips and carrots, and several other kinds of roots, support cold without freezing still longer than apples and pears, but these are less watery, and I believe the vessels in which their fluids are contained are smaller; and both these circumstances ought, according to our assumed principles, to render the passage of their Heat out of them more difficult, and consequently to retard their congelation.

But there is still another circumstance, and a very remarkable one indeed, which, if our conjectures respecting the manner in which Heat is propagated in liquids be true, must act a most important part in the preservation of Heat, and consequently of animal and vegetable life, in cold climates. But as the probability of all these deductions must depend very much on the evidence which is brought to prove the great fundamental fact on which they are established, — that respecting the internal motions among the particles of liquids which *necessarily* take place when they are heated or cooled, — before I proceed any farther in these speculations, I shall endeavour to throw some more light on that curious and interesting subject.

---

## CHAPTER II.

*Farther Investigations of the internal Motions among the Particles of Liquids which necessarily take place when they are heated or cooled. — Description of a mechanical Contrivance, by which these Motions in Water were rendered visible. — An Account of various amusing Experiments which were made with this new-invented Instrument. — They lead to an important Discovery. — Heat cannot be propagated* DOWNWARDS *in Liquids, as long as they continue to be condensed by Cold. — Ice found, by Experiment, to melt more than eighty times slower when boiling-hot Water stood on its Surface, than when the Ice was suffered to swim on the Surface of the hot Water. — The melting of Ice by Water standing on its Surface can be accounted for, even on the supposition that Water is a perfect Non-conductor of Heat. — According to the assumed Hypothesis, Water only eight Degrees of Fahrenheit's Scale above the freezing Point, or at the Temperature of* 40°, *ought to melt as much ice, in any given Time, when standing on its Surface, as an equal Volume of that Fluid at any higher Temperature, even were it boiling hot. — This remarkable Fact is proved by a great Variety of decisive Experiments. — Water at the Temperature of* 41° *is found to melt even* MORE *Ice, when standing on its Surface, than boiling-hot Water. — The Results of all these Experiments tend to prove that Water is, in fact, a perfect Non-conductor of Heat; or that Heat is propagated in it merely in consequence of the Motions it occasions among the insulated or solitary Particles of that Fluid, which, among themselves, have no Communication or Intercourse whatever in this Operation. — The Discovery of this Fact opens to our View one of the grandest and most interesting Scenes in the Economy of Nature.*

AS the particles of water, as also of all other Fluids, are infinitely too small to be seen by human eyes, their motions must of course be imperceptible by us; but we are frequently enabled to judge with the utmost certainty of the motions of invisible Fluids by the motions they occasion in visible bodies. Air is an invisible Fluid, but we acquire very just notions of the motions in air by the dust and other light bodies which are carried along with it in its motions. Nobody who has ever seen a whirlwind sweep over the surface of a ploughed field in dry weather can have any doubt respecting the nature of the motions into which the air is thrown on those occasions, notwithstanding that they are extremely complicated, and would be very difficult to describe.

It was by the motions of the very fine particles of dust, which by accident had been mixed with the spirits of wine in my large thermometer, and which, when strongly illuminated by the direct beams of the sun, became visible, that I first discovered the internal motions in that Fluid which take place when it is cooling; and, availing myself of this kind hint, I contrived to render the internal motions of water equally visible. This I immediately saw could be done with the utmost facility if I could but find any solid body of the same specific gravity as water, which would be proper to mix with it, — that is to say, that would not be liable to be dissolved by it, or to be reduced to such small particles as to become itself invisible; but such a substance was not to be found. On reflection it occurred to me that it is very fortunate that such substances do not abound, for otherwise we should find great difficulty in procuring water in a pure state.

Not being able to find any solid substance fit for my purpose, of the same specific gravity as pure water, I was obliged to have recourse to the following stratagem.

Looking over the tables of specific gravities, I found that the specific gravity of transparent yellow amber was but a little greater than that of water, being 1.078, while that of water is 1.000; and it occurred to me, that, by dissolving a certain quantity of pure alkaline salt, I might augment its specific gravity, or rather bring the specific gravity of the solution to be precisely equal to that of the amber, without impairing the transparency of the liquid, or changing any of its properties, by which the manner of its receiving and transporting Heat could be sensibly affected.

This contrivance was put in execution in the following manner, with complete success. Having provided myself with a number of glass globes of various sizes, with long cylindrical necks, I chose one which was about two inches in diameter, with a cylindrical neck $\frac{3}{4}$ of an inch in diameter and twelve inches long; and putting into it about half a teaspoonful of yellow amber, in the form of a coarse powder (the pieces, which were irregular in their forms, and transparent, being about the size of mustard-seeds), I poured upon it a certain quantity of distilled water, which was at the temperature of the air in my room (about 60° F.).

Finding, as I expected, that the amber remained at the bottom of the globe, I now added to the water as much of a saturated solution of pure vegetable alkali as was sufficient to increase the specific gravity of the water (or rather of the diluted saline solution), till the pieces of amber began to float, and remained apparently motionless in any part of the liquid where they happened to rest.

As the glass body was not yet as full as I wished, I continued to add more of the alkaline solution and of water, in due proportions, till the globe was full, and also till its cylindrical tube was filled to within about three inches of its end, and then closed it well with a clean cork.

Having shaken the contents of this glass body well together, I placed it, with its cylindrical tube in a vertical position, on a wooden stand, and left it to repose in quiet, in order to see how long the solid particles of amber (which appeared to be very equally dispersed about in the whole mass of the liquid) would remain suspended.

Though the greater number of these particles seemed at first to have no tendency either to ascend or to descend, yet some of them soon began to move very slowly upwards, and others to move as slowly downwards; and as these particles were moving at the same time promiscuously in all parts of the same liquid, and even in the same part of it in both directions at the same time, the ascending and descending particles frequently passing each other so near as to touch, I saw that these motions were independent of any internal motion of the liquid, and arose merely from the difference of the specific gravity of the different small pieces of the amber and of that of the liquid. Some of the pieces of amber, being evidently heavier than the liquid, moved downward, while others which were lighter ascended to its surface.

Finding that there was so much difference in the specific gravities of the different pieces of amber, I now added more of this substance to the liquid, and suffering it to subside after I had shaken it well together, I gently poured off what had risen to the top of the liquid, and retaining only that which had settled at the bottom of it, I increased the specific gravity of the liquid by adding a little of the alkaline solution till the small pieces of amber which remained in the glass were just buoyed up and suspended in the different parts of the Fluid, where they seemed to have taken their permanent stations.

I had now an instrument which appeared to me to be well calculated for the very interesting experiments I had projected, and it will easily be imagined that I lost no time in making use of it.

The first experiment I made with this instrument was to plunge it into a tall glass jar nearly filled with water almost boiling hot. The result was just what I expected. Two currents, in opposite directions, began at the same instant to move with great celerity in the liquid in the cylindrical tube, the ascending current occupying the sides of the tube, while that which moved downwards occupied its axis.

As the saline liquor grew warm, the velocity of these currents gradually diminished, and at length, when the liquor had acquired the temperature of the surrounding water in the jar, these motions ceased entirely.

On taking the glass body out of the hot water, the internal motions of the liquor recommenced, but the currents had changed their directions, that which occupied the axis of the tube being now the ascending current.

When the cylindrical tube, instead of being held in a vertical position, was inclined a little, the ascending current occupied that side of it which happened to be uppermost, while the under side of it was occupied by the current which moved (with equal velocity) downwards.

When the contents of the glass body had acquired the temperature of the air of the room, these motions ceased; but they immediately recommenced on exposing the instrument to any change of temperature.

In all cases where the instrument *received Heat*, the current in the axis of its cylindrical tube when it was placed in a vertical position (and that which occupied its *upper side* when it was inclined) moved downwards. When it parted with Heat, its motion was in an opposite direction, that is to say, *upwards*.

A change of temperature amounting only to a few degrees of Fahrenheit's scale was sufficient to set the contents of the instrument in motion; and the motion was more or less rapid as the velocity was greater or less with which it acquired or parted with Heat, and the motion was most rapid in those parts of the instrument where the communication was not rapid.

A partial motion might at any time be produced in any part of the instrument by applying to that part of it any body either hotter or colder than the instrument. If the body so applied were hotter than the instrument, the motion of the saline liquor in it in that part of it immediately in contact with the hot body was *upwards* ; if colder, downwards ; and whenever a hot or cold body produced a current upwards or downwards, this current immediately produced another in some other part of the liquid which flowed in an opposite direction.

On inclining the cylindrical tube of the instrument to an angle of about 45 degrees with the plane of the horizon, and holding the middle of it over the flame of a candle at the distance of three or four inches above the point of the flame, the motion of the Fluid in the upper part of the tube became excessively rapid, while that in the lower end of it where it was united to the globe, as well as that in the globe itself, remained almost perfectly at rest.

I even found that I could make the Fluid in the upper part of the tube *actually boil*, without that in the lower part of it appearing to the hand to be sensibly warmed. But when the flame was directed against the lower part of the tube, all the upper parts of it in contact with the liquid, and especially that side of it which was uppermost as it lay in an inclined position, where the ascending current was most rapid, where it impinged against the glass, were very soon heated very hot.

The motions in opposite directions in the liquid in the tube were exceedingly rapid on this sudden application of a strong Heat, and afforded a very entertaining sight; but to a scientific observer they were much more than amusing. They detected Nature, as it were, in the very act, in one of her most hidden operations, and rendered motions visible in the midst of an invisible medium which never had been seen before, and which most probably had never been suspected.

Encouraged by this success, and confirmed in my opinions respecting the interesting fact I had undertaken to investigate, I now proceeded with confidence to still more direct and decisive experiments.

It is an opinion which, I believe, is generally received among philosophers, that water cannot be heated in contact with ice: reflecting on the subject, I immediately perceived that either this must be a mistake or all my ideas respecting the manner in which Heat is propagated in that Fluid must be erroneous. I saw that as long as the ice floats at the surface of water which is attempted to be warmed over a fire, (or in any other way,) the ice-cold water which results from the melting of the ice must, according to my own hypothesis, descend, and, spreading over the bottom of the containing vessel, and, before it has time to be much heated, being in its turn forced to give place to the ice-cold water which, as long as any ice remains, continues to descend in an uninterrupted stream as long as this operation is going on, the mass of the water cannot be much heated; but on the supposition that water is not a conductor of Heat, according to the common acceptation of that term, or that Heat cannot pass in that Fluid except when it is *carried* by its particles, which, being put in motion by the change it occasions in their specific gravity, *transport* it from place to place, it does not appear how ice, if, instead of being permitted to swim on water, it were confined at the bottom of it or at any given distance below its surface, could in any way affect the temperature of the superincumbent water, or prevent its receiving Heat from other bodies.

Were water a conductor of Heat, there is no doubt but that the influence of the presence of the ice would be propagated in the water in all directions.

The metals are all conductors of Heat, and Professor Pictet found, by an ingenious and decisive experiment,* that in a bar of copper 33 inches in length, placed in a vertical position, Heat passed downwards as well as upwards, and nearly with the same facility in both these directions; and if it can be shown that Heat cannot descend in water, that alone will, I imagine, be thought quite sufficient to prove that water is not a conductor of Heat.

When we meditate profoundly on the nature of Fluidity, it seems to me that we can perceive some faint lights which might lead us to suspect that the *cause*, and I may say the very *essence*, *of fluidity*, is that property which the particles of bodies acquire when they become fluid, by which all farther interchange or communication of Heat among them is prevented. But however this may be, the result of the following experiments will certainly be considered as affording indisputable evidence of one important fact respecting the manner in which Heat is propagated in water.

*Experiment No.* 15.

Into a cylindrical glass jar 4.7 inches in diameter and 14 inches high, I fitted a circular cake of ice nearly as

* Essais de Physique, Tome I. Genève. 1790.

large as the internal diameter of the jar, and $3\frac{1}{2}$ inches thick, weighing $10\frac{1}{8}$ oz.

This cake of ice being ready, I now poured into the jar 6 lb. $1\frac{1}{4}$ oz., Troy, of boiling-hot water, and, putting the ice gently into it, I found that it was entirely melted in 2 minutes and 58 seconds.

Having found by this experiment how long the ice was in melting at the surface of the hot water, I now endeavoured to find out whether it would not require a longer time to melt at the bottom of the water.

### *Experiment No.* 16.

Into the same jar which was used in the foregoing experiment, I now put a cake of ice of the same form and dimensions as that above described, but, instead of letting it swim at the surface of the hot water, I fastened it down on the bottom of the jar and poured the water upon it.

This cake of ice was fastened down in the jar by means of two slender and elastic pieces of deal about $\frac{1}{8}$ of an inch thick, and $\frac{1}{4}$ of an inch wide, which, being a trifle longer than the internal diameter of the jar, were of course a little bent when they were introduced into it in an horizontal position, and, on being put down upon the ice at right angles to each other served to confine the ice, and prevent its rising up to the surface when the water was put into the jar upon it.

To protect the ice while the boiling-hot water was pouring into the jar, its surface was covered with a circular piece of strong writing-paper, which was afterwards removed as gently as possible by means of a string which was fastened to one side of it; and to prevent the glass jar from being cracked by the sudden application of the boiling-hot water, I began by pouring a small quantity of cold water into the jar, just enough to fill up the interstices between the ice and the glass, and to cover the ice to the height of about $\frac{1}{4}$ of an inch ; and in pouring the hot water into the jar, out of a large tea-kettle in which it had been boiled, I took care to direct the stream against the middle of the circular piece of paper which covered the ice.

The jar with the ice and the hot water in it being placed on a table near a window, I drew away as gently as possible the paper which covered the surface of the ice, and prepared myself to observe at my ease the result of this most interesting experiment.

A very few moments were sufficient to show me that my expectation with regard to it would not be disappointed. In the former experiment a similar cake of ice had been entirely melted in less than three minutes ; but in this, after more than twice that time had elapsed, the ice did not show any apparent signs of even *beginning to melt*. Its surface remained smooth and shining, and the water immediately in contact with it appeared to be perfectly at rest, though the internal motions of the hot water above it, which was giving off its heat to the sides of the jar and to the air, were very rapid, as I could distinctly perceive by means of some earthy particles or other impurities which this water happened to contain.

I examined the ice with a very good lens, but it was a long time before I could perceive any signs of its melting. The edges of the cake remained sharp, and the minute particles of dust, which by degrees were precipitated by the hot water as it grew colder, remained motionless as soon as they touched the surface of the ice.

As the hot water had been brought from the kitchen

in a tea-kettle, it was not quite boiling hot when it was poured into the jar. After it had been in the jar one minute I plunged a thermometer into it, and found its temperature to be at 180°.

After 12 minutes had elapsed, its temperature at the depth of one inch under the surface was 170°. At the depth of seven inches, or one inch above the surface of the ice, it was at $169\frac{1}{20}$, while at only $\frac{3}{4}$ of an inch lower, or $\frac{1}{4}$ above the surface of the ice, its temperature was 40°.

When 20 minutes had elapsed, the Heat in the water at different depths was found to be as follows : —

| | |
|---|---|
| Immediately above the surface of the ice | 40° |
| At the distance of ½ an inch above it | 46 |
| At 1 inch | 130 |
| At 3 inches | 159 |
| At 7 inches | 160 |

When 35 minutes had elapsed, the Heat was as follows : —

| | |
|---|---|
| At the surface of the ice | 40° |
| ½ an inch above it | 76 |
| 1 inch above it | 110 |
| 2 inches | 144 |
| 3 inches | 148 |
| 5 inches | 148½ |
| 7 inches | 149 |

At the end of one hour the Heat was as follows : —

| | |
|---|---|
| At the surface of the ice | 40° |
| 1 inch above it | 80 |
| 2 inches | 118 |
| 3 inches | 128 |
| 4 inches | 130 |
| 7 inches | 131 |

After 1 hour and 15 minutes had elapsed, the Heat was found to be as follows: —

| | |
|---|---|
| At the surface of the ice | 40° |
| 1 inch above it | 82 |
| 2 inches | 106 |
| 3 inches | 123 |

The Heat of the water had hitherto been taken near the side of the jar; in the two following trials it was measured in the middle or axis of the jar.

When 1 hour and 30 minutes (reckoning always from the time when the boiling-hot water was poured into the jar) had elapsed, the Heat of the water in the middle of the jar was found to be as follows: —

| | |
|---|---|
| At the surface of the ice | 40° |
| 1 inch above it | 84 |
| 2 inches | 115 |
| 3 inches | 116 |
| 7 inches | 117 |

When 2 hours had elapsed, the Heat in the middle of the jar was found to be as follows: —

| | |
|---|---|
| At the surface of the ice | 40° |
| 1 inch above it | 76 |
| 2 inches | 94 |
| 3 inches | 106 |
| 4 inches | 108 |
| 6 inches | 108¼ |
| 7 inches | 108½ |

An end being now put to the experiment, the hot water was poured off from the ice, and on weighing that which remained, it was found that 5 oz. 6 grains, Troy, (= 2406 grains) of ice had been melted.

Taking the mean temperature of the water at the end

of the experiment at 106°, it appears that the mass of hot water (which weighed $73\frac{1}{4}$ ounces) was cooled 78 degrees, or from the temperature of 184° to that of 106°, during the experiment. Now, as it is known that one ounce of ice absorbs just as much Heat in being changed to water as one ounce of water loses in being cooled 140 degrees, it is evident that one ounce of water which is cooled 78 degrees gives off as much Heat as would be sufficient to melt $\frac{78}{140}$ of an ounce of ice; consequently the $73\frac{1}{4}$ ounces of hot water, which in this experiment were cooled 78 degrees, actually gave off as much Heat as would have been sufficient to have melted $\frac{73\frac{1}{4} \times 78}{140} = 40\frac{8}{10}$ ounces of ice.

But the quantity of ice actually melted was only about five ounces; and hence it appears that *less than one-eighth part of the Heat lost by the water was communicated to the ice*, the rest being carried off by the air.

As the same quantity of hot water was used in this experiment and in that, No. 15, which immediately preceded it, and as this water was contained by the same vessel (the glass jar above described), it appears that ice melts more than *eighty times slower* at the bottom of a mass of boiling-hot water than when it is suffered to swim on its surface. For, as in the experiment No. 15, $10\frac{1}{8}$ oz. of ice were melted in 2 minutes and 58 seconds, 5 ounces at least must have been melted in 1 minute and 29 seconds; but in the experiment No. 16, 2 hours, or 120 minutes, were employed in melting 5 ounces.

The ice however *was melted*, though very slowly, at the bottom of the hot water; and that circumstance alone would have been sufficient to have overturned my hypothesis respecting the manner in which Heat is propagated in liquids, had I not found means to account in a satisfactory manner for that fact without being obliged to abandon my former opinions.

In about half an hour after the hot water had been poured into the jar, in the last experiment, examining the surface of the ice, I discovered an appearance which fixed my attention and excited all my curiosity; I perceived that the ice had been melted and diminished at its surface, excepting only where it had been covered, or, as it were, *shadowed*, by the flat slips of deal by which the cake of ice was fastened down in its place.

Had the ice been protected and prevented from being melted by that piece of the wood *only* which, being undermost of the two, reposed immediately on the surface of the ice, I should not perhaps have been much surprised; but that part of the surface of the ice being likewise protected which was situated immediately under the other piece of wood, — that which, lying across the under piece and resting on it, *did not touch the ice anywhere except just at its edge*, — that circumstance attracted my attention, and I could at first see no way of accounting for these appearances but by supposing that the ice had been melted by the *calorific rays* which had been emitted by the hot water, and that those parts of the ice which had been *shadowed* by the pieces of deal, receiving none of these rays, had, of course, not been melted.

I was so much struck with these appearances that I immediately made the following experiments, with a view merely to the elucidation of this matter.

### *Experiment No.* 17.

Into a cylindrical glass jar, 6½ inches in diameter and 8 inches high, I put a circular cake of ice, as large as could be made to enter the jar, and about 3½ inches thick ; and on the flat and even surface of the ice I placed a circular plate of the thinnest tin I could procure, near $6\frac{1}{2}$ inches in diameter, or sufficiently large just to cover the ice. This plate of tin (which, to preserve its form, or keep it quite flat, was strengthened by a strong wire which went round it at its circumference) had a circular hole in its center, just two inches in diameter, and it was firmly fixed down on the upper surface of the cake of ice by means of several thin wooden wedges which passed between its circumference and the sides of the jar.

A second circular plate of tin, with a circular hole in its center two inches in diameter, and in all other respects exactly like that already described, was now placed over the first, and parallel to it, at the distance of just one inch, and like the first was firmly fixed in its place by wooden wedges.

These perforated circular plates being fixed in their places, the jar was placed in a room where Fahrenheit's thermometer stood at 34°, and ice-cold water was poured into it till the water just covered the upper plate, and then the jar was filled to within half an inch of its brim with boiling water, and, being covered over with a board, was suffered to remain quiet two hours.

At the end of this time the water, which was still warm, was poured off, and, the circular plate being removed, the ice was examined.

A circular excavation just as large as the hole in the tin plate which covered the ice (namely, two inches in diameter), and corresponding with it, perfectly well defined, and about $\frac{2}{10}$ of an inch deep in the center, had been made in the ice.

This was what I expected to find ; but there was something more which I did not expect, and which, for some time, I was quite at a loss to account for. Every part of the surface of the ice which had been covered by the tin plate appeared to be perfect, level, and smooth, and showed no signs of its having been melted or diminished, excepting only in one place, where a channel, about an inch wide and a little more than $\frac{2}{10}$ of an inch deep, which showed evident marks of having been formed by a stream of warm water, led from the excavation just mentioned, in the center of the upper part of the cake of ice, to its circumference. As the edge or vertical side of the cake of ice was evidently worn away where this stream passed, there could be no doubt with respect to its direction. It certainly ran *out of* the circular excavation in the middle of the ice; and though it might at first appear difficult to explain the fact, and to show how this hot water could arrive at that place, yet it was quite evident that the immediate cause of the motion of this stream of water could be no other than its specific gravity being greater than that of the rest of the water at the same depth; and that this greater specific gravity was at the same time accompanied by a higher degree of Heat is evident from the deep channel which this stream had melted in the ice, while other parts of the surface of the ice, at the same level, were not melted by the water which rested on it. To elucidate this point, I made the following experiment: —

### *Experiment No.* 18.

Thinking it probable that if the circular excavation in the ice, which answered to the circular hole in the middle of the tin plate which covered the ice, and also to that in the second plate which was placed an inch higher, had been melted by *radiant Heat* (as it has improperly been called), or by the calorific rays from the hot water; then, in that case, as some of these rays must probably have been reflected downwards at the surface of the water in attempting to pass upwards into the air, I thought that by preventing this part of them from reaching the ice, which I endeavoured to do by causing them to be absorbed by a light black body (a circular piece of deal board covered over with black silk), which I caused to swim on the surface of the water, their effects in melting the ice might perhaps be sensibly diminished. Had this really been the case, it would certainly have afforded strong grounds to suspect that these rays were in fact the cause of the appearances in question; but on making the experiment with the greatest care, I could not perceive that the covering of the surface of the hot water with a black body produced any difference whatever in the result of the experiment as it was first made (experiment No. 17), or when this black covering was not used.

After some meditation on the subject, it occurred to me that this melting of the ice at its upper surface could be accounted for in a manner which appeared to me to be perfectly satisfactory; without supposing either that water is a conductor of Heat or that the effect in question was produced by calorific rays.

Though it is one of the most general laws of nature with which we are acquainted, that all bodies, solids as well as fluids, are condensed by cold, yet in regard to *water* there appears to be a very remarkable exception to this law. Water, like all other known bodies, is indeed condensed by cold at every degree of temperature which is considerably higher than that of freezing, but its condensation, on parting with Heat, does not go on till it is changed to ice; but when, in cooling, its temperature has reached to the 40th degree of Fahrenheit's scale, or eight degrees above freezing, it ceases to be farther condensed; and on being cooled still farther, *it actually expands*, and continues to expand as it goes on to lose more of its Heat, till at last it freezes; and at the moment when it becomes solid, and even after it has become solid, it expands still more on growing colder. This fact, which is noticed by M. de Luc in his excellent treatise on the modifications of the atmosphere, has since been farther investigated and put beyond all doubt by Sir Charles Blagden. See Philosophical Transactions, Vol. LXXVIII.

Now, as water in contact with melting ice is always at the temperature of 32°, it is evident that water at that temperature must be specifically lighter than water which is eight degrees warmer, or at the temperature of 40°; consequently, if two parcels of water at these two temperatures be contained in the same vessel, that which is the coldest and lightest must necessarily give place to that which is warmer and heavier, and currents of the warmer water will *descend* in that which is colder.

In the two last experiments, as the circular tin plate which covered the surface of the ice served to confine the thin sheet of water which was between the plate and the ice, as this water could not rise upwards, being hindered by the plate, and as it had no tendency to descend, it is probable that it remained in its place; and as it was *ice-cold*, it was not capable of melting the ice on which it reposed.

But as the tin plate had a circular hole in its center, the surface of the ice *in that part* was of course naked, and, the ice-cold water in contact with it being displaced by the warmer and heavier water from above, an excavation, in the form of a shallow basin, was formed in the ice by this descending warm current.

The warm water contained in this basin overflowed its banks as soon as the basin began to be formed, and, issuing out on that side which happened to be the lowest, opened itself a passage under the tin plate to the edge of the ice, over which it was precipitated and fell down to the bottom of the jar. The water of this rivulet being warm, it soon formed for itself a deep channel in the ice, and at the end of the experiment it was found to be everywhere *deeper* than the bottom of the basin where it took its rise.

This manner of accounting for the appearances in question seemed to me to be quite satisfactory ; and the more I meditated on the subject, the more I was confirmed in my suspicions that *all liquids* must necessarily be perfect *non-conductors of Heat.*

On these principles I was now enabled to account for the melting of the ice at the bottom of the hot water in the experiment No. 16, as also for the slowness with which that process went on ; and, encouraged by this success, I now proceeded with confidence to plan and to execute still more decisive experiments ; from the results of which, I may venture to say it, the important facts in question have been put beyond all possibility of doubt.

If water be in fact a perfect *non-conductor* of Heat, — that is to say, if there be *no communication whatever of Heat* between neighbouring particles or *molecules* of that fluid (which is what I suppose), then, as Heat cannot be propagated in it but *only* in consequence of the motions occasioned in the fluid by the changes in the specific gravity of those particles which are occasioned by the changes of their temperature, it follows that Heat cannot be propagated *downwards* in water as long as that fluid continues to be condensed with cold; and that it is *only in that direction* (downwards) that it *can be propagated* after the water has arrived at that temperature where it begins to be expanded by cold, which has been found to be at about the 40th degree of Fahrenheit's scale.

Reasoning on these principles, we are led to this remarkable conclusion; namely, that *water which is only eight degrees above the freezing-point, or at the temperature of* 40°, *must be able to melt as much ice in any given time,* WHEN STANDING ON ITS SURFACE, *as an equal volume of water at any higher temperature,* EVEN THOUGH IT WERE BOILING HOT.

My philosophical reader will doubtless think that I must have had no small degree of confidence in the opinion I had formed on this interesting subject, to have had the courage to make, *even in private,* the experiments which were necessary to ascertain that fact.

### *Experiment No.* 19.

Into a cylindrical glass jar, 4.7 inches in diameter and 13.8 inches high, I put 43.87 cubic inches, or 1 lb. 11$\frac{1}{7}$ oz. Troy, in weight, of water, and placing the jar in a freezing mixture, composed of pounded ice and common sea-salt, I caused the water to freeze into one compact mass, which adhered firmly to the bottom and sides of the jar, and which formed a cylinder of ice just three inches high.

Had the bottom of the jar been quite flat, instead of being raised or vaulted, the cylinder of ice would have been no more than 2.67 inches high.

As soon as the water in the jar was completely frozen, the jar was removed from the freezing mixture and placed in a mixture of pounded ice and pure water, where it was suffered to remain four hours, in order that the cake of ice in the jar might be brought to the temperature of 32°.

The jar still standing in a shallow dish in the pounded ice and water, the surface of which cold mixture was just on a level with the surface of the ice in the jar, I covered the top of the cake of ice with a circular piece of strong paper, and poured gently into the jar 73¼ oz., Troy, of boiling-hot water, which filled it to the height of eight inches above the surface of the ice. (See Plate II.)

I then removed very gently the circular piece of paper which covered the surface of the ice, and after leaving the hot water in contact with the ice a certain number of minutes, I poured it off, and, weighing immediately the jar and the unmelted ice which remained in it, I ascertained the *quantity of ice which had been melted* by the hot water during the time it had been suffered to remain in the jar.

This experiment I repeated four times the same day (16th March, 1797), varying at each repetition of it the time the water was permitted to remain on the ice. The results of these experiments were as follows: —

| No. of the Experiment. | Time the Hot Water remained on the Ice. | Temperature of the Hot Water when it was poured on the Ice. | Temperature of the Water 1 Inch below its Surface at the end of the Experiment. | Quantity of Ice melted. |
|---|---|---|---|---|
| | Minutes. | | | Grains. |
| No. 19 | 1 | 186° | Not observed. | 1632 |
| No. 20 | 3¾ | 185 | Not observed. | 1824 |
| No. 21 | 15 | 184 | 170° | 1757 |
| No. 22 | 60 | 186 | 140 | 2573 |

From the results of these experiments it was plain that a very considerable portion of the ice which was melted, was melted in the very beginning of the experiments, or while the hot water was actually *pouring* into the jar, which operation commonly lasted about one minute; and the irregularities in the results of the experiments, and particularly of the three first, showed evidently that the quantity of ice melted in that operation was different in different experiments. I had indeed foreseen that this would be the case, and on that account it was that I covered the surface of the ice with a circular piece of strong paper, and always took care to pour the water very gently into the jar; but I found that all these precautions were not sufficient to prevent very considerable anomalies in the results of the experiments; and as I found reason to suspect that the motion in the mass of the hot water, which was unavoidably occasioned by removing the circular piece of paper which covered the ice, was the principal cause of these inaccuracies, I had recourse to another and a better contrivance.

Having procured a flat, shallow dish, of light wood, half an inch deep, $4\frac{1}{2}$ inches in diameter (or something less than the internal diameter of the jar), and about $\frac{1}{4}$ of an inch thick at its bottom, I bored a great number of very small holes through its bottom, which gave it the appearance of a sieve. This perforated wooden dish, having been previously made *ice-cold*, was placed on the surface of the ice in the jar, and the hot water being gently poured into the dish through a long wooden tube, as this perforated dish floated and remained constantly at the surface of the water, and as the water passing through such a great number (many hundreds) of small holes was not projected downwards with force, it is evident that by this simple contrivance those violent motions in the mass of water in the jar which before took place when the hot water was poured into the ice, and when the paper which covered the ice was removed, were in a great measure prevented.

In order that the water which was poured through the wooden tube (the bore of which was about half an inch in diameter) might not impinge perpendicularly and with force against the bottom of the dish, the lower end of the tube was closed by a fit cork-stopple, and the water was made to issue horizontally through a number of small holes in the sides of this tube, at its lower end.

As soon as the operation of pouring the hot water into the jar was finished, the perforated dish was carefully removed, and the jar was covered with a circular wooden cover, from the center of which a small mercurial thermometer was suspended.

The effects produced by this new arrangement of the machinery will appear by comparing the results of the two following experiments with those just mentioned.

| No. of the Experiment. | Time the Hot Water remained on the Ice. | Temperature of the Hot Water. | | Quantity of Ice melted. |
|---|---|---|---|---|
| | | At the Beginning. | At the End. | |
| | Minutes. | | | Grains. |
| No. 23 | 1 | 196° | 196° | 423 |
| No. 24 | 3 | 190 | 188 | 703 |

In order still more effectually to prevent the inaccuracies arising from the internal motions in the mass of hot water which were occasioned in pouring the water into the jar (and which could not fail to affect, more or less, the results of the experiment), I had recourse to the following contrivance.

I filled a small phial containing $8\frac{1}{4}$ cubic inches with ice-cold water, and then, emptying the phial in the jar, I covered the surface of the ice with this ice-cold water to the height of 0.478 of an inch.

On the surface of this ice-cold water, instead of that of the ice, I now placed the perforated wooden dish, previously made ice-cold, and poured the hot water upon it.

The results of the following experiments show that this contrivance tended much to diminish the apparent irregularities of the experiments.

The air of the room in which these experiments were made was of the temperature of 41°.

| No. of the Experiment. | Time the Hot Water was on the Ice. | Temperature of the Hot Water 1 Inch below its Surface. | | Quantity of Ice melted. |
|---|---|---|---|---|
| | | At the Beginning. | At the End. | |
| | Minutes. | | | Grains. |
| No. 25 | 10 | 192° | 182° | 580 |
| No. 26 | 30 | 190 | 165 | 914 |
| No. 27 | 180 | 190 | 95 | 3200 |

From the results of these last three experiments we can now determine with a very considerable degree of certainty how much ice was melted *in the act of pouring the water into the jar*, and consequently the rate at which it was melted in the ordinary course of the experiment, — supposing equal quantities to be melted in equal times.

As in the 27th experiment 3200 grains were melted in 180 minutes, and in the 25th experiment 580 grains were melted in 10 minutes, we may safely conclude that the same quantity must have been melted in the same time (10 minutes) in the 27th experiment; if, therefore, from 3200 grains — the quantity melted in 180 minutes in this last experiment — we deduct 580 grains for the quantity melted during the first 10 minutes, there will remain 2620 grains for the quantity melted in the succeeding 170 minutes, when, the motions occasioned in the water on its being poured into the jar having subsided, we may suppose the process of melting the ice to have gone on regularly.

But if in the regular course of the experiment no more than 2620 grains were melted in 170 minutes, it is evident that not more than 154 grains could have been melted in the ordinary course of the process in 10 minutes; for 170 minutes : 2620 grains :: 10 minutes : 154 grains. If, therefore, from 580 grains, the quantity of ice actually melted in 10 minutes in the 25th experiment, we deduct 154 grains, there remains 426 for the quantity melted in pouring the water into the jar.

Let us see, now, how far this agrees with the result of the 26th experiment. In this experiment 914 grains of ice were melted in 30 minutes. If from this quantity we deduct 426 grains, the quantity which, according to the foregoing computation, must have been melted *in pouring the hot water into the jar*, there will remain 478 grains for the quantity melted in the ordinary course of the process in 30 minutes; which gives 159 grains for the quantity melted in 10 minutes; which differs very little from the result of the foregoing computation, by which it appeared to be = 154 grains. This difference, however, small as it is, is sufficient to prove an important fact, namely, that the effects produced by the motion into which the hot water had been thrown in being poured into the jar had not ceased entirely in 10 minutes, or when an end was put to the 25th experiment. We shall therefore come nearer the truth, if, in our endeavours to discover the quantity of ice melted in any given time in the ordinary course of the experiments, we found our computation on the results of the two experiments No. 26 and No. 27.

In the latter of these experiments 3200 grains of ice were melted in 180 minutes, and in the former 914 grains were melted in 30 minutes. If, therefore, from 3200 grains, the quantity melted in 180 minutes, we take the quantity melted in the first 30 minutes, = 914 grains, there will remain 2286 grains for the quantity melted in the succeeding 150 minutes, and this gives 152 grains for the quantity melted in 10 minutes. By the former computation it turned out to be 154 grains.

But if 152 grains of ice is the quantity melted in 10 minutes, in the ordinary course of the process three times that quantity, or 456 grains only, could have been melted *in this manner* in the 30 minutes during which the 26th experiment lasted ; and deducting this quantity from 914 grains, the quantity actually melted in that experiment, the remainder, 458 grains, shews how much must have been melted in pouring the hot water on the ice, or in consequence of the motions into which the water was thrown in the performance of that operation. By the preceding computation this quantity turned out to be 426 grains.

From the result of these computations I think we may safely conclude that in the ordinary course of the experiments not more than 152 grains of ice were melted by the hot water in 10 minutes.

I shall now proceed to give an account of several experiments in which the water employed to melt the ice was at a *much lower temperature.*

Having removed a small quantity of ice which remained unmelted in the bottom of the jar, I put a fresh quantity of water into it, and placing the jar in a freezing mixture, caused this water, which filled the jar to the height of four inches, to freeze into one solid mass of ice. I then placed the jar in a shallow earthen dish, and surrounded it to the height of the level of the top of the ice with a mixture of snow and water (see Plate II.), and, placing it in a room in which there had been no fire made for many months, and in which the temperature of the air was at 41°, I let it remain quiet two hours, in order that the ice might acquire the temperature of 32°.

This being done, I took the jar out of the earthen dish, and, wiping the outside of it dry with a cold napkin, I weighed the jar with the ice in it very exactly, and then replaced it in the earthen dish, and surrounded it as before with snow and water to the height of the level of the surface of the ice.

I then poured 73¼ ounces, Troy, (= 15,160 grains) of water, at the temperature of 41°, into the jar, which covered the ice to the same height to which it had been covered in the former experiments, namely, to about 8 inches; and suffering it to stand on the ice a certain number of minutes, I then poured it off, and, wiping the outside of the jar, weighed it, in order to ascertain how much ice had been melted.

In putting this cold water into the jar, the same precautions were used (by pouring it through the wooden tube into the perforated wooden dish, &c.) as were used when the experiment was made with boiling water.

The following table shews the results of six experiments made the same day (the 19th March, 1797), in the order in which they are numbered, and which were all made with the utmost care.

| Number of the Experiment. | Temperature of the Water in the Jar 1 Inch below its Surface. | | Temperature of the Air. | Time the Water remained on the Ice. | Quantity of Ice melted. |
|---|---|---|---|---|---|
| | At the beginning of the Experiment. | At the end of the Experiment. | | | |
| | | | | Minutes. | Grains. |
| No. 28 | 41° | 40° | 41° | 10 | 203 |
| No. 29 | 41 | 40 | 41 | 10 | 220 |
| No. 30 | 41 | 40 | 41 | 10 | 237 |
| No. 31 | 41 | 40 | 41 | 10 | 228 |
| No. 32 | 41 | 38 | 41 | 30 | 617 |
| No. 33 | 41 | 38 | 41 | 30 | 585 |

The agreement in the results of these experiments is not much less extraordinary than the surprising fact which is proved by them, namely, that boiling-hot water does not thaw more ice in any given time *when standing quietly on its surface* than water at the temperature of 41°, or nine degrees only above the point of freezing!

There is reason to conclude that it does not even thaw so much; and this still more remarkable circumstance may, I think, be accounted for in a satisfactory manner on the supposition (which, however, I imagine, will no longer be considered as a bare supposition), that water is a non-conductor of Heat.

It appeared from the results of the experiments made with hot water, that the quantity of ice melted in 10 minutes in the ordinary course of that process amounted to no more than 152 grains; but in these experiments with cold water, the quantity melted in that time was never less than 203 grains, and, taking the mean of four experiments, it amounted to 222 grains.

There is one circumstance, however, respecting these experiments with cold water, which it is necessary to investigate before their results can be admitted as complete proof in the important case in question.

In the experiments which were made with hot water, it was found that a considerable part of the ice which was melted, was melted in consequence of the motions into which the water was thrown upon being poured into the jar, and that the effect of these motions continued to be sensible for a longer time than most of these experiments with cold water lasted. Is it not possible that the results of these experiments with cold water may also have been affected by the same cause? This is what I shall endeavour to find out.

In the 32d experiment 617 grains of ice were melted in 30 minutes, and in the 33d experiment 585 grains were melted in the same time; and taking the mean of these two experiments, it appears that 601 grains were melted in 30 minutes. If now from this quantity we deduct that which, according to the mean result of the four preceding experiments, must have been melted in 10 minutes, namely, 222 grains, there will remain 379 grains for the quantity melted in the last 20 minutes in these two experiments; consequently, half this quantity, or 189½ grains, is what must have been melted in 10 minutes in the ordinary course of the process.

But this quantity (189½ grains), though less than what was actually melted in the experiments which lasted only 10 minutes, is still considerably greater than 152 grains, the quantity which was found to have been melted in the same time in the ordinary course of the process in those experiments in which hot water was used; consequently, the great question, for the decision of which these experiments were contrived, is, I believe I may venture to say, decided.

But, however conclusive the result of these experiments appeared to me to be, I felt myself too much interested in the subject to rest my inquiries here.

Having found, as well from the results of the experiments made with cold water as from those made with hot water, that a considerable quantity of ice was melted in the act of pouring the water into the jar, and in consequence of those undulatory motions into which the water was thrown in that operation, notwithstanding all the pains I had taken to diminish those motions and prevent their effects, I now doubled my precautions in guarding against those sources of error and uncertainty.

Before I poured the water into the jar I covered the surface of the ice to the height of 0.956 of an inch with ice-cold water, and this I did when water at the temperature of 41° was used, as well as in those experiments in which boiling-hot water was employed. In the former experiments I had covered the surface of the ice with ice-cold water only in those experiments in which hot water was used, and even in those I used only half as much ice-cold water as I now employed for that purpose.

I also now poured the water into the jar in a smaller stream, employing no less than three minutes in filling it up to the height of eight inches above the surface of the ice; and I endeavoured to ascertain how far the results of the experiments were influenced by the temperature of the air, and also by wrapping up the jar in warm covering.

The same jar was used in all the experiments, and it was always placed in the same earthen dish, and surrounded, to the level of the top of the ice, with melting snow. This jar is very regular in its form, being very nearly a perfect cylinder, and is on that account peculiarly well calculated for the use for which I selected it.

In each of the three first experiments which are entered in the following table, the jar was well covered up with a very warm covering of cotton-wool. This covering (which was above an inch thick) reached from the surface of the melting snow in which the jar stood quite to the top of the jar. The mouth of the jar was first covered with a round wooden cover (from the center of which a thermometer, the bulb of which reached one inch below the surface of the water, was suspended), and on the top of this wooden cover there was put a thick covering of cotton.

In all the experiments in the following table, except the three first, the jar was exposed naked to the air, except the lower part of it, which, as I have already more than once observed, was always covered, as high as the ice in the jar reached, with melting snow or with pounded ice and water.

In the two experiments No. 37 and No. 38, which are marked with asterisks, the surface of the ice was covered with ice-cold water to the depth of 0.478 of an inch only; in all the other experiments it was covered to the depth of 0.956 of an inch.

| Number of the Experiment. | Temperature of the Water in the Jar 1 Inch below its Surface. | | Temperature of the Air. | Time the Water remained on the Ice. | Quantity of Ice melted. |
|---|---|---|---|---|---|
| | In the beginning of the Experiment. | At the end of the Experiment. | | Minutes. | Grains. |
| No. 34 | 188° | 179° | 41° | 30 | 634 |
| No. 35 | 189 | 180 | 41 | 30 | 747 |
| No. 36 | 190 | 147 | 41 | 180 | 3963 |
| No. 37 | 41 | 38 | 41 | 30 | 592* |
| No. 38 | 41 | 43 | 61 | 30 | 676* |
| No. 39 | 186 | 157 | 61 | 30 | 559 |
| No. 40 | 188 | 156 | 61 | 30 | 575 |
| No. 41 | 190 | 156 | 61 | 30 | 542 |
| No. 42 | 41 | 43 | 61 | 30 | 573 |
| No. 43 | 42 | 44 | 61 | 30 | 575 |
| No. 44 | 42 | 35 | 61 | 120 | 2151 |

The results of these experiments afford matter for much curious speculation; but I shall content myself for the present with making only two or three observations respecting them. And, in the first place, it is remarkable, that, although in the experiments No. 34 and No. 35, of 30 minutes each, considerably less ice was melted than in that No. 26, which lasted the same time, yet in that No. 36, of 180 minutes, more was melted than in that No. 27, of the same duration. This difference in the two last-mentioned experiments will be accounted for hereafter.

With regard to the difference in the results of the experiments of 30 minutes, there is no doubt but that it arose from the precautions which had been taken in this last set of experiments to prevent the effect of the violent motions into which the hot water was thrown in being poured into the jar, that less ice was melted in the experiments No. 34 and No. 35 than in that No. 26.

*Secondly*, It appears that more ice was melted in the same time in the experiments in which the jar was covered up with warm covering than in those in which it was left naked and exposed to the air of the room.

The difference is even considerable. The quantity melted in 30 minutes when the jar was covered, at a mean of two experiments (No. 34 and No.35), was $690\frac{1}{2}$ grains; but when the jar was naked, the quantity at a mean of three experiments (No. 39, No. 40, and No. 41) was only $558\frac{2}{3}$ grains.

*Thirdly*, The quantity of ice melted under similar circumstances — that is to say, when the jar was naked — was sensibly greater when the water was at the temperature of about 41° than when it was nearly boiling hot. In the experiment No. 41, when the water which was poured on the ice was at the temperature of 190 deg., 542 grains only of ice were melted in 30 minutes; whereas in the next experiment (No. 42), when the water was at 41°, or 149 degrees colder, 573 grains were melted in the same time.

Finding that covering up the jar with a thick and warm covering of cotton caused more ice to be melted by the hot water, I was curious to see what effects would be produced by keeping the jar plunged *quite up to its brim* in a mixture of snow and water, instead of merely surrounding that part of it which was occupied by the cake of ice by this cold mixture.

I was likewise desirous of finding out — and, if possible, at the same time — whether water at a temperature something above that at which that Fluid ceases to be condensed with cold would not melt more ice in any given time than an equal quantity of that Fluid, either colder or much hotter. The result of the 43d experiment had shewn me, — what indeed a very simple computation would have pointed out, — namely, that, when the temperature of the water is but a few degrees above the point of freezing, if its quantity or depth is not very considerable, it will soon be so much cooled as very sensibly to retard the process of melting the ice; and with respect to hot water, the increased quantity of ice which was melted by it when the jar was covered up with a warm covering convinced me that the real cause which prevented the hot water from melting as much ice as the cold water in my experiments was the embarrassments in the process of melting the ice, which were occasioned by the descending currents formed in the hot water on its being cooled by the air at its surface, and by the sides of the jar.

These descending currents meeting, in the region of the constant temperature of 40°, with those cold currents which ascended from the surface of the ice, it seems very probable that the ascending currents, on the motion of which the melting of ice depends, were checked by this collision.

By retarding the cooling of the hot water above by wrapping up the jar in a warm covering, the velocity of the descending currents was of course diminished; but when this was done, the results of the experiment shewed that the melting of the ice was accelerated.

When, the jar being naked, the cooling of the hot water, and consequently the motions of the descending currents, were rapid, no more than about 542 grains, or at most 575 grains, were melted in 30 minutes; but when the jar was covered with a warm covering, 634 grains, and in one experiment (that No. 35) 747 grains, were melted in the same time.

As plunging the jar into a cold mixture of snow and water could not fail to accelerate the cooling of the hot water in the jar, and consequently to increase the rapidity of the descending currents in it, ought not this to embarrass, in an extraordinary degree, the ascending currents of ice-cold water from the surface of the ice, and to diminish the quantity of ice melted? This is what the following experiments, compared with the results of those No. 39, No. 40, and No. 41, will shew.

| Number of the Experiment. | Temperature of the Water in the Jar 1 Inch below its Surface. In the beginning of the Experiment. | At the end of the Experiment. | Temperature of the cold Mixture in which the Jar was kept plunged to its brim. | Time the Water remained on the Ice. | Quantity of Ice melted. |
|---|---|---|---|---|---|
| | | | | Minutes. | Grains. |
| No. 45 | 188° | 68° | 32° | 30 | 406 |
| No. 46 | 186 | 67 | 32 | 30 | 440 |
| No. 47 | 189 | 68 | 32 | 30 | 432 |
| No. 48 | 187 | 67 | 32 | 30 | 355 |
| No. 49 | 188 | 68 | 32 | 30 | 364 |
| Quantity of ice melted in these 5 experiments, | | | | | 1997 |

| | Grains. |
|---|---|
| Mean quantity melted by hot water when the jar was kept plunged to its brim in melting ice and water . . . | $399\frac{2}{5}$ |
| Mean quantity melted by hot water in 30 minutes, in the two experiments, No. 26 and No. 27, when the part of the jar occupied by the water was surrounded by air, at the temperature of 41° . . . . . . . . . . | 456 |
| Mean quantity melted by hot water in 30 minutes, in the three experiments, No. 39, No. 40, and No. 41, when the part of the jar occupied by the water was surrounded by air, at the temperature of 61° . . . . . . . . . . | $558\frac{1}{3}$ |
| Mean quantity melted by hot water in 30 minutes, in the two experiments, No. 34 and No. 35, when the part of the jar occupied by the water was kept covered up by a thick and warm covering of cotton . . . . . . . . | $690\frac{1}{2}$ |

As all the experiments were made in the same manner, and with equal care, and differed only in respect to the manner in which the outside of the jar, above the surface of the ice in it, was covered, their results shew the effects produced by those differences.

I should perhaps have suspected that the greater quantity of ice which was melted when the heat of the water in the jar was confined for the longest time had been occasioned, at least in part, by the Heat communicated downwards by the medium of the glass; but that this could not have been the case was evident, not only from the manner in which the ice was always found to have been melted, but also from the results of similar experiments made with much colder water.

Had it been melted by Heat communicated by the glass, it would undoubtedly have been most melted in those parts of its surface where it was in contact with the glass; but this I never once found to be the case.

The results of the following experiments will shew—what indeed might easily have been foreseen—that the temperature of the medium by which the upper part of the jar was surrounded does not always affect the result of the experiment in the same degree, nor even always in *the same manner*, in different experiments in which the temperature of the water in the jar is very different.

To facilitate the comparison of these experiments, and that of the foregoing, which are similar to them, I shall here place them together.

| Number of the Experiment. | Temperature of the Water in the Jar 1 Inch below its Surface. | | Temperature of the Medium by which the upper part of the Jar was surrounded. | Time the Water remained on the Ice. | Quantity of Ice melted. |
|---|---|---|---|---|---|
| | In the beginning of the Experiment. | At the end of the Experiment. | | | |
| | | | | Minutes. | Grains. |
| No. 50 | 41° | 36° | 32° | 30 | 542 |
| No. 37 | 41 | 38 | 41 | 30 | 592 |
| No. 42 | 41 | 43 | 61 | 30 | 576 |

It is certainly very remarkable indeed that so much more ice should be melted by water at the temperature of 41°, when the jar which contained it was surrounded by a cold mixture of pounded ice and water, than by an equal quantity of boiling-hot water in the same circumstances. In the experiment No. 50, the quantity melted by the cold water was 542 grains, while that melted by the boiling-hot water, taking the mean of five experiments (those No. 45, 46, 47, 48, and 49), was no more than $399\frac{2}{5}$ grains. But the results of the four following experiments are, if possible, still more surprising.

These experiments were made with water at the temperature of 61°, the temperature of the air of the room being at the same time 61°; in the two first of these experiments the jar was kept plunged to its brim in a mixture of snow and water; in the two last its lower part only, namely, as high as the level of the surface of the ice, was surrounded by this cold mixture, its upper part being naked, and surrounded by the air of the room.

In each of the experiments (as in those which preceded them), before the water was poured into the jar the surface of the ice was covered to the height of 0.956 of an inch with ice-cold water, in order more effectually to defend it against the effects of the temporary motions into which the water employed to melt the ice was unavoidably thrown in the performance of this operation; and the same quantity of water was always used, namely, $73\frac{1}{4}$ ounces, Troy, or as much as was sufficient to fill the jar to the height of 8 inches.

| Number of the Experiment. | Temperature of the Water in the Jar 1 Inch below its Surface. | | Temperature of the Medium by which the upper part of the Jar was surrounded. | Time the Water remained on the Ice. | Quantity of Ice melted. |
|---|---|---|---|---|---|
| | In the beginning of the Experiment. | At the end of the Experiment. | | Minutes. | Grains. |
| No. 51 | 61° | 49° | 32° | 30 | 660 |
| No. 52 | 61 | 50 | 32 | 30 | 662 |
| No. 53 | 61 | 60 | 61 | 30 | 642 |
| No. 54 | 61 | 60 | 61 | 30 | 650 |

These experiments are remarkable, not only on account of the very small difference in the quantities of ice melted which resulted from the cooling of the sides of the jar, but also, and more especially, as that difference was directly contrary to the effects produced by the same means in the experiments with hot water. More ice was melted when the outside of the jar was kept ice-cold than when it was surrounded by air at the temperature of 61 .

All these appearances might, I think, be accounted for in a satisfactory manner on the principles we have assumed respecting the manner in which Heat is propagated in liquids ; but without engaging ourselves at present too far in these abstruse speculations, let us take a retrospective view of all our experiments, and see what general results may with certainty be drawn from them.

One of the experiments in which the greatest quantity of ice was melted by *hot water* is that No. 36, in which 3963 grains were melted in three hours, or 180 minutes. If now from this quantity we deduct that which, according to the results of the two preceding experiments, must have been melted in the first 30 minutes, namely, 690½ grains, there will remain 3272½ grains for the quantity melted in the last 150 minutes, which gives 654½ grains for the quantity melted in 30 minutes *in the ordinary course of the experiment.*

This quantity, 654½ grains, deducted from that which at a mean of two experiments (those No. 34 and No. 35) was found to be actually melted in 30 minutes, namely, 690½ grains, leaves 36 grains for the quantity which in these two experiments was melted in consequence of the temporary motions into which the hot water was thrown in the operation of pouring it into the jar. The difference between these two quantities ( = 36 grains) is very inconsiderable, and shews that the means used for diminishing the effects produced by those motions had been very efficacious.

As the results of the three experiments No. 34, No. 35, and No. 36, were exceedingly regular and satisfactory, — as the Heat of the water appears to have been so completely confined by the warm covering which surrounded the jar, and as the process of melting the ice went on regularly or equally for so great a length of time (three hours) in the 36th experiment, we may venture to conclude that more ice could not possibly have been melted by boiling-hot water — *standing on it* — than was melted in these experiments.

This quantity was found to be at the rate of 654½ grains in 30 minutes.

But as in these experiments extraordinary means were used, by which an uncommonly large quantity of ice was melted, they cannot be considered as similar to those which were made with cold water, and consequently cannot with propriety be compared with them.

When the experiments were similar, the mean results of those which were made with water at different temperatures were as follows.

| | | Ice melted in 30 minutes. Grains. |
|---|---|---|
| In the experiments in which the part of the jar which was occupied by the water was exposed uncovered to the air at the temperature of 61° | With boiling-hot water (*experiments* No. 39, 40, *and* 41) | $558\frac{2}{3}$ |
| | With water at the temperature of 61° (*experiments* No. 53 *and* No. 54) | 646 |
| | With water at the temperature of 41° (*experiments* No. 42 *and* No. 43) | 574 |
| In the experiments in which the part of the jar which was occupied by the water was surrounded by pounded ice and water, and consequently was at the temperature of 32° | With boiling-hot water (*experiments* No. 45, 46, 47, 48, *and* 49) | $399\frac{2}{5}$ |
| | With water at the temperature of 61° (*experiments* No. 51 *and* No. 52) | 661 |
| | With water at the temperature of 41° (*experiment* No. 50) | 542 |

From the results of all these experiments we may certainly venture to conclude that boiling-hot water is not capable of melting more ice, *when standing on its surface*, than an equal quantity of water at the temperature of 41°, or when it is only *nine degrees* above the temperature of freezing!

This fact will, I flatter myself, be considered as affording the most unquestionable proof that could well be imagined, that water is a perfect *non-conductor of Heat*, and that Heat is propagated in it *only* in consequence of the motions which the Heat occasions in the insulated and solitary particles of that fluid.*

* The insight which this discovery gives us in regard to the nature of the mechanical process which takes place in chemical solutions is too evident to require illustration; and it appears to me that it will enable us to account in a satisfactory manner for all the various phænomena of chemical affinities and vegetation. Perhaps all the motions among inanimate bodies on the surface of the globe may be traced to the same cause, — namely, to the non-conducting power of Fluids with regard to Heat.

The discovery of this fact opens to our view one of the most interesting scenes in the economy of Nature: but in order to prepare our minds for the contemplation of it, it will be not amiss to refresh our memory by recapitulating what has already been said on the Propagation of Heat in Fluids, and particularly in water; and adding such occasional observations as may tend to elucidate that abstruse subject.

Those who enter into the spirit of these investigations will not consider these repetitions and illustrations as either superfluous or tiresome.

---

## CHAPTER III.

*Recapitulation, and farther Investigation of the Subject. — All Bodies are condensed by Cold without Limitation,* WATER ONLY EXCEPTED. — *Wonderful Effects produced in the World in consequence of the particular Law which obtains in the Condensation of Water. — This Exception to one of the most general Laws of Nature, a striking Proof of Contrivance in the Arrangement of the Universe; a Proof which comes home to the Feelings of every ingenuous and grateful Mind. — This particular Law does not obtain in the Condensation of* SALT WATER. — *Final Cause of the Saltness of the Sea. — The Ocean probably designed by the Creator to serve as an Equalizer of Heat. — Could not have answered that Purpose had its Waters been fresh. — Final Cause of the Freshness of Lakes and inland Seas in high Latitudes. — Usefulness of these Speculations.*

AS the immediate cause of the motions in a liquid, which take place on its undergoing a change of temperature, is evidently the change in the specific gravity of those particles of the liquid which become either hotter or colder than the rest of the mass, and as the specific gravities of some liquids are much more changed by any given change of temperature than those of others, ought not this circumstance (independent of the more or less perfect fluidity of the liquid) to make a sensible difference in the conducting power of liquids?

The more a liquid is expanded by any given change of temperature, the more rapid will be the ascent of the particles which first receive the Heat; and as these are immediately replaced by other colder particles, which, in their turns, come to be heated, this must of course produce a rapid communication of Heat from the hot body of the liquid.

But when, on the other hand, the specific gravity of a liquid is but little changed by any given change of temperature, the motions among the particles of the liquid occasioned by this change must be very sluggish, and the communication of Heat of course very slow.

Let us stop here for one moment just to ask ourselves a very interesting question. Suppose that in the general arrangement of things it had been necessary to contrive matters so that water should not freeze in winter, or that it should not freeze *but with the greatest difficulty*, — very slowly, *and in the smallest quantity possible*. How could this have been most readily effected?

Those who are acquainted with the law of the condensation of Water on parting with its Heat have already anticipated me in these speculations; and it does not appear to me that there is anything which human sagacity can fathom, within the wide-extended bounds of the visible creation, which affords a more striking or more palpable proof of the wisdom of the Creator, and of the special care he has taken in the general arrangement of the universe to preserve animal life, than this wonderful contrivance; for though the extensiveness and immutability of the general laws of Nature impress our minds with awe and reverence for the Creator of the universe, yet *exceptions to those laws*, or particular modifications of them, from which we are able to trace effects evidently *salutary* or advantageous to ourselves and our fellow-creatures, afford still more striking proofs of contrivance, and ought certainly to awaken in us the most lively sentiments of admiration, love, and gratitude.

Though in temperatures above blood-heat the expansion of water with Heat is very considerable, yet in the neighbourhood of the freezing point it is almost nothing. And what is still more remarkable, as it is an exception to one of the most general laws of Nature with which we are acquainted, when in cooling it comes within eight or nine degrees of Fahrenheit's scale of the freezing point, instead of going on to be farther condensed as it loses more of its Heat, it *actually expands* as it grows colder, and continues to expand more and more as it is more cooled.

If the whole amount of the condensation of any given quantity of boiling-hot water, on being cooled to the point of freezing, be divided into any given number of equal parts, the condensations corresponding to equal changes of temperature will be very unequal in different temperatures.

In cooling $22\frac{1}{2}$ degrees of Fahrenheit's scale (or one-eighth part of the interval between the boiling and the freezing points) the condensation will be, —

| | | | | Condensation. |
|---|---|---|---|---|
| In cooling $22\frac{1}{2}$°, viz. from | 212° | to | $189\frac{1}{2}$° | 18 parts. |
| | $189\frac{1}{2}$ | " | 167 | 16.2 " |
| | 167 | " | $144\frac{1}{2}$ | 13.8 " |
| | $144\frac{1}{2}$ | " | 122 | 11.5 " |
| | 122 | " | $99\frac{1}{2}$ | 9.3 " |
| | $99\frac{1}{2}$ | " | 77 | 7.1 " |
| | 77 | " | $54\frac{1}{2}$ | 3.9 " |
| | $54\frac{1}{2}$ | " | 32 | 0.2 " |

Hence it appears that the condensation of water, or increase of its specific gravity in being cooled $22\frac{1}{2}$ degrees of Fahrenheit's scale, is at least *ninety times greater* when the water is boiling-hot, than when it is at the mean temperature of the atmosphere in England ($54\frac{1}{2}$°), or within $22\frac{1}{2}$ degrees of freezing, (for 18 is to 0.2 as 90 to 1.)

All liquids, it is true, in cooling, are more condensed by any given change of temperature when they are very hot than when they are colder; but these differences are nothing compared to those we observe in water.

The ratio of the condensation in cooling from 212° to $189\frac{1}{2}$° to that in cooling from $54\frac{1}{2}$° to 32° in each of the under-mentioned fluids has been shown, by the experiments of M. de Luc, to be as follows: —

| | | | | |
|---|---|---|---|---|
| Olive-oil . . . . . . . | as | $1\frac{14}{100}$ | to | 1 |
| Strong spirits of wine . . . . | as | $1\frac{29}{100}$ | to | 1 |
| A saturated solution of sea-salt in water . | as | $1\frac{38}{100}$ | to | 1 |
| Pure water . . . . . . | as | 90 | to | 1 |

The difference between the laws of the condensation of pure water and of the same fluid when it holds in solution a portion of salt is striking; but when we trace *the effects* which are produced in the world by that arrangement, we shall be lost in wonder and admiration.

Let me beg the attention of my reader while I endeavour to investigate this most interesting subject, and let me at the same time bespeak his candour and indulgence. I feel the danger to which a mortal exposes himself who has the temerity to undertake to explain the designs of Infinite Wisdom. The enterprise is adventurous, but it cannot surely be improper.

The wonderful simplicity of the means employed by the Creator of the world to produce the changes of the seasons, with all the innumerable advantages to the inhabitants of the earth which flow from them, cannot fail to make a very deep and a lasting impression on every human being whose mind is not degraded, and quite callous to every ingenuous and noble sentiment; but the farther we pursue our inquiries respecting the constitution of the universe, and the more attentively we examine the effects produced by the various modifications of the active powers which we perceive, the more we shall be disposed to admire, adore, and love that great First Cause which brought all things into existence.

Though winter and summer, spring and autumn, and all the variety of the seasons, are produced in a manner at the same time the most simple and the most stupendous (by the inclination of the axis of the earth to the plane of the ecliptic), yet this mechanical contrivance alone would not have been sufficient (as I shall endeavour to show) to produce that gradual change of temperature in the various climates which we find to exist, and which doubtless is indispensably necessary to the preservation of animal and vegetable life.

Though change of temperature seems necessary to the growth and perfection of most vegetables, yet these changes must be within certain limits. Some plants can support greater changes of temperature than others, but the extremes of Heat and of Cold are alike fatal to all.

As the rays of the sun are the immediate cause of the Heat on the surface of the globe, and as the length of the days in high latitudes is so very different in summer and in winter, it is evident that, in order to render those regions habitable, some contrivance was necessary to prevent the consequences which this great inequality of the Heat generated by the sun in summer and in winter would naturally tend to produce; or, in other words, to equalize the Heat, and moderate its extremes in these two seasons.

Let us see how far *Water* is concerned in this operation, and then let us examine how far the remarkable law which has been found to obtain in its condensation by cold tends to render it well adapted to answer that most important purpose.

The vast extent of the ocean, and its great depth, but still more its numerous currents, and the power of water to absorb a vast quantity of Heat, render it peculiarly well adapted to serve as an equalizer of Heat.

On the retreat of the sun after the solstice, it is closely followed by the cold winds from the regions of eternal frost, which are continually endeavouring to press in towards the equator. As the power of the sun to warm the surface of the earth and the air diminishes very fast in high latitudes on the days growing shorter, it soon becomes too weak to keep back the dense atmosphere which presses on from the polar regions, and the cold increases very fast.

There is, however, a circumstance by which these rapid advances of winter are in some measure moderated. The earth, but more especially the *water*, having imbibed a vast quantity of Heat during the long summer days, while they receive the influence of the sun's vivifying beams; this Heat, being given off to the cold air which rushes in from the polar region, serves to warm it and soften it, and consequently to diminish the impetuosity of its motion, and take off the keenness of its blast. But as the cold air still continues to flow in as the sun retires, the accumulated Heat of summer is soon exhausted, and all solid and fluid bodies are reduced to the temperature of freezing water. In this stage the cold in the atmosphere increases very fast, and would probably increase still faster, were it not for the vast quantity of Heat which is communicated to the air by the watery vapours which are first condensed, and then congealed, in the atmosphere, and which afterwards fall upon the earth in the form of snow; and by that still larger quantity which is given off by the water in the rivers and lakes, and in the ground upon its being frozen.

But in very cold countries the ground is frozen and covered with snow, and all the lakes and rivers are frozen over in the very beginning of winter. The cold then first begins to be extreme, and there appears to be no source of Heat left, which is sufficient to moderate it in any sensible degree.

Let us see what must have happened if things had been left to what might be called their natural course, — if the condensation of water on being deprived of its Heat had followed the law which we find obtains in other fluids, and even in water itself in some cases, namely, when it is mixed with certain bodies.

Had not Providence interfered on this occasion in a manner which may well be considered as *miraculous*, all the fresh water within the polar circle must inevitably have been frozen to a very great depth in one winter, and every plant and tree destroyed ; and it is more than probable that the regions of eternal frost would have spread on every side from the poles, and, advancing towards the equator, would have extended its dreary and solitary reign over a great part of what are now the most fertile and most inhabited climates of the world !

In latitudes where now the return of spring is hailed by the voice of gladness, where the earth decks herself in her gayest attire, and millions of living beings pour forth their songs of joy and gladness, nothing would have been heard but the whistling of the rude winds, and nothing seen but ice and snow, and flying clouds charged with wintry tempests.

Let us, with becoming diffidence and awe, endeavour to see what the means are which have been employed by an almighty and benevolent God to protect his fair creation.

As nourishment and life are conveyed to all living creatures through the medium of water, — *liquid, living* water, — to preserve life, it was absolutely necessary to preserve a great quantity of water in a fluid state in winter as well as in summer.

But in cold climates the temperature of the atmosphere, during many months in the year, is so much below the freezing point, that, had not measures been taken to prevent so fatal an accident, all the water must inevitably have been changed to ice, which would infallibly have caused the destruction of every living thing.

Extraordinary measures were therefore necessary for preserving in a liquid state as much of the water existing in those climates as is indispensably necessary for the preservation of vegetable and animal life; and this could only be done by contriving matters so as to prevent this water from parting with its Heat to the cold atmosphere.

It has been shown, I believe I may venture to say proved, in the most satisfactory manner, that liquids part with their Heat ONLY in consequence of their internal motions; and that the more rapid these motions are, the more rapid is the communication of the Heat; that these motions are produced by the change in the specific gravity of the liquid, occasioned by the change of temperature; and of course that they are more rapid, as the specific gravity of the liquid is the more changed by any given change of temperature.

But it has been shown that the change in the specific gravity of water is extremely small, which takes place in any given change of temperature, *below the mean temperature of the atmosphere*, and particularly when the temperature of the water is very near the freezing point; and hence it follows that water must give off its Heat very slowly when it is near freezing.

But this is not all. There is a still more extraordinary, and in its consequences more wonderful, circumstance which remains to be noticed. When water is cooled to within eight or nine degrees of the freezing point, it not only ceases to be farther condensed, but is actually expanded by farther diminutions of its Heat; and this expansion goes on as the Heat is diminished, as long as the water can be kept fluid; and when it is changed to ice it expands even still more, and the ice floats on the surface of the uncongealed part of the Fluid.

Let us see how very powerfully this wonderful contrivance tends to retard the cooling of water when it is exposed in a cold atmosphere.

It is well known that there is no communication of Heat between two bodies as long as they are both at the same temperature ; and it is likewise known that the *tendency* of Heat to pass from a hot body into one which is colder, with which it is in contact, is greater, as the difference is greater in the temperature of the two bodies.

Suppose now that a mass of very cold air reposes on the quiet surface of a large lake of fresh water at the temperature of 55° of Fahrenheit's thermometer. The particles of water at the surface, on giving off a part of their Heat to the cold air with which they are in contact, and in consequence of this loss of Heat becoming specifically heavier than those hotter particles on which they repose, must of course descend. This descent of the particles which have been cooled necessarily forces other hotter particles to the surface, and these being cooled in their turns bend their course downwards ; and the whole mass of water is put into motion, and continues in motion as long as the process of cooling goes on.

Before I proceed to trace this operation through all its various stages, I must endeavour to remove an objection which may perhaps be made to my explanation of this phænomenon. As I have supposed the mass of air which rests on the surface of the water to be *very cold*, and as I have taken it for granted that there is no communication whatever of Heat between the particles of water in contact with this very cold air and the neighbouring warmer particles of water, it may be asked how it happens that these particles at the surface are not so much cooled as to be immediately changed to ice. To this I answer, that there are two causes which conspire to prevent the *immediate* formation of ice at the surface of the water: *First,* the specific gravity of the particle of water at the surface being increased at the same moment when it parts with Heat, it begins to descend as soon as it begins to be cooled, and before the air has had time to rob it of all its Heat, it escapes and gets out of its reach; and, *secondly,* air being a bad conductor of Heat, it cannot receive and transmit or *transport it* with sufficient celerity to cool the surface of water so suddenly as to embarrass the motions of the particles of that liquid in the operation of giving it off.

But to return to our lake. As soon as the water in cooling has arrived at the temperature of about 40°, as at that temperature it ceases to be farther condensed, its internal motion ceases, and those of its particles which happen to be at its surface remain there; and, after being cooled down to the freezing point, they give off their latent Heat, and ice begins to be formed.

As soon as the surface of the water is covered with ice, the communication of Heat from the water to the atmosphere is rendered extremely slow and difficult; for ice being *a bad conductor of Heat* forms a very warm covering to the water, and moreover it prevents the water from being agitated by the wind. Farther, as the temperature of the ice at its lower surface is always very nearly the same as that of the particles of liquid water with which it is in contact (the warmer particles of this Fluid, in consequence of their greater specific gravity, taking their places below), the communication of Heat between the water and the ice is necessarily very slow on that account.

As soon as the upper surface of the ice is covered

with snow (which commonly happens soon after the ice is formed), this is an additional and very powerful obstacle to prevent the escape of the Heat out of the water; and though the most intense cold may reign in the atmosphere, the increase of the thickness of the ice will be very slow.

During this time the mass of water which remains unfrozen will lose *no part of its Heat;* on the contrary, it will continually be receiving Heat from the ground. This Heat, which is accumulated in the earth during the summer, will not only serve, in some measure, to replace that which is communicated to the atmosphere through the ice, and prevent its being furnished at the expence of the latent Heat of the water in contact with its surface, but when the temperature of the air is not much below that of freezing, this supply of Heat from below will be quite sufficient to replace that which the air carries off; and the thickness of the ice will not increase.

Whenever the temperature of the air is not actually *colder* than freezing water, the Heat which rises from the bottom of the lake will be all employed in melting the ice at its under surface, and diminishing its thickness.

It will indeed frequently happen, when the ice is very thick, and especially when its upper surface is covered with deep snow, that the melting of the ice at its under surface will be going on, when the temperature of the atmosphere is considerably below the freezing point.

As the particles of water which, receiving Heat from the ground at the bottom of the lake, acquire a higher temperature than that of 40°, and being *expanded*, and becoming specifically lighter by this additional Heat, rise up to the upper surface of the fluid water, and give off their sensible Heat to the under surface of the ice, never return to the bottom, this communication of the Heat which exhales from the earth produces very little motion in the mass of the water; and this circumstance is, no doubt, very favourable to the preservation of the Heat of the water.

When a strong wind prevails, and the surface of the water is much agitated, ice is not formed, even though the whole mass of water should, by a long continuance of cold weather, have been previously cooled down to that point to which it is necessary that it should be brought, in order that its internal motions may cease, and it may be disposed to congeal; for though the particles at and near the surface may no longer have any tendency to descend, on being farther cooled, yet, as they have so considerable a quantity of sensible Heat (eight or ten degrees) to dispose of, after their condensation with cold ceases, and as the agitation into which the water is thrown by the wind does not permit any particle to remain long enough in contact with the cold air to give off all its Heat at once, there is a continual succession of fresh particles at the surface, all of which give off Heat to the air; but none of them have time to be cooled sufficiently to be disposed to form ice. The water will lose a vast quantity of Heat, and as soon as the wind ceases, if the cold should continue, ice will be formed very rapidly.

But it is not merely the agitation of the water which renders the communication of the Heat very rapid, the agitation of the wind also tends to produce the same effect.

On the return of spring, the snow melting before the sun as he advances and his rays become more powerful, all the Heat which the earth exhales is employed in dissolving the ice at its under surface, while the sun on the other side acts still more powerfully to produce the same effect.

Though ice is transparent, yet it is not perfectly so; and as the light which is stopped in its passage through it cannot fail to generate Heat *when* and *where* it is stopped or absorbed, it is by no means surprising that snow should be found to melt when exposed in the sun's rays, even when the temperature of the air in the shade is considerably below the point of freezing. Snow exposed to the sun melts long before the even surface of ice begins to be sensibly softened by its beams, and it is not till some time after all the hills are bare that the ice on the lakes and rivers breaks up.

The rays which penetrate a bank of snow, being often reflected and refracted, descend deep into it, and the Heat is deposited in a place where it is not exposed to be carried off by the cold air of the atmosphere; but the rays which fall upon the horizontal and smooth surface of ice are mostly reflected upwards into the atmosphere; and if any part of them are stopped at the surface of the ice, the Heat generated by them *there* is instantaneously carried off by the cold air, and a particle of water is no sooner made fluid than it is again frozen.

Hence we see that the snow which in cold countries covers the ice that is formed on the surface of fresh water not only prevents the Heat of the water from being carried off by the air during the winter, but also assists very powerfully in thawing the ice early in the spring.

Should the waters of a lake be so deep, or so imperfectly transparent as to intercept a great proportion of rays of the sun before they reach the bottom, in that case, the temperature of the water at the bottom of the lake will be *nearly the same all the year round;* and in countries where there is *any* frost in winter, and particularly in those lakes which lie near high mountains, and are fed by torrents which proceed from *Glaciers*, and melting snow, this *constant temperature* at the bottom can never be much above or below 41° F., whatever may be the Heat to which the *surface* of the lake is exposed in summer, or however long and intensely hot the summer may be.*

Let us now see what the consequences would have been, had the condensation of water with cold followed the law which obtains in regard to all other fluids.

As the internal motion of the water could not have failed to continue as long as its specific gravity continued to be increased by parting with Heat, ice would not have begun to be formed till the whole mass of water had arrived at the temperature of 32° of Fahrenheit's thermometer.

To see what an enormous quantity of Heat would be

* In a letter from Professor Pictet, of Geneva, to the Author, of the 7th July, 1797, accompanying the 36th number of the BIBLIOTHÈQUE BRITANNIQUE (in which an account, or rather translation, of the first Edition of this Essay is published in the French language), there is the following paragraph : —

"I took the liberty to throw in, as usual," (in the translation,) "some occasional notes ; one of which will, I hope, deserve your attention. It points out the near coincidence of the mean temperature of the bottom, observed in ten different lakes, by M. de Saussure and myself, viz. 4⅓° R. (equal to 41¾° F.) with the temperature where the *minimum* of volume, or *maximum* of density, of water takes place. We vainly strove to this day to explain the uniformity we observed in that particular in several lakes very differently situated in many respects, but your reflections seem to me fully to resolve the problem."

The following is the note in the *Bibliothèque Britannique*, alluded to by Professor Pictet in the foregoing paragraph of his letter : —

"Ce n'est pas seulement dans le lac de Genève que M. de Saussure, notre savant ami, a fait les expériences curieuses qui sont ici rappellées, et à quelques-unes des quelles nous avons eu le plaisir d'assister ; il les a répétées dans la Méditerranée, et dans dix lacs qui bordent de part et d'autre la chaine des Alpes. Nous tirons de lost when the water is deep in consequence of its whole mass being cooled in this manner, we have only to compute how much ice this Heat would melt, or how much water it would heat from the point of freezing to that of boiling.

It has been shown by experiment, that any given quantity of ice requires as much Heat to melt it as an equal quantity of fluid water loses in cooling 140 degrees; consequently, the quantity of ice which might be melted by the Heat given off bv any given quantity of water in cooling any given number of degrees is to the given quantity of water as the number of degrees which it is cooled to 140 degrees.

Hence it follows that when the temperature of the water is 8 degrees above the freezing point, it gives off

son grand ouvrage sur les montagnes les températures observées au fond de ces lacs comme suit : —

| Noms des Lacs. | Profondeurs en pieds de France. | Températures du fond Degrés de Reamur. |
|---|---|---|
| "Lac de Genève | 950 | 4.3 |
| " de Neuchâtel | 325 | 4.1 |
| " de Bienne | 217 | 5.5 |
| " du Bourget | 240 | 4.5 |
| " d'Annecy | 163 | 4.5 |
| " de Thun | 350 | 4.0 |
| " de Brientz | 500 | 3.8 |
| " de Lucerne | 600 | 3.9 |
| " de Constance | 370 | 3.4 |
| Lac Majeur | 335 | 5.4 |
| Température moyenne du fond de dix lacs | | 4.34, ou 4⅓° R." |

"Il n'est peutêtre aucun de nos lecteurs qui, plein des idées que notre auteur vient de discuter, ne soit frappé de la coincidence entre cette température du fond des lacs dans nos latitudes moyennes et celle à laquelle l'eau atteint son *minimum* de volume ou *maximum* de densité ! La permanence de cette température, et son identité dans des lacs d'ailleurs très-diversement situés, paroissent intimément liées avec cette circonstance du *minimum* de volume. Mais ce n'est pas ici le lieu de donner cours aux idées que peut suggerer ce rapprochement ; nous l'indiquons à l'auteur comme un objet digne de ses méditations."

The Author of this Essay feels himself very much obliged to his ingenious and respectable friend, Professor Pictet, for these interesting observations.

in cooling down to that temperature as much Heat as would melt $\frac{8}{140}$ or $\frac{2}{35}$ of its weight of ice; the water, therefore, which is cooled from the temperature of 40° to that of 32°, if it be 35 feet deep, will give off as much Heat in being so cooled as would melt a covering of ice 2 feet thick.

But this even is not all; for as the particles of water on being cooled at the surface would, in consequence of the increase of their specific gravity on parting with a portion of their Heat, immediately descend to the bottom, the greatest part of the Heat accumulated during the summer in the earth on which the water reposes would be carried off and lost before the water began to freeze; and when ice was once formed, its thickness would increase with great rapidity, and would continue increasing during the whole winter; and it seems very probable, that, in climates which are now temperate, the water in the large lakes would be frozen to such a depth in the course of a severe winter that the Heat of the ensuing summer would not be sufficient to thaw them; and should this once happen, the following winter could hardly fail to change the whole mass of its waters to one solid body of ice, which never more could recover its liquid form, but must remain immovable till the end of time.

In the month of February, after a frost which had lasted a month, the temperature of the air being 38°, M. de Saussure found the temperature of the water of the Lake of Geneva, at the surface, at 41°, and at the depth of 1000 feet at 40°. Had the frost continued but a little longer, ice would have been formed; but had the constitution of water been such that the whole mass of that fluid in the lake must have been cooled down to the temperature of 32° before ice could have been formed, this event could not have happened till the water had given off as much Heat as would be sufficient to melt a covering of ice above 57 feet thick!

This quantity of Heat would be sufficient to heat, to the point of boiling, a quantity of ice-cold water as large as the lake, and 49 feet deep.

We cannot sufficiently admire the simplicity of the contrivance by which all this Heat is saved. It well deserves to be compared with that by which the seasons are produced; and I must think that every candid enquirer who will begin by divesting himself of all unreasonable prejudices will agree with me in attributing them both TO THE SAME AUTHOR.

When we trace still farther the astonishing effects which are produced in the world by the operations of that simple law which has been found to obtain in the condensation of water on its being deprived of Heat, we shall find more and more reason to admire the wisdom of the contrivance.

That high latitudes might be habitable, it was necessary that vegetables should be protected from the effects of the chilling frosts of a long and severe winter; but if it be true that watery liquids do not part with their Heat but in consequence of their internal motions, and if these motions are occasioned merely by the change produced in the specific gravity of those particles of the liquid which receive Heat, or which part with it, who does not see how very powerfully the sudden diminution and final cessation of the condensation of water in cooling, as soon as its temperature approaches to the freezing point, operates to prevent the sap in vegetables from being frozen?

But if, for the purposes of life and vegetation, it be necessary that the ground, the rivers, the lakes, and the trees be defended from the cold winds from the poles, it may be asked how this inundation of cold air is to be warmed? I answer by the waters of the ocean, which there is the greatest reason to think were not only designed principally for that use, but particularly *prepared* for it.

Sea water contains a large proportion of salt in solution; and we have seen that the condensation of a saline solution, on its being cooled, follows a law which is extremely different from that observed in regard to pure water; and which (as may easily be shown) renders it peculiarly well adapted for communicating Heat to the cold winds which blow over its surface.

As sea water continues to be condensed as it goes on to cool, even after it has passed the point at which fresh water freezes, the particles at the surface, instead of remaining there after the mass of the water had been cooled to about 40°, and preventing the other warmer particles below from coming in their turns and giving off their Heat to the cold air (as we have seen always happens when fresh or pure water is so cooled), these cooled particles of *salt water* descend as soon as they have parted with their Heat, and in moving downward force other warmer particles to move upwards; and in consequence of this continual succession of warm particles which come to the surface of the sea, a vast deal of Heat is communicated to the air, — incomparably more than could possibly be communicated to it by an equal quantity of fresh water at the same temperature, as will appear by the following computation.

Without taking into the account that very great advantage which sea water possesses over fresh water, considered as an equalizer of the temperature of the atmosphere, which arises from the comparative *lowness of the point of its congelation;* supposing even sea water to freeze at as high a temperature as fresh water, namely, at 32°; and supposing (what is strictly true) that as soon as either sea water or fresh water is frozen at its surface, and this ice covered with snow, the communication of Heat from the water to the atmosphere ceases almost entirely, — we will endeavour to determine how much more Heat would, even on this supposition, be communicated to the air by salt water than by fresh water, after both have arrived at the temperature of 40°.

When fresh water, in cooling, has arrived at this temperature, it ceases to be farther condensed with cold, and its internal motions (which, as we have already more than once observed, are caused *solely* by the changes produced in the specific gravity of its particles) cease, of course, and ice immediately begins to be formed on its surface; but as the condensation of salt water goes on as its Heat goes on to be diminished, its internal motions will continue; and it is evidently impossible for ice to be formed at its surface till the whole mass of the water has become ice-cold, or till its temperature is brought down to 32°. It would therefore give off a quantity of Heat equal to 8 degrees, at least, of Fahrenheit's thermometer, *more than the fresh water* would part with before ice could be formed on its surface.

To be able to form an idea of this enormous quantity of Heat, we have only to recollect what has already been said, and we shall find reason to conclude that it would be sufficient to melt a covering of ice equal in thickness to $\frac{2}{36}$ of the depth of the sea. It would therefore be sufficient in that part of the North Sea (lat. 67°) where Lord Mulgrave sounded at the depth of 4680 feet, to melt a cake of ice 265 feet thick!

But the Heat evolved in the formation of each superficial foot of ice would be sufficient to raise the temperature of a stratum of incumbent air 2220 times as thick as the ice (consequently, in the case in question, 265 × 2220 feet, or 869 miles thick) 28 degrees, or from the temperature of freezing water to that of 50° of Fahrenheit's thermometer, or to the mean annual temperature of the northern parts of Germany!

The Heat given off to the air by each superficial foot of water in cooling *one degree* is sufficient to heat an incumbent stratum of air 44 times as thick as the depth of the water 10 degrees. Hence we see how very powerfully the water of the ocean, which is never frozen over, except in very high latitudes, must contribute to warm the cold air which flows in from the polar regions.

But the ocean is not more useful in moderating the extreme cold of the polar regions than it is in tempering the excessive heats of the torrid zone; and what is very remarkable, the fitness of the sea water to serve this last important purpose is owing to the very same cause which renders it so peculiarly well adapted for communicating Heat to the cold atmosphere in high latitudes, namely, *to the salt which it holds in solution.*

As the condensation of salt water with cold continues to go on even long after it has been cooled to the temperature at which fresh water freezes, those particles at the surface which are cooled by an immediate contact with the cold winds must descend, and take their places at the bottom of the sea, where they must remain, till, by acquiring an additional quantity of Heat, their specific gravity is again diminished. But this Heat *they never can regain in the polar regions*; for innumerable experiments have proved, beyond all possibility of doubt, that there is no *principle of Heat* in the *interior parts of the globe*, which, by exhaling through the bottom of the ocean, could communicate Heat to the water which rests upon it.

It has been found that the temperature of the earth at great depths under the surface is different in different latitudes, and there is no doubt but this is also the case with respect to the temperature at the bottom of the sea, in as far as it is not influenced by the currents which flow over it; and this proves to a demonstration that the Heat which we find to exist, without any sensible change during summer and winter, at great depths, is owing to the action of the sun, and not to *central fires*, as some have too hastily concluded.

But if the water of the ocean, which, on being deprived of a great part of its Heat by cold winds, descends to the bottom of the sea, cannot be warmed *where it descends*, as its specific gravity is greater than that of water at the same depth in warmer latitudes, it will immediately begin to spread on the bottom of the sea, and to flow towards the equator, and this must necessarily produce a current at the surface in an opposite direction; and there are the most indubitable proofs of the existence of both these currents.

The proof of the existence of one of them would indeed have been quite sufficient to have proved the existence of both, for one of them could not possibly exist without the other; but there are several direct proofs of the existence of each of them.

What has been called the Gulf Stream in the Atlantic

Ocean is no other than one of these currents, — that at the surface, which moves from the equator towards the north pole, modified by the trade winds and by the form of the continent of North America; and the progress of the lower current may be considered as proved directly by the cold which has been found to exist in the sea at great depths in warm latitudes, — a degree of temperature much below the mean annual temperature of the earth in the latitudes where it has been found, and which of course must have been *brought from colder latitudes.*

The mean annual temperature in the latitude of 67° has been determined by Mr. Kirwan, in his excellent treatise on the temperature of different latitudes, to be 39°; but Lord Mulgrave found on the 20th of June, when the temperature of the air was $48\frac{1}{2}°$, that the temperature of the sea at the depth of 4680 feet was 6 degrees below freezing, or 26° of Fahrenheit's thermometer.

On the 31st of August, in the latitude of 69°, where the annual temperature is about 38°, the temperature of the sea at the depth of 4038 feet was 32°; the temperature of the atmosphere (and probably that of the water at the surface of the sea) being at the same time at $59\frac{1}{2}°$.

But a still more striking, and I might, I believe, say an incontrovertible proof of the existence of currents of cold water at the bottom of the sea, setting from the poles towards the equator, is the very remarkable difference that has been found to subsist between the temperature of the sea at the surface and at great depths, at the tropic; though the temperature of the atmosphere there is so constant that the greatest change produced in it by the seasons seldom amounts to more than five or six degrees, yet the difference between the Heat of the water at the surface of the sea, and that at the depth of 3600 feet has been found to amount to no less than 31 degrees; the temperature above or at the surface being 84°, and at the given depth below no more than 53°.*

It appears to me to be extremely difficult, if not quite impossible, to account for this degree of cold at the bottom of the sea in the torrid zone on any other supposition than that of cold currents from the poles; and the utility of these currents in tempering the excessive heats of those climates is too evident to require any illustration.

These currents are produced, as we have already seen, in consequence of the difference in the specific gravity of the sea water at different temperatures; their velocities must therefore be in proportion to the change produced in the specific gravity of water by any given change of temperature; and hence we see how much greater they must be in salt water than they could possibly have been had the ocean been composed of fresh water.

It is not a little remarkable that the water of all great lakes is fresh, and nearly so in all inland seas (like the Baltic) in cold climates, and which communicate with the ocean by narrow channels. We shall find reason to conclude that this did not happen without design, when we consider what consequences would probably ensue should the waters of a large lake in an inland situation, in a cold country (such as the lake Superior, for instance, in North America), become as salt as the sea.

Though the cold winds which blow over the lake in the beginning of winter would be more warmed, and the temperature of the air on the side of the lake opposite to the quarter from whence these winds arrive would be rendered somewhat milder than it now is; yet, as the

* Phil. Transactions, 1752.

water of the lake would give off an immense quantity of Heat before a covering of ice could be formed on its surface for its protection, it would, on the return of spring, be found to be *extremely cold*; and as it would require a long time to regain from the influence of the returning sun the enormous quantity of Heat lost during the winter, it would remain very cold during the spring, and probably during the greatest part of the summer; and this could not fail to chill the atmosphere, and check vegetation in the surrounding country to a very considerable distance. And though a large lake of salt water in a cold country would tend to render the winter *somewhat milder* on one side of it, namely, on the side opposite to the quarter from whence the cold winds came; yet this advantage would not only be confined to a small tract of country, but would not anywhere be very important, and would by no means counterbalance the extensive and fatal consequences which would be produced in summer by so large a collection of very cold water.

When the winter is once fairly set in, — when the earth is well covered with snow, and the rivers and lakes with ice, and more especially when the ice as well as the land is covered with that warm winter garment, a few degrees more of cold in the air cannot produce any lasting bad consequences. It may oblige the inhabitants to use additional precautions to guard themselves, their domestic animals, and their provisions from the uncommon severity of the weather; but it can have very little influence in the temperature of the ensuing summer; and even it is probable, if it influences it at all, that it tends rather to make it *warmer* than *colder*. Lakes of salt water could therefore be of no real use *in winter* in cold countries, and in summer they could not fail to be very hurtful; while fresh lakes, as they are frozen over almost as soon as the winter sets in, and long before the whole mass of their water is cooled down to the temperature of freezing, preserve the greater part of their Heat through the winter, and if they are of no use during the cold season, they probably do little or no harm in summer.

But I must take care not to tire my reader by pursuing these speculations too far. If I have persisted in them, if I have dwelt on them with peculiar satisfaction and complacency, it is because I think them uncommonly interesting, and also because I conceived that they might be of real use in this age of *refinement* and *scepticism.*

If, among barbarous nations, the *fear of a God*, and the practice of religious duties, tend to soften savage dispositions, and to prepare the mind for all those sweet enjoyments which result from peace, order, industry, and friendly intercourse, — a *belief in the existence of a Supreme Intelligence*, who rules and governs the universe with wisdom and goodness, is not less essential to the happiness of those who, by cultivating their mental powers, HAVE LEARNED TO KNOW HOW LITTLE CAN BE KNOWN.

## DESCRIPTION OF THE PLATES.

### PLATE I.

THIS Plate represents the cylindrical Passage Thermometer used in the experiments on the conducting power of liquids with regard to Heat.

Fig. 1. *a*, *b*, is a section of the brass tube in which the Thermometer *c*, with an oblong copper bulb, is placed.

*e*, *f*, is the glass tube of the thermometer, which, for want of room in the Plate, is represented as broken off at *f*.

*g*, is a stopple of cork by which the end of the brass tube, *a*, *b*, is closed; and

*h*, is a circular disk of the same substance.

The space in the brass tube below this disk *h*, surrounding the bulb of the thermometer, was occupied by the liquid whose conducting power was determined. The space between the disk and the cork-stopper *g*, was filled with eider-down.

Between the inside of the brass tube and the lower part of the bulb of the thermometer are seen the wooden pins which served to confine the thermometer in its place.

Fig. 2. This is an horizontal section of the brass tube, and a bird's-eye view of the thermometer in its place.

### PLATE II.

Fig. 3. This Figure shows the manner in which the experiments were made, in which a cake of ice at the

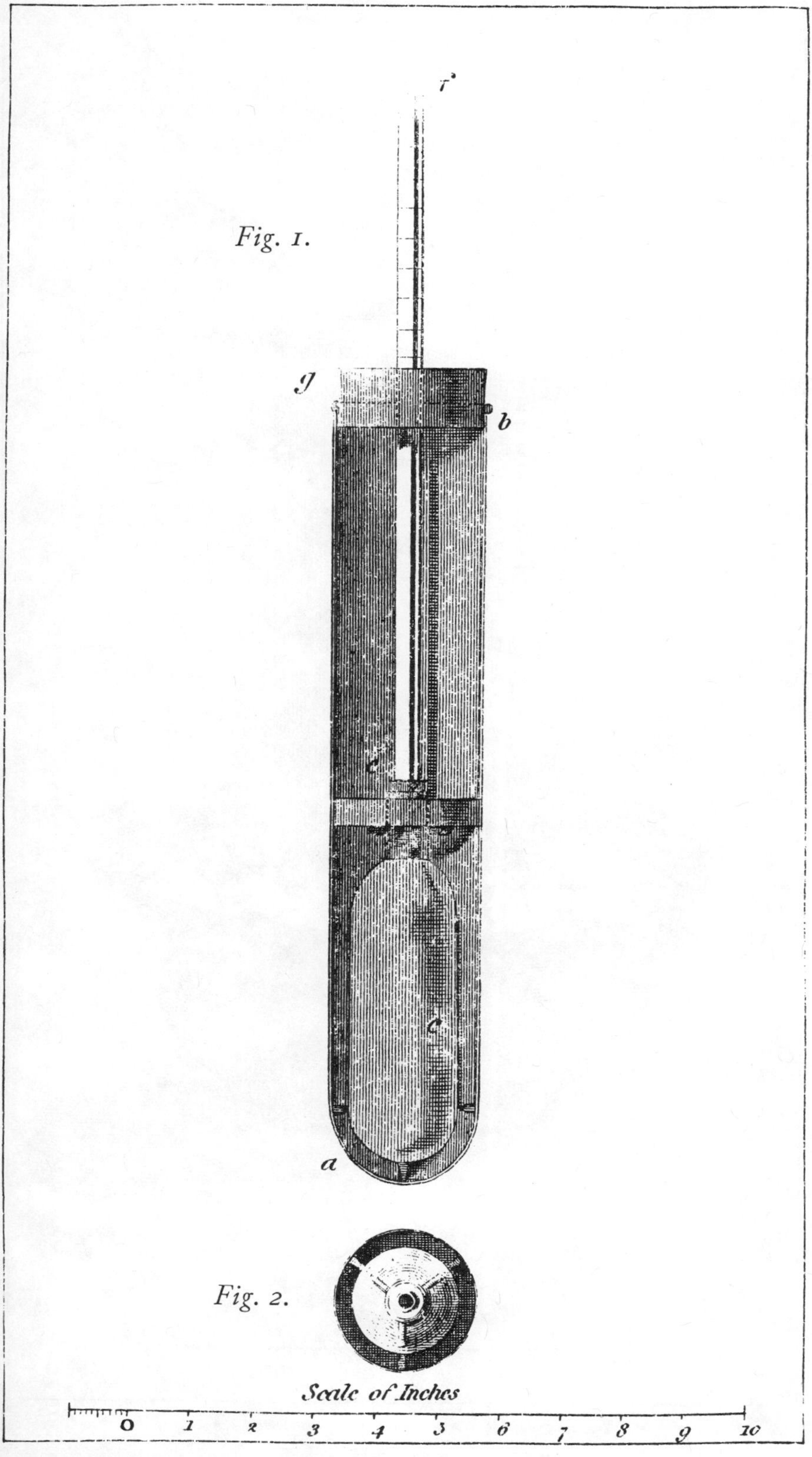
r
Fig. 1.
g
b
d
c
a
Fig. 2.
Scale of Inches
0 1 2 3 4 5 6 7 8 9 10

Plate II.

Fig. 3.

Scale of Inches

bottom of a tall glass jar was thawed by hot water standing on its surface.

*a*, is an earthen bowl filled with pounded ice and water, in which the glass jar, *b*, was placed.

*c*, *d*, is the level of the upper surface of the ice in the jar.

*e*, *f*, is the level of the surface of the water standing on the ice in the jar.

# PART II.

## AN ACCOUNT

OF

## SEVERAL NEW EXPERIMENTS,

WITH

## OCCASIONAL REMARKS AND OBSERVATIONS,

AND

## CONJECTURES

RESPECTING CHEMICAL AFFINITY AND SOLUTION, AND THE MECHANICAL PRINCIPLE OF ANIMAL LIFE.

## CHAPTER I.

*Account of a Circumstance of a private Nature, by which the Author has been induced to add this and the following Chapters to the Second Edition of this Essay. — Experimental Investigation of the Subject continued. —* OIL *found by Experiment to be a Non-Conductor of Heat. —* MERCURY *is likewise a Non-Conductor. — Probability that all* FLUIDS *are* NON-CONDUCTORS, *and that this Property is* ESSENTIAL TO FLUIDITY. *— The Knowledge of that Fact may be of great Use in enabling us to form more just Ideas with regard to the Nature of those mechanical Operations which take place in chemical Solutions and Combinations, in the Process of Vegetation, and in the various Changes effected by the Powers of Life in the Animal Economy. — Rapidity of Solution no Proof of the Existence of an Attraction of Affinity. — Strata of fresh Water and of salt Water may be made to repose on each other in actual Contact without mixing. — Probability that* *the Water at the Bottom of fresh Lakes, that are very deep, may be actually salt.*

AT the end of a French translation of the First Edition of this Essay, published at Geneva, Professor Pictet (the translator) has added the following extract of one of my private letters to him (of the 9th of June, 1797), written in answer to one from him to me, acknowledging the receipt of a manuscript copy of the Essay which I had sent him.

"I should have been much surprised if my Seventh Essay had not interested you;[8] for in my life I never felt pleasure equal to that I enjoyed in making the experiments of which I have given an account in that performance. You will perhaps be surprised when I tell you, that I have suppressed a whole Chapter of interesting speculation, merely with a view of leaving to others a tempting field of curious investigation *untouched*, and to give more effect to my concluding reflection, which I consider as being by far the most important of any I have ever published."

As these assertions (which were not originally intended for the public eye) are liable to several interpretations, I think it my duty, not only to explain them, but also to let the Public know precisely how far I have pushed my inquiries in the investigation of the subject under consideration: This is an act of justice which I owe to those who may be engaged in the same pursuits; for it would be very unfair, by *obscure hints of important information kept back*, to keep others in doubt with respect to the originality of the discoveries they may make in the prosecution of *their* investigations. This would tend to *damp* the spirit of inquiry, instead of *exciting* it; and throwing out such hints looks so much like lying in wait to seize on the fair fruits of the labours of others, that I cannot rest till I have shewn that I do not deserve to be suspected of such pitiful views.

My worthy friend, Professor Pictet, certainly did not suspect any unhandsome design in any thing I said to him in my (private) letter; but those who are less acquainted with my character may not be disposed to give me credit for candour and disinterestedness without proofs.

With regard to the assertion in my letter, "that I had suppressed a whole Chapter of interesting speculation, with a view to leaving to others a tempting field, *untouched*, for curious investigation,"—this is perfectly true in fact, as will, I flatter myself, appear by what I shall now lay before the Public; and I am confident that those who will take the trouble to consider with attention the reasons which induced me to do this, will find them such as will deserve their approbation.

Having, as I flattered myself, laid open a new and most enticing prospect to those who are fond of philosophical pursuits, I was afraid, if I advanced too far, that others, instead of striking out roads for themselves, might perhaps content themselves with following in my footsteps; and consequently that many, and probably the most interesting, parts of the new field of inquiry would remain a long time unexplored. And with regard to the reputation of being a *discoverer*, though I rejoice—I might say, exult and triumph—in the progress of human knowledge, and enjoy the sweetest delight in contemplating the advantages to mankind which are derived from the introduction of useful improvements; yet I can truly say, that I set no very high value on the honour of being the first to stumble on those treasures which everywhere lie so slightly covered.

In respect to the "concluding reflection" of the First Edition of this Essay, — though some may smile in pity, and others frown at it, I am neither ashamed nor afraid to own, that I consider the subject as being of the *utmost importance* to the peace, order, and happiness of mankind, *in our present advanced state of society.* But to return from these digressions —

Though it appeared to me that the important fact I undertook to investigate, relative to the *manner* in which heat is propagated in Fluids, is fully established by the experiments, of which an account has been given in the preceding Chapters of this Essay; yet, as a thorough examination of the subject is a matter of much importance in many respects, I did not rest my inquiries here, but made a number of experiments with a view to throwing still more light upon it, and enabling us to form more clear and distinct ideas respecting those curious mechanical operations which appear to take place in Fluids, when Heat is propagated in them.

Having frequently observed when a quantity of water in one of my glass jars was frozen to a cake of ice, by placing the jar in a freezing mixture, that, as the ice first began to be formed at the sides of the jar, and increased gradually in thickness, the portion of water in the axis of the jar (which last retained its fluidity), being compressed by the expansion of the ice, was always forced upwards towards the end of the process, and formed a pointed projection of ice in the form of a nipple (*papilla*), which was sometimes above half an inch high in the middle of the upper side of the cake, — I was led by that circumstance to make the following interesting experiments.

*Experiment No.* 55.

A cake of ice, 3 inches thick, which had a pointed projection, ½ an inch high, which arose from the center of its upper surface, being frozen fast in the bottom of a tall cylindrical glass jar, 4¾ inches in diameter, this jar, standing in an earthen pan, and being surrounded by pounded ice and water, to the height of an inch above the level of the upper surface of the cake of ice, was placed on a table, near a window, in a room where the air was at the temperature of 31° of Fahrenheit's thermometer; and fine *olive oil*, which had previously been cooled down to the temperature of 32°, was poured into the jar till it stood at the height of 3 inches above the surface of the cake of ice.

Having ready a solid cylinder of wrought iron, 1¼ inch in diameter, and 12 inches long, with a small hook at one end of it, by means of which it could occasionally be suspended in a vertical position, and furnished with a fit hollow cylindrical sheath of thick paper, into which it just passed, — open at both ends, and about $\frac{1}{10}$ of an inch longer than the solid cylinder of iron, to which it served as a covering for keeping it warm, — this iron cylinder, being heated to the temperature of 210° in boiling water, and being suddenly introduced into its sheath, was suspended by an iron wire which descended from the ceiling of the room, in such a manner that its lower end entering the jar (in the direction of its axis) was immersed in the oil to such a depth that the middle of the flat surface of this end of the hot iron, which was directly above the point of the conical projection of ice, was distant from it only $\frac{2}{10}$ of an inch. The end of the sheath descended $\frac{1}{10}$ of an inch lower than the end of the hot metallic cylinder.

As the oil was very transparent, and the jar placed in a favourable light, the conical projection of ice was perfectly visible, even after the hot cylinder was introduced into the jar; and had *any Heat* DESCENDED through the thin stratum of fluid oil which remained interposed between the hot surface of the iron and the pointed projection of ice which was under it, there is no doubt but this Heat must have been apparent, by the melting of the ice; which event would have been discovered, either by the diminution of the height of this projection, or by an alteration of its form. But this was not the case: the ice did not appear in the smallest degree diminished, or otherwise affected by the vicinity of the hot iron.

My reader will naturally suppose, without my mentioning the circumstance, that due care was taken, in introducing the cylinder into the jar, to do it in the most gentle manner possible, to prevent the oil from being thrown into undulatory motions; and that proper means were used for confining the cylinder, motionless, in its place, when it had arrived there.

As this experiment appears to me to be unexceptionable, and its result unequivocal and decisive, in order that a perfect idea may the more easily be formed of it, I have added the Figure 4, where a section of the whole of the apparatus used in making it may be seen, expressed in a clear and distinct manner.

If the general result of the experiments, of which an account has been given in the two first Chapters of this Essay, afforded reason to conclude that *water* is a *non-conductor* of Heat, the result of that here described certainly proves, in a manner quite as satisfactory, that *oil* is also a *non-conductor;* and serves to give an additional degree of probability to the conjecture, that all Fluids are *necessarily* non-conductors of heat.

As *mercury*, which is a metal in fusion, is different in many respects from all other Fluids, I was very impatient to know if it agreed with them in that essential property, from which they have been denominated non-conductors of Heat, and this I found to be actually the case, by the result of the following decisive experiment.

*Experiment No.* 56.

Having emptied and cleaned out the cylindrical glass jar used in the last-mentioned experiment, and replenished it with a fresh cake of ice, with a conical projection in the middle of its upper side, I placed the jar, surrounded by pounded ice and water, on the table, in the cold room, where the foregoing experiment had been made; and poured over the cake of ice as much ice-cold *mercury* as covered it to the height of about an inch. Having cleaned the surface of the mercury in the jar with blotting-paper, I suffered the whole to remain quiet about an hour; and then very gently introduced the end of the hot cylinder of iron (inclosed in its paper sheath) into the mercury, and fixed it immoveably in such a position, that its flat end, which was naked, was immediately over the point of the conical projection of ice, and distant from it about $\frac{1}{4}$ of an inch; where I suffered it to remain several minutes.

It is necessary that I should mention, that, in order to prevent the internal motions in the mass of mercury, which would otherwise have been occasioned by the rising and spreading out on its surface of those particles of that fluid, which, having touched the flat end of the hot iron, became specifically lighter in consequence of their increase of temperature, the end of the hollow cylindrical sheath, in which the solid cylinder of iron was placed, was made to project about $\frac{1}{10}$ of an inch below the flat end of the iron. This precaution was likewise used, and for a similar reason, in the preceding experiment, when oil was used in the place of the mercury; as was mentioned, though without being explained, in giving an account of that experiment.

As the cake of ice, on which the mercury reposed, was at that temperature precisely at which ice is disposed to melt with the smallest additional quantity of Heat, if *any Heat* had found its way *downwards* through the mercury to the ice in this experiment, water would most undoubtedly have been formed, and this water would as undoubtedly have appeared on the surface of the mercury on taking away the iron; but there was not the smallest appearance of any ice having been melted.

To find out whether the cake of ice was *really* at that temperature at which it was disposed to melt with any additional Heat, I thrust down the end of my finger through the mercury, and touched the ice; and this experiment removed all my doubts, for I found that, however expeditiously I performed that operation, it was hardly possible for me to touch the ice without evident signs of water having been produced being left behind, on the clean and bright surface of the mercury, on taking away my finger.

From the results of all these experimental investigations it appears to me that we may safely conclude that *water*, *oil*, and *mercury* are perfect *non-conductors* of Heat; or, that when either of those substances takes the form of a Fluid, all interchange and communication of Heat *among its particles*, or from one of them to the other, directly, becomes from that moment *absolutely impossible.*

That this is also the case with respect to the particles

of *air*, has been rendered extremely probable — I believe I might say proved — by the experiments of which I gave an account in one of my papers on Heat published in the Transactions of the Royal Society; and I have shewn elsewhere (in my Sixth Essay) how much reason there is to conclude that the particles of *Steam* and of *Flame* are in the same predicament.

But if all interchange and communication of Heat, from particle to particle, *immediately*, or *de proche en proche*, be absolutely impossible in so many *elastic* and *unelastic Fluids*, and in Fluids so essentially different in many other respects, are there not sufficient grounds to conclude from hence, that this property is common to all Fluids, and that it is even *essential to fluidity?*

It is easy to conceive that the discovery of so important a circumstance must necessarily occasion a considerable change in the ideas we have formed in respect to the mechanical operations which take place in many of the great phenomena of Nature; as well as in many of those still more interesting chemical operations, which we are able to direct, but which we find, alas! very difficult to explain.

In my paper on Heat, above mentioned, published in the Philosophical Transactions for the year 1792,[7] I endeavoured to apply the discovery of the non-conducting power of *air* in accounting for the warmth of the hair of beasts, of the feathers of birds, of artificial clothing, and of snow, the winter garment of the earth; and also, in explaining the causes of the cold winds from the polar regions, and of their different directions in different countries, which prevail at the end of winter, and early in the spring.

In my Sixth Essay (on the Management of Heat and

the Economy of Fuel)[6] I availed myself of the knowledge of the non-conducting power of *steam* and of *flame*, in explaining the effects of a blow-pipe in increasing the action of pure flame, and in investigating the most advantageous forms for boilers; and in the Third Chapter of this Essay I have endeavoured to apply the discoveries which have been made, respecting the manner in which Heat is propagated in *water*, in explaining the means which appear to have been used by the Creator of the world for equalizing the temperatures of the different climates, and preventing the fatal effects of the extremes of heat and of cold on the surface of the globe. But a most interesting application remains to be made of these discoveries, to *chemistry*, *vegetation*, and the *animal economy;* and to the learned in those branches of science I beg leave most earnestly to recommend them. If I am not much mistaken, they will throw a new light on many of those mysterious operations of Nature, in which *inanimate bodies* are put in motion, their forms changed, their component parts separated, and new combinations formed; and it is possible that they may even enable us to account, on mechanical principles, for those surprising appearances of preference and predilection among bodies, which, without ever having been attempted to be explained, have been distinguished by the appellation of *chemical affinity*.

Perhaps it will be found that every change of form, in every kind of substance, is owing to Heat, and to Heat alone; that every concretion is a true *congelation*, effected by cold or a diminution of Heat; and that every change from a solid to a fluid form is a real *fusion;* that the difference between calcination in the *wet* and in the *dry* way is, in fact, much less than has hitherto been generally imagined; and that no metal is ever dissolved till it has *first been melted.*

Perhaps it will be found, that the apparent violence with which solid bodies of some kinds are attacked by their liquid solvents — and which has, I believe, been considered as a proof of a strong chemical affinity — is not owing to any particular attraction, or election, but to the considerable degree of heat, or of cold, which is produced in their union with their menstrua, — or to a great difference in the specific gravity of the menstruum in its natural state, and that of the same fluid after it has been changed to a saturated solution.

If Fluids are non-conductors of Heat, it is evident that, if any change of temperature takes place in chemical solution, it must necessarily produce *currents* in the solvent, and that these currents must be the more rapid, as the change of temperature is greater; and as they necessarily cause a succession of fresh particles of the solvent to come into contact with the solid, it is evident — all other things being equal — that the rapidity of the process of solution will be as the rapidity of these currents, or as the change of temperature.

But the currents produced by the difference in the specific gravity of the fluid menstruum and of the saturated solution, have perhaps, in general, a still greater effect in bringing a rapid succession of fresh particles of the menstruum into contact with the solid body that is dissolved in it, than those produced by the change of temperature.

When these two causes conspire to accelerate the motion of the same current, or when their tendencies are *in the same direction*, as is the case in the solution of common sea-salt in water, — the solution ought to be most rapid.

When common salt is dissolved in water, the specific gravity of the saturated solution is greater than that of pure water, and will therefore descend in it; and cold being produced in the process, and water being a non-conductor of Heat, the specific gravity of the saturated solution will be *still farther increased*, in consequence of its condensation with this cold, by which its descent in the water will be still farther accelerated.

A curious question here presents itself, which, could it be resolved, might greatly tend to elucidate this abstruse subject of philosophical investigation. Supposing that, in a case where Heat is generated in the solution of a solid in a fluid menstruum, the *addition* to the specific gravity of the menstruum, arising from its chemical union with the solid, should so precisely counterbalance the *diminution* of the specific gravity of the Fluid, by the Heat generated in the process, that the *hot* saturated solution should be precisely of the same specific gravity as the *cold* menstruum, — would or would not the process of solution be possible under such circumstances?

If the *apparent* tendency to approach each other, which we sometimes perceive in solids and their fluid menstrua, were real; if that peculiar kind of attraction of predilection which has been called chemical affinity has a real existence, and if its influence reaches *beyond the point of actual contact* (as has, I believe, been generally supposed), as there is no appearance of any attraction whatever, or affinity, between any solid body and a saturated solution of the same body in its proper menstruum, it seems probable that the solution would take place, under the circumstances described; but should the attraction of affinity, according to the definition of it here given, have no existence in fact (which is what I very much suspect), in that case it is evident that the solution, though it would not be absolutely impossible, would be so very slow as hardly to be perceptible.

It would not be *impossible*, because the particles of the menstruum in immediate contact with the solid, though in the moment of their saturation they would have no tendency to move out of their places, yet, as they would by degrees necessarily give off to the undissolved part of the solid a part of the Heat acquired in the chemical process by which they were saturated, being condensed by this loss of Heat, they would, at length, begin to descend, and give place to other particles of the menstruum; which, in their turns, would follow them, but with velocities, however, continually decreasing, on account of the gradual augmentation of temperature of the undissolved part of the solid, and of the Heat communicated by that solid substance to the whole mass of the liquid menstruum.

Though it would, probably, be extremely difficult to contrive any single experiment, from the result of which a satisfactory decision of this question could be obtained, yet it does not appear to be impossible to discover by *indirect means* the principal fact on which its decision must depend.

It is a well-known fact, that, when water which holds sea-salt in solution is mixed, in any vessel, with fresh water, the salt will, after a short time, be found to be very equally distributed in every part of the whole mass; and I believe that it has been generally considered that this equal distribution of the salt is owing to the affinity which is supposed to exist between sea-salt and water.

Having doubts with respect to the existence of this

supposed attraction, and suspecting that the equal distribution of the salt was owing to a very different cause, — the internal motions among the particles of the water, occasioned by accidental changes of temperature, — I made the following experiment, which, I fancy, will be considered as decisive.

*Experiment No.* 57.

I took a cylindrical glass jar, $4\frac{1}{4}$ inches in diameter, and $7\frac{3}{4}$ inches high, and placing it in the middle of another cylindrical glass jar, $7\frac{1}{2}$ inches in diameter and 8 inches high, which stood in a very shallow earthen dish, nearly filled with pounded ice and water, I placed the dish, with its contents, on a strong table, in an uninhabited room in a retired part of the house, where the temperature of the air, which was the same, with very little variation, day and night, was at about 36° F. Having prepared, and at hand, a quantity of the strongest *brine* I could make with sea-salt, which was very clear, transparent, perfectly colourless, and ice-cold, —and also a quantity of fresh or pure water, ice-cold, lightly tinged of a red colour with turnsol, — and some ice-cold *olive oil*, I first poured as much of the fresh water into the small cylindrical jar as was necessary to fill it up to the height of above 2 inches; and then, by means of a glass funnel, which ended in a long and narrow tube, by introducing this tube into the fresh water, and resting it on the bottom of the jar, I poured a quantity of the brine, equal to that of the fresh water, into the jar; and in performing this operation I took so much care to do it gently, and without disturbing the fresh water already in the jar, that, when it was finished, the fresh water, which, as it was coloured red, could easily be distinguished from the brine, remained perfectly separated from this heavier saline liquor, on which it reposed quietly, without the smallest appearance of any tendency to mix with it.

I now filled to the height of about 5 inches the void space between the outside of the small jar and the inside of the large jar in which it was placed with ice-cold water, mixed with a quantity of ice, in pieces as large as walnuts (pounded ice would have obstructed the view in observing, through the sides of the large jar, what passed in the smaller), and when this was done, I very carefully poured ice-cold *olive oil* * into the smaller jar till it covered the surface of the (tinged) fresh water to the height of about an inch (see Fig. 5, Plate IV.); and placing myself near the table, in a situation where I had a distinct view of the contents of the small jar, I set myself to observe the result of the experiment.

After waiting above an hour without being able to perceive the smallest appearance of any motion, either in the brine or in the fresh water (the one continuing to repose on the other with the most perfect tranquillity, and without the smallest disposition to mix together), I left the room.

When I returned to it the next day, I found things precisely in the state in which I had left them; and they continued in this state, without the smallest appearance of any change, or of any disposition to change, during *four days.*

At the end of that time, thinking that any farther prolongation of the experiment would be quite useless, I removed the small jar, taking care not to agitate its

* This oil served not only to keep the water on which it reposed quiet, but also to prevent any communication of heat between it and the air of the atmosphere.

contents, and placed it in the window of a room heated by a German stove.

In less than an hour I perceived that the brine and the (tinged) fresh water began to mix, and at the end of 24 hours they were intimately mixed throughout, as was evident by the colour of the aqueous fluid on which the oil reposed; which now appeared to the eye to form one uniform mass of a light red tint.

I shall leave it to philosophers to draw their own conclusions from the results of this experiment. In the mean time there is one fact which it seems to point out that I shall just mention, which is not only curious in itself, but may lead to very important discoveries. It appears to me to afford strong reasons to conclude that, were a lake but *very deep*, its waters, near the surface, would necessarily be fresh, even though its bottom should be one solid mass of rock salt!

Would it be ridiculous to make experiments to determine whether the water at the bottom of some very deep lakes is not impregnated with salt? Should it be found to be actually the case, it might prove an unexhaustible treasure in an inland country, where salt is scarce.

As mines of rock-salt are often found in the neighbourhood of fresh lakes, it seems reasonable to suppose that the waters of such lakes should *sometimes* be in contact with *strata* of these mines; and when I first began to meditate on the subject, I was much surprised, not that the salt water which may lie at the bottom of fresh lakes should not already have been discovered, — for from the first I plainly perceived that nothing could happen in the ordinary course of things that could bring it to the light, or even afford any grounds to suspect its existence, — but, as *strata* of salt mines frequently lie higher than the mean level of the country, I was surprised that lakes *of salt water* should not more frequently be found ; and as these reflections occurred to me *after* I had discovered what appeared to me to be an evident proof of the wisdom and goodness of the Creator in *making* all lakes in cold countries *fresh*, I began to be alarmed for the fatal consequences that might ensue, if, by chance, the side of a lake should come into contact with a mountain of salt, as I saw might easily happen.

Shall I, or shall I not, attempt to give my reader an idea of what I felt, when, meditating on the subject, and almost beginning to repent of what many, no doubt, have already condemned as the foolish dream of an enthusiastic imagination, I saw, all at once, that the most effectual care had been taken to prevent the evils I apprehended, — that from the very constitution of things, and the ordinary and uniform operation of the known laws of Nature, the permanent *existence of a lake*, SALT AT THE SURFACE, is *absolutely impossible*, even though it should be surrounded on every side by mountains of salt? *

Though the explosion of a volcano, an earthquake, or any other great convulsion, by which the shores of a lake might be brought into contact with a vast mine of salt, might cause the whole mass of its water to be salt for a time, yet the evil would soon effect its own remedy: the falling in of the crust of earth and stones by which mines of salt are everywhere found to be covered (and without which they could not exist) would very soon cover the naked salt, and the water *at the surface of the*

* By the word *Lake* I mean, as is easy to perceive, a collection of water, in a high inland situation, from which there is a constant efflux.

*lake* would again become perfectly fresh. Should, however, the lake be so deep that the temperature at the bottom should remain the same summer and winter, without any sensible variation, it is most certain that its waters *there* (at the bottom of the lake) would remain perfectly saturated with salt forever.

But are there not some reasons to conclude that the water at the bottoms of *all very deep lakes* ought necessarily to be salt, even in situations where there are no mines of salt near?

The sea-shells that are frequently found in high inland situations, as well as many other appearances noticed by naturalists, strongly indicate that most of our continents have been covered by the waters of the ocean. Now if that event ever happened, — however remote the period may be at which it took place, — it seems highly probable that the salt water left at the bottoms of all deep lakes, by the sea, on its retiring, *must be there now.*

I cannot take my leave of this subject without just observing, that the discovery of the *impossibility* of the permanent existence of what we can plainly perceive would be an evil certainly ought not to *diminish* our admiration of the wisdom of the great Architect of the Universe.

## CHAPTER II.

*Water made to congeal at its under Surface. — Observation respecting the Formation of Ice at the Bottoms of Rivers. — Reasons for concluding that Heat can never be equally distributed in any Fluid. — Perpetual Motions occasioned in Fluids by the unequal Distribution of Heat. — An inconceivably rapid Succession of Collisions among the integrant Particles of Fluids is occasioned by the internal Motions into which Fluids are thrown in the Propagation of Heat. — An Attempt to estimate the Number of those Collisions which take place in a given Time. — These Investigations will greatly change our Ideas respecting the real State of Fluids apparently at rest. —* FLUIDITY *may be called the* LIFE OF INANIMATE BODIES. *— Conjectures respecting the* VITAL PRINCIPLE *in living animals; and the Nature of Physical* STIMULATION.

WHATEVER the mechanical operation may in fact be, by which those effects are produced that have given rise to the idea of the existence of an attraction of affinity (a power different from gravitation) between solid bodies and their liquid menstrua, and between different portions of the same menstruum differently saturated, the result of the foregoing experiment (No. 57) proves that two particles of water in combination with very different quantities of sea-salt, — or a particle of water *saturated* with salt, and another perfectly free from salt, *may be* in contact with each other for any length of time without showing any appearance of a disposition to equalize the salt between them.

But should we even admit as a fact, what this experiment seems to indicate, namely, that there is no such thing as an *attraction of predilection* between solids and their solvents, and that all those motions which have been attributed to the action of that supposed power (as well as all other motions which take place in Fluids) are the immediate effects of *gravitation* acting according to immutable laws, and *changes of specific gravity by Heat;* yet there would still remain one great difficulty in explaining chemical solution. As all mechanical operations require a *certain time* for their performance; and as the motion which is occasioned in a Fluid by a change of specific gravity in any individual particles of it *begins* as soon as the change begins to take place, if there be no attraction between the particles of solid bodies and the particles of their menstrua; as Heat is supposed to be generated or absorbed, or — to speak more properly — both generated and absorbed, in the *contact* of those particles, and previous to the completion of their chemical union, — how does it happen that the particle of the menstruum whose specific gravity is necessarily changed by this change of temperature does not *immediately* quit the solid, in consequence of this change, and before the process of solution has *had time to be completed?*

A consideration of the effects of the *vis inertiæ* of the particle of the menstruum whose specific gravity is thus changed, and also of the *vis inertiæ* of the rest of the Fluid, and the resistance it must oppose to the motion of its individual solitary particles, would furnish us with arguments that might be employed with advantage in removing this difficulty; but I fancy that the result of the experiment of which I shall presently give an account will be more satisfactory than any reasoning, unsupported by facts, that I could offer on the subject.

When a doubt arises with regard to the *possibility* of any operation of a peculiar kind, which is *supposed* to take place, in any process of nature among those infinitely small integrant particles of bodies which escape, and must ever escape, the cognizance of our gross organs, however they may be assisted by art, the shortest way of deciding the question is to put the known powers of nature in action under such circumstances that the effects produced by them must show, unequivocally, whether the supposed operation be possible or not; and if it be found to be possible in one case, we may then argue with less diffidence on the probability of its actually taking place in the specific case in question.

It has been abundantly proved by the experiments of M. de Luc, and by those of my friend Sir Charles Blagden, that when water, in cooling, has arrived at the temperature of about 41° F., its condensation with cold ceases, and it begins to expand, and continues to expand gradually as its temperature goes on to be farther diminished, till it is changed to ice. Availing myself of that most important discovery, I made the following experiment.

### *Experiment No.* 58.

Having poured *mercury*, at the temperature of 60°, into a common glass tumbler, till this Fluid stood at the height of about an inch, I then poured about twice as much water (at the same temperature) upon it; and, placing the tumbler in a shallow earthen dish, surrounded it to the height of the level of the surface of the mercury with a freezing mixture of snow and common salt. Having done this, I was very curious indeed to see in what part of the water ice would first make its appearance. Could it be at the upper surface of it? That appeared to me to be impossible; for, the experiment being made in a room warmed by a German stove, the temperature of the air which reposed on that surface was considerably above the point at which water freezes.

Could it be at its lower surface, where it rested on the upper surface of the mercury? If that should happen, it would show that, notwithstanding the diminution of the specific gravity of the water in passing from the temperature of 41° to that of 32°, and the tendency which this diminution gave it to quit the service of the mercury from the instant when, in being cooled by a contact with it, it had passed the point of 41°, yet there was time sufficient for the congelation to be completed *before the particle of water so cooled could make its escape.*

The reader will naturally conclude from what was said in the preceding page, that it was merely with a view to the determination of that single fact that this experiment was contrived; and he will perceive by the result of it that my expectations with regard to it were fully answered.

Ice was not only formed *at the bottom of the water*, at its under surface, where it was in contact with the cold mercury, but I found on repeating the experiment, and varying it, by previously cooling the mercury in the tumbler to about 10°, that *boiling hot water*, poured gently upon it, was instantly frozen, and gradually formed a thick cake of ice, covering the mercury; though almost the whole of the mass of the unfrozen water which rested on this ice remained nearly boiling hot.

This experiment not only determines the point for the decision of which it was undertaken, but also enables us to form a just opinion respecting a matter of fact which has been the subject of a good deal of dispute.

Though many accounts have been published of ice found at the bottom of rivers, yet doubts have been entertained of the possibility of its being *formed* in that situation. From the result of the foregoing experiment it appears to me that we may safely conclude, that, if after a very long and a very severe frost, by which the surface of the ground has not only been frozen to a considerable depth, but also cooled several degrees below the freezing-point, a river should overflow its banks, and cover the surface of ground *previously so cooled*, ice would be formed at the bottom of the water; but all the experiments that have been made on the congelation of water show the absolute impossibility of ice being ever formed, in any country, at the bottom of a river which constantly fills its banks, or which never leaves its bed exposed, dry, to the cold air of the atmosphere.

By reflecting on the various consequences that ought to follow from the peculiar manner in which Heat appears to be propagated in Fluids, we are led to conclude, that it is almost impossible that any Fluid exposed to the action of light should ever be throughout of the same temperature, though its mass be ever so small; and that the difference in the Heat of its different particles must occasion perpetual motions among them.

Suppose any open vessel, — as a common glass tumbler, for instance, — containing a piece of money, a small pebble, or any other small solid opaque body, to be filled with water, and exposed in a window, or elsewhere, to the action of the sun's rays. As a ray of light cannot fail to generate Heat when and where it is stopped or absorbed, the rays, which, entering the water, and passing through it, impinge against the small solid opaque body at the bottom of the vessel, and are *there absorbed*, must necessarily generate a certain quantity of Heat; a part of which will penetrate into the interior parts of the solid, and a part of it will be communicated to those colder particles of the water which repose on its surface.

Let us suppose the quantity of Heat so communicated to one of the integrant particles of the water to be so small, that its effect in diminishing the specific gravity of the particle is but just sufficient to cause it to move upwards in the mass of the liquid with the very smallest degree of velocity that would be perceptible by our organs of sight, were the particle in motion large enough to be visible. This would be at the rate of about *one hundredth part of an inch* in a second.

This velocity, though it appears to us to be slow in the extreme, when we compare it with those motions that we perceive among the various bodies by which we are surrounded, yet we shall be surprised when we find what a rapid succession of events it is capable of producing.

If we suppose the diameter of the integrant particles, or *molecules* of water, to be *one millionth part of an inch* (and it is highly probable that they are even less *), in that case, it is most certain that an individual particle, moving on in a quiescent mass of that Fluid with the velocity in question, namely, at the rate of $\frac{1}{100}$ part of an inch in 1 second, would run through a space equal to *ten thousand times the length of its diameter* in *one second*,

* Leaf gold, such as is prepared and sold by the gold-beaters, is not *four times* as thick as the diameter here assumed for the integrant particles of water. These leaves of solid metal have been found by computation to be no more than $\frac{1}{282020}$ of an inch in thickness. How much less must be the diameter of the integrant particles of gold?

and, consequently, would come into contact with at least *six hundred thousand* different particles of water in that time.

Hence it appears how inconceivably short the time must be that an individual particle, in motion, of any Fluid, can remain in contact with any other individual particle, not in motion, against which it strikes in its progress, however slow that progress may appear to us to be through the quiescent mass of the Fluid!

Supposing the contact to last as long as the moving particle employs in passing through a space equal to the length of its diameter, — which is evidently all that is possible, and more than is probable; then, in the case just stated, the contact could not possibly last longer than $\frac{1}{10000}$ part of a second! This is the time which a cannon bullet, flying with its greatest velocity (that of 1600 feet in a second) would employ in advancing 2 inches.

If the cannon bullet be a *nine pounder*, its diameter will be four inches; and if it move with a velocity of 1600 feet (= 19200 inches) in a second, it will pass through a space just equal to 4800 times the length of its diameter in 1 second. But we have seen that a particle of water moving $\frac{1}{100}$ of an inch in a second actually passes through a space equal to 10000 times the length of its diameter in that time. Hence it appears that *the velocity with which the moving body quits the spaces it occupies* is more than twice as great in the particle of water as in the cannon bullet!

There is one more computation which may be of use in enabling us to form more just ideas of the subject under consideration; and surely too much cannot be done to enlighten the mind, and assist the imagination, in our attempts to contemplate those invisible operations of nature which nothing but the sharpest ken of the intellectual eye will ever be able to detect and seize.

As succeeding events which fall under the cognizance of our senses cannot be distinguished if they happen oftener than about *ten times in a second,** it appears that when a particle of water moves in a quiescent mass of that fluid at the rate of $\frac{1}{100}$ part of an inch only, in one second, its succeeding collisions with the different particles, at rest, of that fluid, against which it strikes as it moves on, must be so inconceivably rapid that no less than *one thousand* of them must actually take place, *one after the other*, in the shortest space of time that is perceptible by the human mind. †

* This assertion, in as far, at least, as it relates to objects of sight, may be proved by the following easy experiment: Let a wheel, with any known number of spokes, be turned round its axis with such a velocity as shall be found necessary, in order that the spokes may disappear or become invisible. From the velocity of the wheel, and the number of spokes in it, the fact will be decided.

† It probably will not escape the observation of my learned readers, that the velocity which I have here assigned to the single particle of water, moving upwards in that fluid in consequence of a change of its specific gravity by Heat, though apparently very small ($\frac{1}{100}$ part of an inch in a second), is however, most probably, considerably greater, in fact, than any individual *solitary* particle of that fluid could possibly acquire, in the supposed circumstances, by any change of temperature, however great, owing to the resistance which would necessarily be opposed to its motion by the quiescent particles of the fluid. Aware of this objection, and being desirous of being prepared to meet it, I took some pains to compute, by the rules laid down by Sir Isaac Newton in his *Principia*, Book II. Sect. vii., what the greatest velocity is that a solitary particle of water (supposed to be $\frac{1}{1000000}$ of an inch in diameter) could possibly acquire by a given change of its specific gravity. And I found that if the specific gravity of water at the temperature of 32° F. be taken at 1.00082, and its specific gravity at 80° at 0.99759, as lately determined by accurate experiments, then a single particle of water at the temperature of 80°, situated in a quiescent mass of that fluid at 32°, the greatest velocity this hot particle could acquire in moving upwards in consequence of its comparative levity would be that of $\frac{1}{2638}$ part of an inch in 1 second. This is at the rate of about one inch and an half in 1 hour. But it is evident, that when great numbers of particles unite and form currents, they will make their way through the quiescent fluid with greater facility, and consequently will move faster.

After we have patiently examined the result of these investigations, and the imagination has become *familiarized* with the contemplation of the interesting facts they present to it, how much will our ideas be changed with regard to the real state of fluids apparently at rest! They will then appear to us to be, what no doubt they really are in fact, an assemblage of an infinite number of infinitely small particles of matter moving continually, or without ceasing, and with inconceivable velocities.

We shall then consider fluidity as the *life of inanimate bodies*, and congelation as the *sleep of death*; and we shall cease to ascribe active powers, or exertions of any kind, to dead *motionless* matter.

But what shall we think of the *vital principle* in living animals? Does not their life also depend on the internal motions in *their* fluids, occasioned by an *unequal* distribution of heat? And is not *stimulation*, in all cases, the mere mechanical effect of the communication of Heat?

It is an opinion which we know to be as old as the days of Moses, that *the life of an animal resides in its blood*; and it is highly probable that it dates from a period still more remote. It was lately revived by an anatomist and physiologist (now no more), who was eminently distinguished for sagacity; and it appears to me that the late discoveries respecting the manner in which Heat is propagated in Fluids tend greatly to elucidate the subject, and to give to the hypothesis a high degree of probability.

According to this hypothesis (as it may now be explained), everything that increases the *inequality of the distribution* of the Heat in the mass of the blood (even though it should not immediately augment its quantity) ought to increase the intensity of those *actions* in which life consists. But are there not many striking proofs that this is the case in fact?

Do not *respiration*, *digestion*, and *insensible perspiration*, all tend evidently (that is to say, according to our assumed principles with regard to the manner in which Heat is propagated in Fluids) to *produce* and to *perpetuate* this inequality of Heat in the animal fluids? And do we not see what an immediate and powerful effect they have in increasing the intensity of the action of the powers of life?

If animal life depends essentially on those *internal* motions in the animal fluids, which, as has been shown, are occasioned by the difference of the *specific gravities* of their integrant particles, or *molecules*, arising from their different temperatures, — in that case it is evident that the *vital powers* would be strengthened, or their action increased, either by *heat* or by *cold* properly applied. But is not this found to be the case in fact? Does not the *dram of brandy* at St. Petersburgh produce the same effects as the *draught of iced lemonade* at Naples, and by the same mechanical operation, but acting in opposite directions? And does not the *loss of Heat*, by insensible perspiration, contribute as efficaciously to the preservation of that *inequality of temperature* which is essential to life, as the *introduction of Heat* into the system in respiration?

Is not the sudden coagulation of blood, when drawn from a living animal, and are not all the other rapid changes that take place in it, evident proofs of an unequal distribution of Heat? And does not the *viscosity* of blood, as well as its perpetual motions in the vascular system, contribute very powerfully to the preservation of that inequality?

Are not the livid spots on the surface of the body, which indicate a beginning of mortification, produced in consequence of a separation or *precipitation* of the heterogeneous particles of the animal Fluids, according to their specific gravities and individual temperatures, occasioned by rest or an interruption of circulation? And may we not emphatically pronounce such Fluids to be dead?

Would not any liquid in which Heat were *equally distributed* be a *fatal poison* if injected into the veins of a living animal? And would not this be the case even were the liquid so injected a portion of the animal's own blood, or of the lymph or any other of its component parts, and were it at the mean temperature precisely of the healthy Fluids circulating in the veins and arteries of the animal?

Is not glandular secretion a true precipitation? and is it not possible that the formation of the solids and the growth of an animal body may be effected by a process exactly similar to congelation? And are there not even circumstances from which we might conclude, with a considerable degree of probability, that most of these congelations are formed at or about the temperature of boiling water?

But I forbear to enlarge on this subject. I find I have unawares entered a province, where, if I advance farther, I shall certainly be exposed to the danger of being considered and treated as an intruder; and I must hasten to make my retreat, which I shall endeavour to effect by abruptly putting an end to this Chapter.

## CHAPTER III.

*Probability that intense Heat frequently exists in the solitary Particles of Fluids, which neither the Feeling nor the Thermometer can detect.— The Evaporation of Ice during the severest Frost explained on that Supposition.— Probability that the Metals would evaporate when exposed to the Action of the Sun's Rays were they not good Conductors of Heat.— Mercury is actually found to evaporate under the mean Temperature of the Atmosphere.— This Fact is a striking Proof that* FLUID MERCURY *is a Non-conductor of Heat.— Probability that the Heat generated by the Rays of Light is always the same in intensity; and that those Effects which have been attributed to Light ought perhaps in all Cases to be ascribed to the Action of the Heat generated by them.— A striking Proof that the most intense Heat does sometimes exist where we should not expect to find it.— Gold actually melted by the Heat which exists in the Air of the Atmosphere, where there is no Appearance of Fire, or of anything red-hot.— We ought to be cautious in attributing to the Action of unknown Powers, Effects similar to those produced by the Agency of Heat.— The most intense Heat may exist without leaving any visible Traces of its Existence behind it.— This important Fact illustrated by the necessary Result of an imaginary Experiment.*

HOW far the possibility of the communication of Heat between the integrant particles of a Fluid may or may not be owing to the extreme mobility of those particles, and to the infinitely short time that two of them, of different specific gravities (owing to a difference of temperature), can remain in contact, I leave others to determine; in the mean time, it is most certain that the existence of this impossibility of any immediate communication of Heat among the particles of a Fluid renders the distribution of Heat very unequal; and it seems highly probable that many appearances which have been attributed to very different causes are in fact owing to *intense Heat* existing and producing the effects proper to it in situations where its existence has not even been suspected.

If Fluids are non-conductors of Heat, no situation can possibly be more favourable to its preservation than when it exists in them; and it is not only evident, *a priori*, that the most intense Heat *may exist* in a few solitary particles of some Fluids without its being possible for us to detect it, or to discover the fact, either by our feeling or by the thermometer; but there are many appearances that strongly indicate, and others that prove, that intense Heat actually does exist in that concealed or imperceptible state very often.

There is no reason to suppose that it is possible for ice to be reduced to steam without being previously melted; and it is well known that ice cannot be melted with a lower degree of Heat than that of 32° of Fahrenheit's scale: but in the midst of winter, in the coldest climates, and when the temperature of air of the atmosphere, as shown by the thermometer, has been much below 32°, ice, exposed to the air, has been found to evaporate.

How can we account for this event, except it be by supposing that some of the particles of air which accidentally (as we express it) come into contact with the ice are so hot, as not only to melt the small particles of ice which they happen to touch, but also to reduce a part of the generated water to steam, before it has time to freeze again; or by supposing that this is effected by intense Heat generated by light absorbed by small projecting points of the ice? As ice is a very bad conductor of Heat, that circumstance renders it more likely that the event in question should actually take place in either of these ways.

If the metals were very bad conductors of Heat, instead of being very good conductors of it, I think it more than probable that even they would be found to evaporate when exposed to the action of the direct rays of the sun; and perhaps also in situations in which such an event would appear still more extraordinary.

MERCURY has been actually found to *evaporate* under the mean temperature of the atmosphere! What a striking proof is this that *fluid mercury* is a non-conductor of Heat, and also, that very intense Heat may be generated, or exist where it would not naturally be expected to be found! And does not the evaporation of water under the mean temperature of the atmosphere afford another proof of this last fact?

That the most intense heat is often excited in very small particles of solid bodies dispersed about in the midst of masses of cold liquids is not to be doubted. It is well known what an intense Heat the rays of the sun are capable of exciting; and it seems to be highly probable that Heat actually excited by them is always the same, that is to say, *intense in the extreme:* but when the rays are few, and when circumstances are not favourable to the *accumulation* of the Heat they generate, it is often so soon dispersed that it escapes the cognizance of our senses and of our instruments, and sometimes leaves no visible traces of its existence behind it.

Why should we not suppose that the Heat generated by a ray of light, which, entering a mass of cold water, accidentally meets with an infinitely small particle of any solid and opaque substance which happens to be floating in the liquid, and is absorbed by it, is not just as intense as that generated in the focus of the most powerful burning mirror or lens?

Mr. Senebier has given us an account of a great number of interesting experiments on the effects produced on different bodies by exposure to the direct rays of the sun; but why may we not attribute all those effects to the intense *local* Heat, generated by the light absorbed by the infinitely small, and, if I may use the expression, *insulated* particles of the bodies which were found to be affected by it?

The surface of wood of various kinds was turned brown. The same appearances might be produced in a shorter time by the rays which proceed from a red-hot iron, which change the surface of the wood to an imperfect coal. But were not the surfaces of the woods which were turned brown by the light of the sun in Mr. Senebier's experiments changed to an imperfect coal? And is it possible for a Heat less intense than that of *incandescence* to produce that effect?

Among the many facts that might be adduced to prove that the most intense Heat *may*, and frequently *does exist* where we should not expect to find it, the following appears to me to be very striking and convincing. It is, I believe, generally imagined that the intensity of the heat generated in the combustion of fuel is much less in a small fire than in a great one; but there is reason to think that this is an erroneous opinion, founded on appearances that are not conclusive; at least, it is certain that the intense Heat of a large smelting furnace, such as is necessary for melting the most refractory metals, actually exists in the feeble flame of the smallest candle; and what may appear still more extraordinary, this intense degree of Heat often exists in the air of the atmosphere, *where no visible signs of Heat appear*, as I shall presently show.

Iron is fully *red-hot* by daylight at the temperature of about 1000° of Fahrenheit's scale; brass melts at 3807°, copper at 4587°, silver at 4717°, and gold at 5237°; and nothing is more certain than that the Heat must be at that intensity which corresponds to the 5237th degree of Fahrenheit's scale, *where gold is found to melt.* But very fine gold, silver, or copper wire, flatted, (such as is used to cover thread to make lace,) melts instantaneously on being held in the flame of a candle. It will even be melted if it be held a few seconds *over* the flame of a candle, *at the distance of more than an inch from the top of the flame*, in a place where there is no appearance of fire, or of anything red-hot.

From the important information which we acquire from the result of these experiments, we see how much we ought to be on our guard in forming an opinion with respect to the *intensity* of the Heat which *may exist* in the invisible insulated particles of matter of any kind that may be scattered about in a given space, or which may float in any Fluid, where neither our feeling nor our thermometers can possibly be sensibly affected by it.

A thermometer can do no more than indicate the *mean of the different temperatures of all those bodies or particles of matter which happen to come into contact with it.* If it be suspended in air, it will indicate the mean of the temperatures of those particles of air *which happen to touch it;* but it can never give us any information respecting the *relative* temperatures of those particles of air.

If, during the most intense frost, a thermometer were suspended in the neighbourhood of a burning candle, — in the same room, for instance, — if it were placed over the candle, or nearly so, though it should be distant from it several feet, as air is a non-conductor of Heat, there is not the smallest doubt but that some solitary particles of air, heated by the candle to the intense Heat of melting gold, would reach the thermometer; but neither the thermometer, nor the hand held in the same place, could give any indication of such an event.

As it appears, from all that has been said, that intense heat *may exist* even under the form of *sensible Heat*, where its presence cannot be discovered or detected by us; and as it seems highly probable that in many cases, where its existence may escape our observation, it may nevertheless be capable of producing very visible effects, I think we ought always to be much on our guard in accounting for effects similar to those which are known to be produced by Heat, and never, without very sufficient reasons, attribute them to the agency of any other *unknown* power; and this caution appears to me to be peculiarly necessary in accounting for those effects which have been found to be produced in various bodies when they are exposed to the action of the sun's rays.

If the solar rays concentrated in the focus of a lens, when they are made to fall on a piece of wood, instantly change its surface to a black colour, and reduce it to charcoal, why may we not conclude that the change of colour which is gradually or more slowly produced in the same kind of wood when it is simply exposed in the sunbeams is produced in the same manner?

The difference in the *times* necessary to produce similar effects in these two cases is no proof that they are not produced *in the same manner;* for if they are effected merely by the agency of Heat (which I suppose), then the effects produced in any given time will not be as the density of the light or as the number of rays, but as that part of the Heat generated which, not being immediately dispersed or carried off by the air, has time to produce the action proper to it in the wood; and consequently must be incomparably greater, in proportion, when the rays are concentrated, than when they are not.

*Luna cornea* exposed to the action of light changes colour; but why should we not attribute this change to the expulsion of the oxygen united with the metal, by the agency of the Heat generated by the light? To remove every possible objection to this explanation of the phenomenon nothing more appears to be necessary than to admit what is well known, — that this metallic oxyd may be reduced, without addition, *with some degree of Heat*, and that this substance is a bad conductor of Heat.

Will not the admission of our hypothesis respecting the *intensity* of the Heat which is supposed to be generated where light is stopped, and of that respecting the non-conducting power of Fluids with regard to Heat, enable us to account, in a manner more satisfactory than has hitherto been done, for the effects of the sun's light in bleaching linen, when it is exposed wet to the action of his direct rays? as also for the reduction of those metallic oxyds which have been found to be revived by exposure to light? And will it not also assist us in accounting for the production of pure air in the beautiful experiment of Dr. Ingen-Housz, in which the green leaves of living vegetables are exposed, immersed in water, to the sun's rays?

Mr. Senebier has shown that the colouring matter of healthy green leaves of vegetables, which is extracted from them by spirits of wine, and which tinges the spirits of a beautiful green colour, is destroyed, or rather changed to a dirty brown colour, in a few minutes, on exposing this tincture in a transparent phial, and *in contact with pure air*, to the direct rays of a bright sun; but why should we not consider this process as a real combustion?

The Heat acquired by the liquid, — which, as I have often perceived in repeating the experiment, is very considerable, — and the necessity there is for the presence of *pure air*, that the experiment may succeed, seem to indicate that something very like combustion must take place in it.

If liquids are non-conductors of Heat, they ought, certainly, *on that account*, to be peculiarly well calculated for confining and consequently furthering the operations of that Heat which is generated by light, or by any other means, in their integrant particles, or in the infinitely small and insulated particles of other bodies that are dispersed about, or held in solution in them; as I have already more than once had occasion to observe.

If this supposition be admitted, a very great difficulty will be removed in accounting for chemical solution on the hypothesis that the change of form from a solid to a fluid state is in all cases a real fusion; or that it is effected by the sole agency of Heat; and that concretion, or crystallization, is a process in all respects perfectly analogous to freezing.

There are but three forms under which sensible bodies

are found to exist, — namely, that of a *solid*, that of a *fluid*, and that of an *elastic fluid*, or *gas;* and it is well known that every substance with which we are acquainted — all ponderable matter without exception — is capable of existing alternately under all those forms indifferently; and that the form under which it appears *at any given time* depends on its *temperature at that time.*

We know farther that every identical substance undergoes these different changes of form at certain fixed temperatures; and when we consider the subject with attention we shall find that, had not these temperatures been fixed, and had they not been different in different bodies, it would have been utterly impossible for us to have identified any substance whatever.

Perhaps this is the only essential difference that really exists among bodies that appear to us to be different.

But not only the degrees of Heat, or points in the scale of temperature at which the forms of different bodies are changed, are various, but the *extent of the variation of temperature* under which a substance can persevere, or continue to maintain its form in its *middle state*, — that of *fluidity*, or rather *liquidity*, — is very different in different bodies; and this last circumstance has a wonderful effect in increasing the variety of the compositions and decompositions which are continually taking place in the various operations of nature on the surface of the globe.

Another circumstance, not less prolific in events, is the union which takes place between bodies of *different kinds;* and those most important changes in regard to the degrees of Heat which the bodies so united can support without having their forms changed, which are found to result from such union.

When, to the established laws which have been discovered in the operations of nature in the change of form in substances that appear to us to be *simple*, we add those which have been found to obtain in the changes of form of bodies that are known to be *compounded*, we shall perhaps be able to conceive some more distinct ideas with regard to the nature of those mechanical operations which take place in chemical processes. I call them *mechanical*, — for mechanical they must of necessity be, according to the most rigid interpretation of that expression.

But the hypothesis of the existence of *intense Heat* in the midst of cold liquids is so new, and seems to be so contrary to the result of all our experience and observation, that I feel it to be necessary to take some pains to illustrate the matter.

And first, we must not expect always to find traces remaining of the existence of intense Heat, even where there are the strongest reasons to think it has actually existed; for as often as Heat is dispersed or carried off, before it has had time to produce any changes of form, or chemical changes or combinations in the bodies to which it is communicated, it leaves no marks behind it.

Fire-arms are often found to miss fire, even when many live sparks from the flint and steel actually fall into the pan among the priming; but nobody, surely, will pretend that the small particles of *red-hot iron* which fall among the grains of the gunpowder, and cool in contact with them, are not intensely hot, — incomparably more so than would be necessary to inflame the powder were their Heat of sufficient *duration* to produce that effect. Had these small sparks been invisible, it is highly probable that their existence would never have been suspected, and that the fact which they prove would not have been believed.

That gunpowder may be inflamed, it is necessary that the sulphur which constitutes one of its component parts should be first *melted* and then *boiled;* for it is the vapour of boiling sulphur which always takes fire when gunpowder is kindled.

Were melted sulphur a conductor of Heat, there is reason to think that gunpowder would be very far from being so inflammable as we find it to be.

As those who have not been much accustomed to meditate on the subject under consideration may find some difficulty in conceiving how it is possible for intense Heat to be *excited* in or to *exist* in the midst of a mass of any cold liquid, — as of water, for instance, — without immediately producing visible effects, I feel it to be my duty to put that matter in the clearest light possible, and to show that what I have considered as being *probable* is most undoubtedly very far from being impossible.

The best method of proceeding in inquiries of this kind, where the principal object is to discover whether a supposed event, which, from its nature, cannot fall under the cognizance of our senses, is or is not possible, seems to me to be, to begin by supposing the event to have actually taken place, and then to trace its necessary consequences, and compare them with those appearances which are actually found to take place.

Adopting this method, we will suppose a quantity of pure water, at the mean temperature of the atmosphere in England, that of 55° F., to be put into a clean and very transparent glass tumbler, placed in a window and exposed to the direct rays of the sun. If the glass and the water are both *perfectly transparent*, it is evident that no Heat will be generated in either of them by the sun's light.

If now a small particle of any opaque solid body be suspended in the midst of the water in the tumbler, those rays of light, which, impinging against it, are absorbed by it, must necessarily generate Heat in the very moment when they are stopped. This is an incontrovertible fact, which nobody will dispute.

In order to render this imaginary experiment more interesting, we will suppose the solid body put into the water to be a small particle of yellow amber; and that its specific gravity is so exactly equal to that of the water that it has no tendency to move in it, either upwards or downwards, and consequently will remain in the situation where it is placed, without being suspended; and we will suppose, farther, that this solid particle of amber is nearly globular, and $\frac{1}{1500}$ of an inch in diameter, which is just equal to the diameter of a single thread of silk, as spun by the worm, and is probably one of the smallest objects that is perceptible by the human eye, unassisted by art.

As it is evident that Heat must be generated or excited in this small particle of amber by the light it stops or absorbs, the points which remain to be discussed are, therefore, what *its intensity* is at the moment of its existence; and what are the effects which it ought to produce in consequence of that intensity.

The reasons have already been mentioned which render it probable that when Heat is generated by the rays of light, its intensity, *where it is generated*, and before it has been diminished in consequence of its dispersion, is always the same; and, taking it for granted that this is the case in fact, we will endeavour to trace the operations of that Heat — extreme in its intensity, or degree, but small in regard to its quantity, or to the space it occupies — which is generated in the particle of amber in the experiment under consideration.

As this Heat must first exist where it is generated, it is evident that it must exist at the surface of the particle of amber; and as all solid bodies are, in a greater or less degree, conductors of Heat, a part of this Heat will penetrate the substance of the solid particle, while another part of it will be carried off by the cold particles of water in contact with the surface thus heated by the light.

It remains, therefore, to be determined what the effects are which this Heat — so absorbed, on the one hand, by the solid particle of amber, and communicated to the water, on the other — ought necessarily to produce. And first, if the dispersion of the Heat by both these means should be sufficiently rapid to prevent its accumulation to such a degree as to melt the amber, it is evident that no visible effects by which its existence could be discovered would be produced in that substance; and this event (the fusion of the amber) will depend on three circumstances, namely, — First, on the temperature at which amber melts; Secondly, on the facility with which Heat expands and is dispersed in a solid mass of that substance, or on its conducting power; and Thirdly, on the rapidity with which the Heat generated at the surface of the amber is carried off by the cold Fluid in which it is immersed.

Though I do not think there would be any reason for surprise, even admitting the existence of the supposed intense Heat, should the amber be found not to be melted under the circumstances described, yet it appears to me to be extremely probable, that if amber, in a very fine powder, were mixed with any transparent oil, capable of supporting a great degree of Heat without being reduced to vapour, and exposed in it to the direct rays of a very bright sun, the amber would melt and be dissolved, though perhaps very slowly.

But if amber does not melt when exposed in water to the action of the sun's beams, and consequently suffers no visible change by which the existence of the Heat supposed to be generated at its surface by the light can be detected, ought not this Heat, were it, in fact, as intense as it is supposed to be, to produce some visible effects in the water, by which its existence would necessarily be discovered?

To resolve this doubt, we must inquire what visible effects it would be possible for the Heat in question to produce in the water. Now if we suppose the water not to be decomposed by this Heat, which, as no chemical change is supposed to take place in the amber, cannot happen, the only effect this Heat can possibly produce on the water is an increase of its temperature, which increase must, however, be much too small to be detected, either by the feeling or by the thermometer.

It might, perhaps, be expected that *steam* would be found at the heated surface of the particle of amber, and become visible; but when we consider the matter for a moment, we shall see that it is quite impossible that such an event should happen, for even on the supposition (which, however, is far from being probable) that the same individual particles of water which come into contact with the hot surface of the amber should remain in contact with it till their temperatures should gradually be raised to that point at which water is changed to steam, yet, from the extreme rapidity with which steam condenses when in contact with cold water, it is evident that it could not exist an instant under the circumstances here supposed. Indeed, we have direct proofs that steam cannot exist under such circumstances, by what is found to happen when large masses of iron or steel, raised to a most intense heat, in a blast furnace, are suddenly plunged into cold water by smiths, in tempering edge-tools; for these masses of red-hot metal may be distinctly seen to be in actual contact with the cold water, and did not a part of the water which is decomposed by the hot iron make its escape in the form of inflammable air, it is not probable that there would be any visible appearance from which the formation of steam could be suspected.

Hence we see the possibility of the existence of *intense Heat* in the midst of a mass of cold water, or of any other transparent liquid, without producing any visible effects, or leaving behind it any traces by which its existence could be suspected.

Let us now consider a case in which this intense Heat, though perfectly imperceptible on account of the extreme minuteness of the particles of matter in which it exists, is capable, nevertheless, of producing very visible effects. Let us suppose a solution of nitro-muriate of gold in water to be exposed to the action of the sun's rays. If this solution were *perfectly* transparent, no Heat could possibly be generated in it by light; but as it is not so, Heat, in the highest degree of intensity, must necessarily be generated by those opaque particles (of the oxyd of gold) by which it is stopped. Now as gold is a very heavy substance, it is evident that it must be reduced to extremely small particles in order that, when changed to an oxyd by its union with oxygen, it may be dissolved in and continue suspended in water; and it is clear that the smaller any insulated particle of matter is, at the surface of which Heat is generated in consequence of the absorption of light, the more suddenly must the Heat so generated be dispersed through the whole substance of the particle, and the more equally and more intensely must that particle be heated; from hence it appears evidently, that, if the particles of the oxyd dispersed about in the water are but *small enough*, the Heat generated in them by the sun's rays will be sufficient to expel the oxygen united to the gold, and revive that metal.

There is one very obvious objection that will doubtless be made to this conclusion, which, however, may easily be removed. The particle of the metallic oxyd which is supposed to be heated is in contact with the water; how does it happen that a great part of this Heat does not immediately pass off into that cold Fluid? I might answer, because both water and steam are non-conductors of Heat, and might adduce in support of this reason the well-known fact that a drop of water dropped on a piece of iron, heated to most intense white Heat, will remain some time on the iron without being evaporated, even considerably longer than if the iron were much less hot; but a circumstance attending the beautiful experiment in which iron is burned in oxygen gas affords a more direct proof of the fact in question.

As this experiment is commonly made, the iron, which is a piece of small wire, a few inches long, is introduced into a bottle, with a narrow neck, which contains the oxygen gas; the wire being fixed in its place, by causing its upper end to pass through a cork stopple, which is fitted to the mouth of the bottle. The lower end of the wire is pointed; and it is set on fire by being first heated in the flame of a candle, and then plunged suddenly, while red-hot, into the bottle. The combustion begins the moment the end of the wire enters the oxygen gas; and the metal continues to burn with the utmost violence, and with a copious emission of intense white light, till the wire, or till all the gas is consumed, affording one of the most brilliant and most interesting sights that can be imagined.

The product of this combustion is the oxygenation of the iron; and this metallic oxyd, in a state of fusion, and heated to the most intense white Heat, falls to the bottom of the bottle in globules of different sizes.

To protect the glass against these drops of calx of iron in fusion, it is usual to leave a quantity of cold water in the bottle, — enough, for instance, to cover its bottom to the height of about an inch; but I have frequently seen numbers of these globules, much smaller than peas, which have not only descended *red-hot* through the water, but have remained red-hot at the bottom of the bottle, surrounded by the water, at least two or three seconds, and actually melted the glass on which they reposed (and as far as I can recollect) without producing the smallest appearance of steam.

The water could not be decomposed, for the iron was already saturated with oxygen.

This experiment will, I fancy, be considered as affording an indisputable proof that *intense Heat* may exist, at least for a short time, in a small particle of matter surrounded by a cold Fluid.

Now, as it has been found by actual experiment, that when a solution of nitro-muriate of gold in water is exposed to the action of the sun's rays, the gold is revived;

and as it is known that an oxyd of gold may be reduced in the dry way without addition, or merely by intense Heat, why should we not conclude that it is merely by *Heat* that that metal is revived in the case under consideration, and that the *intensity* of the Heat by which this oxygenation is effected is precisely the same in both cases?

Should this supposition be admitted, we might, perhaps, venture to proceed one step farther, and consider the nature and progress of the mechanical operations which take place in disoxygenation of metals, or their precipitation from a solution of their oxyds, when that operation is effected by means of Heat generated, not by light, but by the contact or union of infinitely small particles of bodies, different in kind, and disposed to generate or to absorb sensible Heat on coming together; which particles being dispersed about in the liquid solution, and in the substance added to it to effect the precipitation, are by this mixture brought into contact.

This would naturally lead us to an examination of the phenomena of solution; and those clearly understood would, no doubt, give us a distinct view of the mechanical operations by which those tendencies to union are effected which have been designated under the name *elective attraction.*

But how arduous an undertaking! what intense study! what efforts of the imagination would be necessary to trace out and form distinct ideas of such a succession of events, all perfectly imperceptible by our organs, though assisted by all the resources of art!

Sensible of my own weakness, I dare not proceed any farther. Perhaps it will be thought that I have already advanced much too far; but it is right that I should acknowledge fairly, that in the present case the temerity I have shown has not been entirely without design.

There are two ways in which philosophers, as well as other men, may be excited to action, and induced to engage zealously in the investigation of any curious subject of inquiry, — they may be *enticed*, and they may be *provoked*.

It will probably not escape the penetration of my reader, that I have endeavoured to use both these methods. I am well aware of the danger that attends the latter of them; but the passionate fondness that I feel for the favourite objects of my pursuits frequently hurries me on far beyond the bounds which prudence would mark to circumscribe my adventurous excursions.

---

## CHAPTER IV.

*An Account of a Variety of Miscellaneous Experiments. — Thermometers with cylindrical Bulbs may be used to show that Liquids are Non-conductors of Heat. — Ice-cold Water may be heated and made to boil standing on Ice. — Remarkable Appearances attending the thawing of Ice, and the melting of Tallow and of Bees-Wax, by means of the radiant Heat projected downwards by a red-hot Bullet. — Beautiful Crystals of Sea-Salt formed in Brine standing on Mercury. — Olive-Oil soon rendered colourless by Exposure to the Air standing on Brine. — An Attempt to cause radiant Heat from a red-hot Iron Bullet to descend in Oil. — Account of an artificial Atmosphere in which horizontal Currents were produced by Heat. — Conjectures respecting the proximate Causes of the Winds.*

THOUGH this Essay is already grown to a much larger size than I originally intended, and even larger than I could have wished (well knowing how great an evil a great book is generally thought to be), I could not bring it to a conclusion without adding one Chapter more. In this Chapter the reader will find accounts of several experiments, some of which he will probably consider as not altogether uninteresting. To take up as little of his time as possible, I shall be very brief in these accounts, and in general shall leave the reader to draw his own conclusions from the results of the experiments I shall describe.

§ 1. *An Account of several simple Experiments, which show that Heat does not descend in Fluids.*

If a thermometer constructed with a long and narrow, naked cylindrical bulb (6 inches long, for instance, and $\frac{1}{2}$ an inch in diameter), and filled with mercury, oil, spirits of wine, or any other Fluid proper for that purpose, with which it is required to make the experiment in question, — such thermometer being at the temperature of the air in summer, or at any temperature above the point of freezing water, — if the lower end, or half, of its bulb be plunged into a glass tumbler filled quite full to the brim with pounded ice and water, the height of the Fluid in the tube of the instrument will show that half the Fluid in the cylindrical bulb of the instrument is ice-cold, while the temperature of the other half of it remains unchanged.

The result will be the same, when, to prevent the communication of Heat from the air during the experiment, that part of the bulb of the thermometer (the superior half of it) which projects above the level of the top of the tumbler is covered with a sheath lined with soft fur.

When more or less than half of the bulb of the thermometer is plunged into the ice and water, the height of the liquid in the tube of the instrument will show that that part only of the Fluid in the bulb is cooled which occupies the part of the bulb that is immersed in the ice and water.

§ 2. *Ice-cold Water, standing on Ice, may be heated and made to boil without melting the Ice, contrary to an Opinion that has generally prevailed.*

Take a thin glass tube 1 inch in diameter, and about 8 or 10 inches long, containing about two or three inches of water, and by plunging the end of the tube into a freezing mixture of pounded ice and sea-salt cause the water in the tube to congeal; this being done, pour two or three inches of ice-cold water on the ice; and wrapping up about two inches of the lower end of the tube with a piece of flannel, and holding it inclined at an angle of about 45°, by that part of it which is so covered, bring that part of the tube which is at the height of the surface of the fluid-water to be just over the point of the flame of a burning candle, and distant from it about two or three inches. When the water in that part of the tube begins to boil, the tube may be advanced slowly over the flame of the candle; and if due care be taken to prevent a too sudden application of the Heat, all the water in the tube to within one quarter of an inch of the ice may be brought into the most violent ebullition before the ice will begin to be melted, and at last will appear to boil even at the very surface of the ice.

§ 3. *The radiant Heat from a red-hot Iron Bullet does not appear to be able to make its Way downwards through liquid Water, nor through melted Tallow, nor melted Wax.*

1*st Experiment.* — A very small mercurial thermometer, with a naked globular bulb, was laid down in an horizontal position on two small projections of wax, in the bottom of a shallow wooden dish, in such a manner that, the engraved scale of the thermometer lying uppermost, the height of the mercury in its tube could be observed. This being done, I poured cold water into the dish till it stood at the height of about $\frac{1}{4}$ of an inch above the bulb of the thermometer, and then presented to the thermometer an iron bullet about $1\frac{1}{2}$ inches in diameter, red-hot, which I held (by means of a fit handle) directly over its bulb at the distance of about an inch.

The thermometer seemed to take very little notice of the vicinity of the red-hot iron.

When its bulb was covered with oil the result of the experiment was much the same, but when it was exposed naked, or uncovered by a liquid, to the rays from the hot iron, it appeared to acquire Heat very rapidly. But the two following experiments were still more decisive and satisfactory.

2*d Experiment.* — A shallow earthen dish, about 3 inches deep and 12 inches in diameter at its brim, was filled with water, and, being exposed in a cold room in winter, the water was frozen, and formed a cake of ice at its surface about an inch thick. Letting the dish remain in its place, in order that the surface of the ice might remain perfectly horizontal (which was necessary to the complete success of the experiment, as will presently be seen), I entered the room with a chafing-dish filled with live coals, in the midst of which was my iron bullet, perfectly red-hot; and taking out the bullet from among those burning coals, I held it over the center of this horizontal sheet of ice, and distant from it about $\frac{1}{10}$ of an inch.

The ice directly under the red-hot bullet was soon thawed, but the depth to which it was thawed was very inconsiderable; the water, however, extended itself slowly from the center towards the circumference, and at length a circular spot 2 or 3 inches in diameter in the center of the surface of the ice was covered with it, though but to a very inconsiderable depth.

This little spreading sea appeared to prey on the wall of ice by which it was surrounded on every side.

The particles of water in contact with this wall being rendered specifically lighter on becoming ice-cold, they move upwards, and, making way for other warmer particles to advance from below, cause currents in opposite directions to set between the center (where the hot iron remains) and the circumference. As a current at the temperature of 41° must necessarily set downwards at the middle of the circle, this current, striking against the middle of the excavation formed in the ice, ought to deepen it gradually in that part, though but slowly,— and this is what was actually found to be the case; for the bottom of this excavation was not perfectly flat, but was deeper at and near its center than at its sides.

*3d Experiment.* — When this experiment was varied by using a flat cake of tallow instead of a cake of ice, a very extraordinary appearance indeed presented itself, which at first surprised me very much, but which I soon perceived was a new and very striking proof that Fluids are non-conductors of Heat.

The bottom of the circular cavity in the cake of tallow which was occupied by that part of the tallow that had been melted in the experiment, instead of being concave, as I had found that in the ice to be, or flat, as I expected to find this, was *convex* in the middle, or rather rose up in the form of a protuberance, or very blunt point, the extremity of which reached almost to the surface of the melted tallow! As the iron bullet was held as near as possible to the tallow, the end of this projection, which remained unmelted, was certainly not more than $\frac{2}{10}$ of an inch distant from this red-hot ball! Reflecting on the unexpected result of this experiment I was much struck, and not a little humiliated, with the proof it seemed to me to afford of the impossibility of predicting with certainty any event, however inevitable it may appear, which has not actually been seen to happen.

Though I well knew how the Heat must be communicated under the given circumstances, and could foretel with certainty the directions of the currents it must necessarily occasion in the melting tallow; yet the utmost efforts of my intellectual powers, exercised as they were by much meditation, were not sufficient to enable me to foresee that the point where least Heat would be communicated was that precisely which was nearest to the red-hot bullet, and that a protuberance of unmelted tallow would be left in that place.

Let those be very cautious who speculate on the supposed results of experiments they have never made!

On repeating this experiment, and varying it by using a cake of fine bleached *bees-wax*, instead of tallow, the result was much the same; the protuberance, however, in the middle of the circular cavity occupied by the melted wax, though perfectly perceptible, was less considerable in height than that in the cake of tallow.

### § 4. *Beautiful Crystals of Sea-Salt formed in Brine standing on Mercury.*

A small quantity of strong brine, standing on mercury in an open glass tumbler, having by accident been left in a room in a retired part of the house, I observed at the end of about six months that two beautiful crystals of salt, perfectly quadrangular, had been formed in it, one of which was $\frac{14}{40}$ of an inch long, $\frac{11}{40}$ of an inch wide, and $\frac{5}{40}$ of an inch in thickness; and the other $\frac{12}{40}$ of an inch long, $\frac{10}{40}$ of an inch wide, and $\frac{11}{80}$ of an inch thick.

Did the Fluid mercury on which this brine reposed contribute — and how — to the regularity of the form and the uncommon size of these crystals? And might not beautiful crystals of other salts be procured by similar means?

### § 5. *Olive-Oil rendered colourless by Exposure to the Air standing on Brine.*

A quantity of *olive-oil*, about $\frac{3}{4}$ of an inch in depth, having by accident been left standing in an open glass jar, about four inches in diameter, on about a quart of brine, moderately strong, in a retired room where the sun's rays never enter, — at the end of about six months I observed that the oil had become perfectly colourless, and appeared to me to be nearly as transparent as the purest water. On the approach of winter I found that this oil was much more liable to be congealed with cold than oil of the same kind which had stood near it many months in a large glass bottle closed with a cork.

§ 6. *An unsuccessful Attempt to cause radiant Heat from a red-hot Iron Bullet to descend in Oil.*

Having poured a quantity of this colourless oil into a glass tumbler, and caused it to congeal throughout, I presented to its upper surface a red-hot iron bullet, $1\frac{1}{2}$ inches in diameter, and held it quite close to the oil several minutes, till the bullet ceased to be red-hot. As the oil seemed rather to be merely thickened by the cold, and to have lost its transparency in consequence of the presence of a number of opaque particles, which were everywhere dispersed about in it, than to be congealed into a solid mass, I thought that if it were possible for radiant Heat to descend in any Fluid it might perhaps be in this; and if this should happen I was certain to make the discovery by the manner in which the oil recovered its transparency; for should radiant heat descend, the form of the mass of oil first restored to its transparency must necessarily have been *hemispherical*, or some section of a sphere, or at least of some convex figure; but the under part of that part of the oil which was restored to its transparency in this experiment was, to all appearance, as perfectly flat and horizontal as the upper surface of it, which proves that the Heat, by which the congealed oil was thawed, was communicated to it, not immediately by the red-hot bullet, but *me-*

*diately* by means of the Heat absorbed by or generated in the sides of the tumbler. This experiment appears to me to be important in many respects ; but it would be foreign to my present purpose to engage in an investigation of the subject with which it is most intimately connected.

I cannot finish this Essay without giving my reader an account of one more experiment, the result of which was not only quite unexpected, but uncommonly interesting.

Happening accidentally to place in a window the little instrument I had contrived for rendering visible the internal motions which are occasioned in water when Heat is propagated in that fluid,* as it was winter, and the room was warmed by a German stove, that side of the instrument which happened to be nearest the window being exposed to a current of cold air, while the instrument received Heat continually on the other side from the warmer air of the room, the liquid in the instrument was thrown into motions which never ceased, and afforded a very interesting sight.

With a view merely to amuse myself, and the friends who should happen to call in to visit me, and without the smallest expectation of making any new discoveries, I contrived, and caused to be executed, the instrument I am now about to describe, which I thought could not fail to render these motions perpetual, and exhibit them in a striking manner.

A flat box was formed of two equal panes, each 13 inches high, and 10½ inches wide, of fine ground glass, fitted into a square frame of brass in such a manner that these two panes (which are parallel to each other) are at

* For a description of this instrument see Chapter II. of this Essay.

the distance of 1 inch from each other. In the middle of the top of this brass frame there is a circular opening about ½ an inch in diameter, into which a projecting cylindrical brass tube, about half an inch in length, is soldered; and in the middle of the bottom of the frame there is a similar tube which projects downward. The first of these openings serves for introducing into the flat box the liquid with which it is filled, and the other for drawing it off; and they are both well closed with fit stopples of cork.

On both sides of this brass frame there are deep grooves into which the panes of glass are fitted, and the box was made water-tight by luting the joinings of the glass with the frame with glazier's putty. On the outside of the frame there are thin projections of sheet brass, by means of which the box was fixed in one of the sashes of a window in my room, where it occupied the place of a pane of glass, which was removed to make way for it. This window fronts the southeast, and consequently is exposed to the sun a great part of the day.

Having provided a sufficient quantity of the saline solution (of the same kind as was used in constructing the instrument above mentioned, contrived for rendering visible the internal motions in Fluids), and having mixed with it a due proportion of pulverized yellow amber, I now filled the box half full with this mixture, and as the air in the room was considerably warmer than that without, I expected that the motions in the liquid occasioned by the passage of the Heat would immediately commence.

This actually happened; but how great was my surprise when, instead of the vertical currents I expected, I discovered horizontal currents running in opposite directions, one above another, — or regular WINDS, which, springing up in the different regions of this artificial atmosphere, prevailed for a long time with the utmost regularity, while the small particles of the amber collecting themselves together formed clouds of the most fantastic forms, which, being carried by the winds, rendered the scene perfectly fascinating!

It would be impossible to describe the avidity with which I gazed on these enchanting appearances.

In the state of enthusiasm in which I then was, it really seemed to me that Nature had for a moment drawn back the veil with which she hides from mortal eyes her most secret and most interesting operations, and that I now saw the machinery at work by which winds and storms are raised in the atmosphere!

Nothing seemed to be wanting to complete this bewitching scene, and give it the air of perfect enchantment, but that lightning, in miniature, should burst from these little clouds; and they were frequently so thickened up, and had so much the appearance of preparing for a storm, that had that event actually taken place, it could hardly have increased my wonder and ecstasy.

There were several accidental circumstances attending this experiment which contributed to render it more interesting. The sun, which happened to be remarkably bright, shone full upon the window where the apparatus was placed; and as the grooves in the frame in which the plates of glass were fixed were not deep, that part of this frame which formed the narrow bottom of the box being exposed to the sun's rays, a considerable quantity of Heat was generated by them in that place, as appeared by the motions of the particles of pulverized amber which lay on the bottom of the box, or those which were brought there by the currents.

When these particles, on being heated by the sunbeams, began to move, they first arose up nearly perpendicularly ; but before they had risen to any considerable height, they were carried away obliquely and nearly in an horizontal direction by the lower current, answering to the wind which in the atmosphere prevails at the surface of the earth.

The perpendicular rise of these particles from the bottom of the box, and the subsequent change of their direction, called to my remembrance an appearance very common in hot countries, which I recollected to have often seen, and by which I had often been amused in my youth : in very hot and dry weather, when the wind is still and the sun very powerful, the air which lies on the ground often appears in the most violent agitation, resembling that of a boiling liquid; which motion is most rapid at the surface of the earth, and appears to cease at the height of five or six feet above the ground.

Is not this violent agitation occasioned by the conflict which takes place between the hot and the comparatively cold air moving *vertically*, and in opposite directions, very near the surface of the ground ? And are not the winds which prevail above occasioned by the efforts of whole *strata* of air to ascend or descend obliquely ?

The currents I observed to prevail in my artificial atmosphere were never perfectly horizontal; and if my suspicions with respect to the cause of the winds are well founded, neither can those winds be horizontal which prevail in the superior regions of the atmosphere of the earth, though they may be very nearly so.

The greatest velocity of the currents in the saline liquid in this experiment was nearly two inches in a minute, but their motions were in general much slower.

As the windows in the room in which this experiment was made are double (as are all those both in summer and winter in the apartment I inhabit), and as the apparatus above described occupied the place of a pane of glass belonging to the inside window, it was in my power, by opening either the inside window or the outside window, to cause the Heat on the two opposite sides of the box to be either equal or unequal at pleasure; and by variations which that arrangement enabled me to make in the experiments, I produced several interesting appearances.

There was one very striking appearance indeed which never failed to present itself regularly every day during the three weeks that the experiment was continued.* The clouds, after having been driven about all day by the different currents in the liquid (of which there were sometimes as many as six or seven running in opposite directions at the same time), never failed to collect themselves together in the evening into large masses; sometimes forming only one, and sometimes two or three *strata* at different heights, where they remained to all appearance perfectly motionless during the night.

There can be no question with respect to the *proximate* cause of this phenomenon, for it was undoubtedly owing to a diminution or total cessation of the operation of that cause — of those causes, or of some of them — by which an inequality of temperature in the liquid was produced and continued; but it would be highly curious to investigate the more remote causes of this appearance, and see how far *light*, or rather the absence of it, was concerned in producing it: but that discussion would lead me into a very abstruse inquiry, — that re-

* An end was put to the experiment by an accident; the box being broken by the carelessness of a servant in shutting the window-shutter.

specting radiant Heat, — which would take up more time than I am at present able to bestow on it. Perhaps I may find leisure and courage at some future period to attempt that most difficult investigation. My reader will doubtless have observed that I have hitherto taken pains to avoid it.

I cannot take my leave of the experiment I have been describing without giving my reader a faithful account of everything I can recollect respecting it, and particularly of one accidental circumstance, which it is possible may have had some share in producing the interesting appearances which so powerfully attracted my attention.

The saline liquor and the pulverized amber were mixed in a bottle, and were not put into the flat box till after it had been fixed in the sash or frame of the window, but when I came to pour this mixture into the box I found that I had not provided enough of it. To supply this defect, without the trouble of emptying the box, I added, at several different times, pure water, and a strong solution of potash, in such proportions as I knew to be proper to produce the specific gravity required, and then endeavoured to mix the whole as intimately as possible by agitating the liquor for some considerable time by means of a long and strong quill, the end of which I thrust down into the box through the hole by which the liquor was introduced.

Whether those different portions of liquor were in fact intimately mingled by these means I cannot positively determine. They certainly had every appearance of being so; for the amber was evidently well mixed, and very equally distributed in every part of the Fluid. But even should we grant that the liquid remained divided in different *strata*, arranged according to the specific gravities of the different portions of it that were poured into the box at different times, it does not appear to me that the result of the experiment would be less interesting on that account, or the application of it less satisfactory in explaining the cause of the winds in the atmosphere.

I am, however, far from being desirous that much stress should be laid on this single experiment, being perfectly sensible that others may be contrived, the results of which would be more decisive; in the mean time it appears to me that the hint given us is too plain not to deserve some attention. If it should awaken the curiosity of experimental philosophers, and excite them to farther investigation, the end I had principally in view in publishing this account of it will be completely answered.

# DESCRIPTION OF THE PLATES.

## PLATE III.

FIG. 4. This figure represents a vertical section of the apparatus used in the experiment No. 55 (see page 223), in which an attempt was made to melt the top of a projecting point of ice by Heat transmitted *downwards* through olive-oil communicated by a solid cylinder of iron, heated in boiling water.

In this figure the tall glass jar (in the bottom of which the cake of ice was frozen) is standing in an earthen pan filled with pounded ice.

The *oil* is also represented standing on the *cake of ice* in the jar; and the iron cylinder in its sheath of paper suspended in the axis of the jar in such a manner that the lower end of this cylinder, which is flat, is directly over the pointed projection of ice, and distant from it $\frac{2}{10}$ of an inch.

## PLATE IV.

Fig. 5. This figure shows the manner in which the experiment No. 57 (see page 232) was made when pure or fresh water in a glass jar was made to repose on brine, or water saturated with sea-salt, without mixing with it.

In this experiment the smaller jar, which contained the brine, the pure water, and a quantity of olive-oil by which the surface of the pure water was covered, stood in a larger glass jar, which last stood in a shallow earthen dish filled with pounded ice and water.

*Plate III.*

*Fig. 4.*

*Scale of Inches*

Plate IV.

*Fig. 5.*

*Scale of Inches*

The space between the outside of the smaller jar and the inside of the larger jar was filled to the height of about an inch above the level of the surface of the oil in the smaller jar with pieces of ice nearly as large as walnuts, and ice-cold water.

# EXPERIMENTAL INVESTIGATIONS CONCERNING HEAT

SECTION I. — *Short Account of a new Experiment on Heat.*

I HAVE lately made a new experiment, the result of which appears to me sufficiently interesting to deserve the attention of the Class.

Having found, by experiments often repeated, that metallic bodies, exposed in the free air of a large apartment, are much more speedily heated and cooled when their surfaces have been blackened (over the flame of a candle, for example) than when they are clean and polished, I was curious to know whether the same phenomena would take place when, instead of exposing these bodies in the open air, they should be placed in close metallic vessels, surrounded by a certain thickness of included air, and these vessels should be then plunged in a large mass of hot or cold water. In order to clear up this important point, I made the following experiment: —

A cylindrical vessel of brass, three inches in diameter and four inches long, was enclosed in another larger cylindrical vessel, in the centre of which it was suspended by its neck, so as to touch it in no other part, leaving on all sides an interval of one inch between the vessels.

The external vessel, as well as the smaller one included within it, is made of thin sheets of brass; its diameter is five inches, and its height six. It is one inch and a half in diameter, and six inches high. Its neck is one inch and a quarter in diameter, and two inches and a half long.

The interior vessel is suspended in the centre of the external one by a stopper of cork. This stopper is adjusted to the neck of the external vessel, and there is a cylindrical hole of three quarters of an inch diameter through the cork, and having the same axis; which perforation receives the neck of the interior vessel, and retains it in its place.

The interior vessel was introduced and fixed in its place before the bottom of the exterior vessel was soldered in.

At the centre of the bottom of the great vessel is a small metallic tube, of three quarters of an inch diameter and one inch and a half long, by means of which this instrument is attached to a solid heavy foot of metal, which supports it in a vertical position when the whole instrument is submerged in a vessel of water.

This instrument, which greatly resembles that described in my seventh Essay on the Propagation of Heat in Fluids, which I have called the *passage thermometer*,[9] may be used to make a number of interesting experiments on the cooling of bodies through different fluids. In the present experiment I employed it in the following manner: —

The interior vessel was entirely filled with hot water to the height of half an inch in its neck, and a good thermometer, having its cylindrical bulb four inches long, was inserted therein. The instrument was then plunged in a mixture of pounded ice and water, and the time was noted by means of the thermometer, during which the hot water in the small vessel became cold.

I was careful to plunge the instrument in this frigorific mixture, so that the large vessel was completely submerged, except the upper extremity of its neck; and I added, from time to time, a sufficient quantity of pounded ice to keep the frigorific mixture constantly and throughout at the temperature of melting ice.

The following were the results afforded by two similar instruments, employed at the same time: —

These two instruments, which I shall distinguish respectively by the letters A and B, are of the same form and dimensions; there is no difference between them but in the state of their surfaces. In the instrument A the exterior surface of the small vessel and the interior surface of the great vessel which encloses it are bright and polished; but in the instrument B the exterior surface of the small vessel and the interior surface of the large vessel are black, having been blackened over the flame of a candle, before the bottom of the great vessel was soldered in its place.

Having filled the interior vessel of each of these instruments with boiling water till the water rose to the height of half an inch in the neck, I placed a thermometer in each; and then plunging both instruments at the same time into a tub filled with cold water, mixed with pounded ice, I observed the course of their refrigeration during several hours.

Each of the instruments was completely submerged in the frigorific mixture, excepting about one inch of the superior extremity of the neck of the exterior vessel, and I was careful to add new quantities of pounded ice, from time to time, in order to keep the frigorific mixture constantly at the precise temperature of melting ice.

As the specific gravity of water at the temperature of three or four degrees of the thermometer of Reaumur is greater than that of melting ice, the water which lies at the bottom of a vessel containing a mixture of water and pounded ice is usually warmer than the fluid which occupies the upper part of the vessel. To remedy this inconvenience, my refrigeratory for the frigorific mixture was a tin vessel, supported on three feet of one inch in length; and I placed this first vessel in a larger one of wood, containing a certain quantity of ice surrounding the bottom and part of the sides of the metallic vessel.

As in the first moments of the experiment the thermometers descended too quickly to be observed with precision, I waited till each of them had arrived at the 55th degree of Reaumur; after which I carefully observed the number of minutes and seconds employed in passing through each interval of five degrees of the lower part of the scale of the thermometer to the fifth degree above zero.

| Degrees of the thermometer. | Time employed in cooling By the instrument A. | | By the instrument B | |
|---|---|---|---|---|
| | m. | s. | m. | s. |
| From 55 to 50 | 11 | 6 | 7 | 50 |
| " 50 " 45 | 13 | 15 | 8 | 10 |
| " 45 " 40 | 15 | 12 | 9 | 5 |
| " 40 " 35 | 19 | 10 | 10 | 50 |
| " 35 " 30 | 22 | 24 | 12 | 18 |
| " 30 " 25 | 27 | 50 | 15 | 10 |
| " 25 " 20 | 37 | 6 | 21 | 15 |
| " 20 " 15 | 54 | 15 | 28 | 15 |
| " 15 " 10 | 80 | 25 | 41 | 25 |
| " 10 " 5 | 183 | 45 | 85 | 15 |
| Time employed in cooling from 55° to 5°, | 478 | 4 | 254 | 5 |

The foregoing table exhibits the depression of the thermometers during eight hours employed in the experiment.

It is evident, from the results of this experiment, that the blackened body is constantly cooled in less time than the polished body; but it appears, by the course of the thermometers, that the difference between the quickness of cooling of these two bodies varies, and that this difference was less considerable in proportion as the temperature of the bodies was more elevated in comparison to that of the medium in which they were exposed to cool.

In cooling from the 55th degree to the 50th above the temperature of the surrounding medium, the polished body employed 11 minutes and 6 seconds, and the blackened body employed 7 minutes and 50 seconds to pass through the same interval. But from the 10th to the 15th degree above the temperature of the medium, the polished body employed 183 minutes and 45 seconds, while the blackened body employed only 85 minutes and 15 seconds; but it is extremely probable that this difference between the proportion of the times employed in cooling the two bodies at different temperatures is only apparent, and that it depends on the greater or less time required for the thermometers in the vessels to arrive at the mean temperatures of the masses of water which surround them.

In order to compare the results of this experiment with those I made last year with metallic vessels, polished and blackened, and left to cool in the undisturbed air of a large chamber, it is necessary to ascertain how much time the two bodies in question employed in cooling from the 50th to the 40th degree of Fahrenheit above the temperature of the medium. Now, I found, by observation, that the polished vessel A employed 39 minutes and 30 seconds to pass over that interval of cooling, while the blackened vessel B employed only 22 minutes. These times are in the proportion of 10,000 to 5810. By one of my experiments, made last year, I found that the times employed in passing through the same interval of cooling in the open air by a clean polished metallic vessel, and another of the same form and capacity, but blackened without, were as 10,000 to 5654.

Reflecting on the consequences which ought to result from the radiations of bodies, on the supposition that the temperatures of bodies are always changing by means of these radiations, I was led to the following conclusion: If the intensity of the action of the rays which proceed from a body be universally as the squares of the distances of bodies inversely, which is extremely probable, a hot body exposed to cool in a close place, or surrounded on all sides by walls, ought to cool with the same celerity, or in the same time, whatever may be the magnitude of this enclosure, provided the temperature of the sides or walls be at a constant given temperature; and the results of the experiment here described, in which the hot body was enclosed in a vessel of a few inches diameter, compared with those of several experiments made last year, in which the heated bodies were exposed to cool between the walls of a large chamber, appear to confirm this conclusion.

As to the effect produced by the air in cooling a heated body exposed to cool in a close place filled with that fluid, I have reason to believe that it is much less considerable than has been supposed.

I have shown, by direct and conclusive experiments, that bodies cool and are heated, and that with considerable celerity, when placed in a space void of air;[10] and by experiments made last year, with the intention of clearing up this point, I found reasons to conclude that when a hot body cools in tranquil air, not agitated by winds, one twenty-seventh only of the heat lost by this body (or, to speak more correctly, which it excites in surrounding bodies) is communicated to the air, all the rest being carried to a distance through the air and communicated by radiation to the surrounding solid bodies.

SECTION IV. — *The Heat produced in a Body by a given Quantity of solar Light is the same whether the Rays be denser or rarer, convergent, parallel, or divergent.*

In all cases where the rays of the sun strike on the surface of an opaque body without being reflected, heat is generated and the temperature of the body is increased; but is the *quantity* of heat thus excited always in proportion to the quantity of light that has disappeared? This is a very interesting question and has not hitherto found a decisive solution.

When we consider the prodigious intensity of the heat excited in the focus of a burning mirror or a lens, we are tempted to believe that the concentration and condensation of the solar rays increase their power of exciting heat; but, if we examine the matter more closely, we are obliged to confess that such an augmentation would be inexplicable. It would be equally so on both the hypotheses which natural philosophers have formed of the nature of light; for if light be analogous to sound, since it has been proved, both by calculation and experiment, that two undulations in an elastic fluid may approach and even cross each other without deranging either their respective directions or velocities, we do not see how the concentration or condensation of these undulations can increase their force of impulse; and if light be a real emanation, as its velocity is not altered either by the change of direction it undergoes in passing through a lens or by its reflection from the surface of a polished body, it seems to me that the power of each of these particles to excite or impart heat must necessarily be the same after refraction or reflection as before, and consequently, that the heat communicated or excited must be, in all cases, as the quantity of light absorbed.

I have just made some experiments which appear to me to establish this fact beyond question.

Having procured from the optician Lerebours two lenses perfectly equal, and of the same kind of glass, 4 inches in diameter, and of 11½ focus, I exposed them at the same time to the sun, side by side, about noon, when the sky was very clear; and by means of two thermometers, or reservoirs of heat, of a peculiar construction, I determined the relative quantities of heat that were excited in given times by the solar rays at different distances from the foci of the lenses.

The two reservoirs of heat are a sort of flat boxes of brass filled with water. Each of these reservoirs is 3 inches 10½ lines in diameter, and 6 lines thick, well polished externally on all sides except one of its two flat faces, which was blackened by the smoke of a candle. On this face the solar rays were received in the experiments.

Each of these reservoirs of heat weighs when empty 6850 grains, *poids de marc* (near a pound troy), and contains 1210 grains of water (about 2 oz. 2 dwts.).

Taking the capacity of brass for heat to be to that of water as 1 to 11, it appears that the capacity of the metallic box weighing 6850 grains is equal to the capacity of 622 grains of water; and, adding this quantity of water to that contained in the box, we shall have the capacity of the reservoir prepared for the experiments equal to that of 1832 grains of water.

Each reservoir is kept in its place by a cylinder of dry wood, one of the extremities of the cylinder being fixed in a socket in the centre of the interior face of the reservoir; and each reservoir has a little neck, through which it is filled with water, and which after receives the bulb of a cylindrical thermometer, that reaches completely across the inside of the box in the direction of its diameter.

The two reservoirs of heat, with their two lenses, are firmly fixed in an open frame, which being movable in all directions by means of a pivot and a hinge, the apparatus is easily directed toward the sun, and made to follow its motion regularly, so as to keep the solar spectra constantly in the centres of the blackened faces of the reservoirs.

In order that the quantities of light passing through the two lenses should be perfectly equal, a circular plate of well-polished brass, in the centre of which is a circular hole $3\frac{1}{2}$ inches in diameter, is placed immediately before each of the lenses.

When the reservoirs of heat are placed at different distances from the focuses of their respective lenses, the diameters of the solar spectra which are formed on the blackened faces of the reservoirs are necessarily different; and as the quantities of light are equal, its density at the surface of each reservoir is inversely as the square of the diameter of the spectrum formed on that surface.

*Experiment No.* 1.—In this experiment the reservoir A was placed so near the focus of the lens, between the lens and the focus, that the diameter of the solar spectrum falling on it was only half an inch, or 6 lines, while the reservoir B was advanced so far before the focus that the spectrum was 2 inches in diameter, or 24 lines.

As the quantities of light falling on both were equal, the density of the light at the surface of the reservoir

A was to the density of that at the surface of the reservoir B as the square of 24 to the square of 6, or as 16 to 1.

I imagined that if the quantity of heat which a given quantity of light is capable of exciting depended any way on its density, as the densities were so different in this experiment, I could not fail to discover the fact by the difference of time which it would require to raise the two thermometers the same number of degrees.

Having continued the experiment more than an hour, on a very fine day, when the sun was near the meridian and shone extremely bright, I did not find that one of the reservoirs was heated perceptibly quicker than the other.

*Experiment No.* 2. — I placed the reservoir of heat A still nearer the focus of the lens, in a situation where the solar spectrum was only $4\frac{3}{4}$ lines in diameter, and where blackened paper caught fire in two or three seconds; and I removed the reservoir B still farther from the focus, advancing it forward till the diameter of the spectrum was 2 inches 3 lines.

The densities of the light at the surfaces of the reservoirs in this experiment were as 32 to 1.

The temperature of the reservoirs as well as that of the atmosphere, at the beginning of the experiment, was $54°$ F., $= 9\frac{7}{9}°$ R.

The reservoir A, after having been exposed to the action of very intense light near the focus of the lens for 24 minutes 40 seconds, was raised to the temperature of $80°$ F., $= 21\frac{1}{3}°$ R.

The reservoir B, which was much farther from the focus of its lens, was raised to the same temperature,

80° F., a little more quickly, or in 23 minutes 40 seconds.

To raise the temperature of the reservoir A to 100° F., $= 30\frac{2}{9}$° R., it was necessary to continue the experiment for 1 hour 15 minutes 10 seconds, reckoning from the commencement of it; but the reservoir B reached the same temperature in 1 hour 12 minutes 10 seconds.

The progress of this experiment from the beginning to the end is exhibited in the following table.

| Increases of Temperature. | Time taken. | |
|---|---|---|
| | By A. | By B. |
| | m. s. | m. s. |
| From 54° to 80° F. | 24 40 | 23 40 |
| 80 85 | 7 45 | 7 30 |
| 85 90 | 9 55 | 9 0 |
| 90 95 | 13 30 | 13 0 |
| 95 100 | 19 20 | 19 0 |
| From 54° to 100° | 75 10 | 72 10 |

This experiment was begun at 7 minutes 30 seconds after 11, and finished at 22 minutes 40 seconds after 12, the sky being perfectly clear during the time.

On comparing all the results of this experiment, we see that the reservoir A, which was placed very near the focus, was more slowly heated than the reservoir B, which was at a considerable distance from it. The differences of time, however, taken to heat them an equal number of degrees were very trifling, and I think may be easily explained without supposing the condensation of light to increase its faculty of exciting heat.

In both the preceding experiments the solar rays striking on the reservoirs of heat were *convergent*, and they were even equally so on both sides. To determine whether *parallel* rays have the same power of exciting heat as convergent rays, I made the following experiment.

*Experiment No.* 3. — Having removed the lens from before the reservoir B, I suffered the direct rays of the sun to fall on the blackened face of the reservoir, through the circular hole, $3\frac{1}{2}$ inches in diameter, in the round brass plate which had been constantly placed before that lens in the preceding experiments.

The reservoir A was placed behind its lens as in the former experiments, and at the place where the solar spectrum had 6 lines diameter.

Having exposed this apparatus to the sun, I found that the reservoir B, on which the direct rays fell, was heated sensibly quicker than the reservoir A, which was exposed to the action of the concentrated rays near the focus of the lens.

The temperature of the apparatus and of the atmosphere at the beginning of the experiment being 53° F., $= 9\frac{1}{3}°$ R, the reservoir A required 23 minutes 30 seconds to raise it to the temperature of 80° F., $= 21\frac{2}{9}°$ R.; but the reservoir B, which was exposed to the direct rays of the sun, acquired the same temperature in 18 minutes 30 seconds.

To reach the temperature of 100° F., $= 30\frac{2}{9}°$ R., took the reservoir A 1 hour and 3 minutes, but the reservoir B 47 minutes 15 seconds only.

The following table will show the progress of this experiment from the beginning to the end.

| Increases of Temperature. | Time taken. By A. | | Time taken. By B. | |
|---|---|---|---|---|
| | m. | s. | m. | s. |
| From 53° to 65° F. | 8 | 26 | 7 | 0 |
| 65 70 | 4 | 10 | 3 | 15 |
| 70 75 | 5 | 10 | 3 | 45 |
| 75 80 | 5 | 40 | 4 | 30 |
| 80 85 | 7 | 0 | 4 | 45 |
| 85 90 | 7 | 30 | 5 | 45 |
| 90 95 | 10 | 30 | 8 | 0 |
| 95 100 | 13 | 10 | 10 | 15 |
| 100 105 | 20 | 0 | 14 | 45 |
| From 53° to 105° | 81 | 36 | 62 | 30 |

As a considerable part of the light that fell on the lens before the reservoir A was lost in passing through it, it is evident that the quantity received by this reservoir was less than that received by the reservoir B, which was exposed to the direct rays of the sun; and we have seen that the latter was heated more rapidly than the former.

As we know not exactly how much light was lost in passing through the lens, we cannot determine from the results of this experiment whether convergent rays be more or less efficacious in exciting heat than parallel rays; but the difference in the times of heating was not greater, as it appears to me, than we might have expected to find it, supposing it to be occasioned solely by the difference between the quantities of light acting on the reservoirs.

The result of the following experiment will establish this point beyond doubt.

*Experiment No.* 4. — Having replaced the lens belonging to the reservoir B, I adjusted this reservoir to such a distance between the lens and its focus that the solar spectrum was one inch in diameter; and I placed the reservoir A at the same distance beyond its focus.

As the quantities of light directed toward both were equal, and as the diameters of the spectra, and consequently the densities of the light that formed them were also equal, there could be no difference between the results of the experiments with the two reservoirs, except what was occasioned by the difference in the *direction* of the rays that formed the spectra. On one hand these rays were *convergent*, and on the other *divergent*; and I had inferred that if parallel rays were in reality less efficacious in exciting heat than convergent rays, as some philosophers have supposed, *divergent* rays must be still less efficacious than parallel rays, and consequently much less than convergent rays.

Having made the experiment with all possible care, I found no sensible difference between the quantities of heat excited in a given time by divergent and convergent rays.

The following are the particulars of the progress and results of this experiment.

| Increases of Heat. | Time taken. | |
|---|---|---|
| | By A, with divergent Rays. | By B, with convergent Rays. |
| | m. s. | m. s. |
| From 60° to 65° F. | 4 50 | 4 50 |
| 65 70 | 4 55 | 5 0 |
| 70 75 | 5 27 | 5 25 |
| 75 80 | 6 13 | 6 15 |
| From 60° to 80° | 21 25 | 21 30 |

From the results of all the experiments of which I have just given an account to the Class, we may conclude that the quantity of heat excited or communicated by the solar rays is always, and under all circumstances, as the quantity of light that disappears.

# REFLECTIONS ON HEAT.

THE most excellent gift which man has received from the Author of his being is the power which he possesses of freeing himself from the prejudices arising from the deceptive testimony of his senses, and of penetrating into the mysteries of Nature.

The animals see as we do, without doubt, that the sun, moon, and stars rise and set; man in a state of nature, when his attention is aroused, discovers irregularities in these movements; the man of genius, however, does not allow himself to be deceived by appearances, but causes to come forth from this confusion that vast and wonderful system of laws which govern the mechanism of the Universe.

The first step in science is to observe facts attentively, and in their proper connection; the second is to learn to doubt. The sublime in science consists in employing it to extend the power and increase the innocent enjoyments of the human race.

There is no branch of the physical sciences which is so intimately connected with all the every-day occupations of man as that of *Heat*, and consequently there is no one of them which interests him so closely.

Fire is the most universal and active agent with which we are acquainted, and it is to the power which he has been able to acquire over this wonderful principle that man owes the supernatural strength which has made him superior to all animals, and master of land and sea.

It is not at all surprising that an agent at once so powerful and so manageable, so beneficent and so terrible, should have become an object of admiration and even of adoration among the nations of the earth; but it is more than surprising that a subject, the investigation of which is of such interest, should have been for so long a time neglected.

This indifference to an object at once so curious and so interesting can only be attributed to that lack of attention with which men always regard those things that they are accustomed to have before them at all times.

A proof that our knowledge on the subject of heat is still extremely limited and imperfect lies in the difference of opinion which exists among the learned on the nature of heat and its mode of action. Some regard it as a *substance*, others as a *vibratory motion* of the particles of matter of which a body is composed.

Those who have adopted the hypothesis of a peculiar calorific substance which they call *caloric* suppose that the heating of a body is always the result of an *accumulation* of this substance in the body; on the other hand, those who regard heat as a vibratory motion which is conceived to exist always with greater or less rapidity among the particles of all bodies, consider heat as an *acceleration* of this motion.

On the hypothesis of vibratory motion, a body which has become cold is thought to have lost nothing except motion; on the other hypothesis it is supposed to have lost some material substance, that is, caloric.

The eminent French philosophers, who proposed, twenty-five years ago, the modern hypothesis of caloric, far from considering the existence of this substance as proved, always speak of it with that modest reserve which characterizes men of superior excellence. They propose the word in order to avoid circumlocutions and to render the language of science more concise, rather than to introduce a new opinion.

One of these philosophers, whom science and mankind still mourn, thus expresses himself in his admirable *Traité Elémentaire de Chimie:* "In the labours which M. de Morveau, M. Berthollet, M. de Fourcroy, and myself have performed in common on the reform of the language of chemistry, we have felt that we ought to banish from it those circumlocutions which render the form of expression longer and more cumbersome, less exact, and less clear, and which not seldom even do not allow of ideas sufficiently well defined. We have, therefore, designated the cause of heat, the eminently elastic fluid which produces it, by the name *caloric*. Independently of the fact that this expression answers our purpose in the system which we have adopted, it has besides another advantage, that of being able to adapt itself to all sorts of opinions, since, strictly speaking, we are not even obliged to suppose that caloric is a material substance."

If the point in question, the existence or non-existence of caloric, were less important we might be content with leaving it undecided; but the use of heat is so general, and the art of exciting and directing it is so intimately connected with the perfecting of all the mechanical arts and with a great number of domestic applications, that we cannot take too much trouble in becoming acquainted with it.

Without entering into the details of the various experiments which have been performed in order to determine the nature of heat, I will limit myself in this memoir to some of the principal results of these researches.

A very remarkable phenomenon, and one which must have been noticed as soon as men had any acquaintance with fire, is the radiation from solid bodies as soon as they become very warm.

When a solid body—a bar of iron, for example—is at about the temperature of the surrounding air, we do not see or perceive anything which indicates that it possesses a radiating surface; but if we heat it strongly in the glowing fire of a forge, the body changes color, becomes at first red, then white, is visible in the dark, lights up surrounding objects, and warms in a sensible degree all objects which are struck by the rays which it emits in all directions.

If we allow it to cool slowly in the undisturbed air of a dark room, we see that it changes color again; from white it becomes red, then a darker red; the light which it gives forth gradually diminishes; the intensity of its calorific rays becomes less at the same time, and soon it ceases to shed light round about it.

It continues, however, to emit from its surface calorific rays for some time after it has ceased to be luminous, as may be perceived by holding the hand near it.

The calorific rays which very warm bodies send off from their surfaces pass through the transparent air without heating it, nor do they heat sensibly those bodies at whose surfaces they are reflected.

These very important facts, which ought not to be forgotten, have been established by the results of a large number of experiments.

We have thus taken one sure step in the investigation of heat. We see that very warm bodies emit from their surfaces rays which, passing (like rays of light) through the air excite, at a distance, heat at the surfaces of the surrounding objects on which they fall *without being reflected.*

The existence of the calorific rays, which we are now discussing, being actually proved, and their manner of acting being evidently as I have described it, it is important to ascertain whether the knowledge of these facts be not sufficient to form a theory of heat which will explain all these phenomena.

A theory which should have the advantage of explaining the communication of heat by a *single* method, at once simple and easily understood, would be preferable, it seems to me, to one which, in order to explain various phenomena, would be obliged to admit *two different* modes of the communication of heat.

In order to form a clear and exact idea of the rays in question and of the effects which they are capable of producing, we must go back to their mechanical cause, and consider them with regard both to their existence and to their operation.

There are two ways of looking at the radiation from an object; the first, by conceiving the rays as emanations of an actual substance thrown off from the surface of the body; the second, by considering these rays as *undulations* which, starting from every point of the surface of the radiating object, are propagated in all directions in straight lines in an elastic fluid which surrounds it on every side.

The system of Newton *supposes* that the rays of light are real emanations.

Sound, with which we are better acquainted than we are with light, affords us an example of radiation or undulation in an elastic fluid which most certainly is not an emanation.

We have sufficiently clear ideas of the mechanical operations by means of which the undulations of the air which constitute sound are excited and propagated; but we have no conception of any possible mechanical operation by means of which a *material substance* could be sent forth continually and *in all directions* from the surface of a body.

In physics, in order that an hypothesis may be admitted, it must be founded on the supposition of a *conceivable* mechanical operation.

In order that the theory of heat which is founded on the vibratory hypothesis may be admitted, it is necessary to show that the vibrations in question can exist, and that they can cause the rays or undulations which objects emit from their surfaces, and by means of which we suppose that bodies of different temperatures influence each other even at a distance, bringing about reciprocal and simultaneous changes of temperature, so that little by little they arrive at a common and intermediate temperature.

If the particles which compose a body do not touch each other (an opinion which is generally received, and which appears very probable), as there is no doubt that these particles are continually drawn one towards another by the recognized force of universal gravitation, it is impossible to conceive how, in an assemblage of particles which form a tangible solid body, these particles can preserve their relative situations without being in motion.

From this course of reasoning we might conclude that the particles which compose a body are of necessity in motion; and if we admit the existence of an eminently elastic fluid, — an ether which fills all space throughout the universe, with the exception of that occupied by the scattered particles of ponderable bodies, — it is easy to conceive that the movements of the particles which make up material objects must cause undulations in this fluid; and, on the other hand, the undulations of this fluid must affect to a sensible degree and modify the motions of the particles of these bodies.

It might perhaps seem that these motions among the particles of solid bodies would be incompatible with the preservation of the forms of those bodies; but by reflecting attentively on this subject it will be found that such motions as are here supposed can well exist without diminishing at all the stability of the external form of the bodies.

It would follow necessarily, from the state of things supposed by the hypothesis in question, first, that the sum of the active forces in the universe must always remain constant, in spite of all actions and reactions taking place among the various bodies; secondly, that the particles of all ponderable bodies must of necessity have the property of producing radiations.

Now, if we admit the existence of the *ether*, it is possible to explain the radiations of bodies in still another manner; it is by supposing that the particles are kept apart from each other, not in consequence of the action of the centrifugal force of those particles, but by atmospheres composed of ether or of some other fluid unknown to us, which is extremely elastic, and that it is by the very rapid vibrations which take place in these atmospheres that those undulations in the surrounding ether are excited by means of which the temperature of objects is altered.

The adoption of this latter hypothesis will reconcile, to a certain extent, the theory of vibrations with that of a calorific substance; but still the heating of a body cannot be regarded, in any respect, as the result of the *accumulation* of this substance, but as the *acceleration* of its motion.

In order to establish on a firm foundation the theory of heat which is based upon the vibratory hypothesis, it is necessary not only to show that the vibrations in question are possible, but also to prove that the undulations which they should cause do really exist.

In the ordinary condition of things, the objects which surround us do not afford any indication of radiation, nor do they produce any effect capable of manifesting itself to any one of our senses in such a way as to lead us to suspect that they possess radiating surfaces. But the philosopher who aims at penetrating into the mysteries of Nature must be continually on his guard that he may not be deceived either by the testimony or by the silence of his senses.

In the first place it is evident that our various organs were formed with reference to the daily wants of life; and that, if they were too sensitive, the pleasure which they afford us would be turned into actual pain.

If our ears had been constructed so as to be sensibly affected by all the vibrations which take place in the air, we should, without doubt, be stunned by the intolerable noise, even in the deepest retirement; and if our eyes took cognizance of all the rays that strike them, we should be dazzled by an insupportable flood of light, even in the darkest night.

It is well known that, if the vibrations of a sonorous body be less frequent than 30 in a second, or more frequent than 3000 in a second, the undulations of the air caused by these vibrations do not perceptibly affect our organs of hearing; and it is very probable that the range of our organs of sight is still more limited.

When we have found strong reasons for suspecting the existence of agents which fail to manifest themselves to our senses, we ought to employ all our skill in devising means for compelling them to discover themselves and to unveil the mysteries of their invisible operations.

By means of an instrument which I have called a *thermoscope*, and which is extraordinarily sensitive, I have found not only that all bodies at all temperatures emit rays, but also that the rays emanating from cold bodies are as effectual in cooling warm bodies as the rays from the latter are effectual in warming cold bodies.

The principal part of the instrument of which I have made use in these delicate experiments consists of a long glass tube bent at both ends, and having at each extremity a very thin glass bulb an inch and a half in diameter. The middle portion of this tube, which is straight, is placed in a horizontal position, while the two end portions, whose extremities are the two bulbs, are turned upwards in such a way as to form right angles with the horizontal portion of the tube. The horizontal portion is from 15 to 16 inches in length, and each of the two end portions, which are vertical, is from 6 to 7 inches long. The internal diameter of the tube should be about half a line.

By means of a little glass *reservoir*, an inch in length and a line in internal diameter, inserted in the tube at one of the elbows, there is introduced into the interior of the instrument a small quantity of coloured spirit of wine (exactly enough to fill the reservoir without interfering with the free passage of the air from one bulb to the other); this being done, the extremity of the reservoir is sealed hermetically, and all communication between the air enclosed in the instrument and the air of the outside atmosphere is forever interrupted.

The instrument is adjusted and prepared for use as follows: —

The bulb which is farthest from the reservoir having been warmed slightly with the hand, the instrument is suddenly turned over, so as to bring the reservoir uppermost, and in this way a small quantity of the spirit of wine passes from the reservoir into the horizontal part of the tube; restoring immediately the instrument to its natural position, the observer withdraws himself from it, and waits for the small quantity of spirit of wine which has passed into the horizontal part of the tube to become stationary; this will be as soon as the two bulbs have acquired the same temperature.

The little bubble of spirit of wine, which serves as the index of the instrument, and which may be about three quarters of an inch long, should become stationary nearly in the middle of the horizontal portion of the tube; if it is too near either of the elbows it must be returned to the reservoir, and the operation performed anew.

When this delicate operation is finished, the instrument is ready for use. The method of employing it is as follows: —

One of the two bulbs is protected from the influence (calorific or frigorific) of the warm or cold bodies presented to the other bulb by means of light screens covered with gilt paper; when the air in this latter bulb is warmed or cooled by a body warmer or colder than the thermoscope to which it is thus presented, the elasticity of the air is affected by this change of temperature, and the little bubble or column of spirit of wine which is in the horizontal portion of the tube is compelled to move and to take a new position.

The direction of the motion of this bubble indicates the nature of the change which has taken place in the temperature of the air which is enclosed in the bulb to which the body is presented, and the distance traversed by the bubble is the measure of the increase or diminution of the elasticity, and, as a consequence, of the temperature of that air.

If the bubble recedes from the bulb to which the object experimented upon is presented, it is evident that the air enclosed in the bulb has been heated by the influence of this body; but when the bubble of spirit of wine advances towards this bulb, we have a proof that the air in the bulb has been cooled.

The *rapidity* with which the bulb moves is proportional to the *intensity* of the action of the object presented to the instrument.

In order to compare the intensity of the calorific or frigorific actions of two different objects, they are presented at the same time to the two bulbs of the instrument, and their respective distances from the bulbs so regulated that the bubble of spirit of wine remains at rest in its proper position.

In this case it is evident that the action of the two

objects, each on the bulb to which it is presented, is of precisely the same amount; hence we can calculate the relative intensity of the radiation of each one of the two objects from the extent of the surface presented to the bulb, and from the square of its distance from the bulb.

If we desire to compare the calorific action of a warm body with the frigorific action of a cold body, we begin by protecting one of the bulbs of the instrument by the screens, and then present to the other bulb the two objects, — regulating their respective distances in such a manner that their actions exerted at the same time produce equal effects, that is, so that one warms the bulb as much as the other cools it.

The equality in the amount of action is denoted by the remaining at rest of the bubble of spirit of wine which serves as the index of the instrument, and when this equality is established, the relative intensity of the radiation from the objects in question is calculated from the amount of surface which they respectively present to the bulb, and from the squares of their distances from it.

The sensibility of this instrument is so great that, when it is at a temperature of 15° or 16° of Reaumur's scale, if the hand be presented to one of the bulbs at a distance of three feet, the heat radiating from the hand is sufficient to cause the bubble of spirit of wine to move forward several lines; and the cooling influence of a blackened metallic disk four inches in diameter, at the temperature of melting ice, is such that, when presented to the bulb at a distance of 18 inches, it causes the bubble to advance in the opposite direction with a rapidity which is very perceptible to the eye.

By means of this instrument I have discovered,

first, that all bodies at all temperatures (cold bodies as well as warm ones) emit continually from their surfaces rays, or rather, as I believe, *undulations*, similar to the undulations which sonorous bodies send out into the air in all directions, and that these rays or undulations influence and change, little by little, the temperature of all bodies upon which they fall without being reflected, in case the bodies upon which they fall are either warmer or colder than the body from the surface of which the rays or undulations proceed; secondly, that the intensity of the rays from different bodies *at the same temperature* is very different, and that it is less in bodies which reflect the rays of light than in those which absorb them, less in the metals than in their oxides, less in opaque and polished bodies than in those which are imperfectly transparent and unpolished, (a surface of brass, for instance, emits four times as large a quantity of rays at a given temperature when it is covered with a coating of oxide, and five times as large a quantity when it is blackened by the flame of a candle, as when the surface of the metal is clean and well polished); thirdly, that the rays which bodies of the same temperature send out to each other have no tendency to bring about any change of temperature in these bodies; fourthly, that the rays which any body whatever, at a given temperature, sends continually from its surface in all directions, are calorific or frigorific with regard to other bodies on which they fall, according as these latter are less warm or warmer than the body from which the rays come; so that the same rays are calorific as regards all bodies less warm than the one from which they proceed, and frigorific as regards all those which are warmer than this body.

From these facts we might conclude *a priori*, that those bodies which, when warm, give off many calorific rays would, when colder than the surrounding objects, give off to them many frigorific rays. This is exactly what my experiments have made evident to me.

In experiments made with bodies of the same size, and of the same material, the intervals of temperature being equal, the frigorific influences of cold bodies have always appeared as real and effective as the calorific influences of warm bodies.

To one of the bulbs of a thermoscope, the temperature of which was 20° of Reaumur's thermometer, were presented at the same time and at equal distances two disks of metal of the same diameter. The temperature of one of these disks was 0° (that of melting ice), that of the other was 40°. The index of the instrument by remaining at rest showed that the bulb was cooled by the rays from the cold body as much as it was heated by the rays from the warm body.

If the surface of one of the disks, it matters not of which, is blackened, the intensity of the radiation from this blackened disk is increased to such an extent that the other can no longer counterbalance it; but if the second one is blackened also, the equality of action is immediately re-established.

If the emanations from warm and cold bodies are really undulations in an extremely rare and elastic fluid which has been called *ether*, the communication of heat and cold ought to be similar to the communication of sound; and all the mechanical contrivances which have been invented to increase the intensity of sound ought to be just as applicable for increasing the effects produced by these emanations from warm and cold bodies;

and, indeed, I found that a speaking-tube (a conical brass tube, well polished on the inside) placed between one of the bulbs of the thermoscope and a hollow ball of thin copper 3 inches in diameter, which, being filled with pounded ice, was presented to it at a distance of 12 inches, increased more than three times the effect of the cold body.

To use a rather strong metaphor, but one which expresses perfectly the idea which I have conceived of the mechanical operation in question, I will say that the cold ball *spoke* at the larger opening of the speaking-tube while the bulb of the thermoscope *listened* at the smaller opening.

If it is true that the particles which make up all material bodies are agitated continually by very rapid vibratory motions, and that, in consequence of these motions, all bodies at all temperatures send continually from every point of their surfaces rays or undulations similar to the undulations caused in the air by the vibration of sonorous bodies; and if bodies of different temperatures act one upon another at a distance, by means of these rays or undulations, working simultaneously an interchange in temperature and gradually bringing about a mean intermediate temperature, — we ought then to regard the cooling of a warm body as the result of the actual and positive operation of the surrounding bodies less warm than itself; and since the rays coming from warm bodies, and, as a consequence, from cold bodies, are reflected in great measure by the polished surfaces of opaque bodies, and since the rays which are reflected produce little or no effect on the bodies at whose surfaces they are reflected, we might conclude *a priori* that opaque polished bodies ought to cool or become warm more slowly than bodies imperfectly transparent and unpolished.

I will now detail the results of a series of experiments made with a design of throwing light on this point, so important in the science of heat.

I had made two cylindrical vessels, four inches in diameter and four inches high, of thin sheet brass, well polished on the outside. Having blackened one of them over the flame of a candle, I filled them both with boiling water, and left them at the same time to cool in the air of a large quiet room. The one which was blackened cooled almost twice as fast as the one whose metallic surface remained bright and clean. When the two vessels had become of the same temperature as that of the room in which they were situated, they were removed into a room warmed by a stove, and I found that the blackened vessel was heated twice as quickly as the other.

The blackened vessel was cleaned and covered with a single covering of fine linen, fitting closely to the body of the instrument. Repeating the experiments with the two vessels, that which was exposed naked to the cold air took up 45 minutes in cooling through an interval of 10 degrees on Fahrenheit's scale, that is, from the 50th to the 40th degree above the temperature of the room; the other vessel, covered with a *coat* of fine linen, took up only 29 minutes in cooling through the same interval.

When the two vessels had become of the same temperature, they were removed into a warm room, and I found that the vessel which was clothed with linen acquired heat faster than the one whose surface was naked.

If the results of these experiments do not furnish a conclusive proof of the radiation from all bodies, and that it is by means of these radiations from surrounding objects that the temperature of a given body is changed, they certainly lend to this conjecture a great degree of probability.

Several other similar experiments were undertaken in order to throw light on this point, and results were invariably obtained which tended to confirm the hypothesis in question.

Of all known bodies the metals are the most *opaque*, and it appears that they are so to an equal degree; it appears also that a naked metallic surface, or one that is free from all dirt, is always polished in spite of those irregularities of form by which the brilliancy of its metallic lustre is broken up and apparently diminished. If these conjectures are well founded, we may conclude that all metals are equally competent to reflect from their surfaces the rays that impinge upon them; and if objects are heated and cooled by rays from surrounding objects, we might conclude not only that of all known bodies the metals ought to acquire heat or become cold the least rapidly, but also that they ought to acquire heat or become cold with the same degree of difficulty or rapidity.

To put these suppositions to the test of experiment, I procured several cylindrical vessels, of the same form and dimensions but of different metals, and I found that they did indeed all cool or acquire heat in the same time. There were vessels of brass, tin, lead, and others covered with thin coatings of gold and silver; each vessel was four inches in diameter and four inches high, and when filled with boiling water and exposed, in winter, to the air of a large quiet room, they all passed, in cooling, through the given interval of 10 degrees in from 45 to 46 minutes.

This equality in the degree of readiness with which all the metals become cool or acquire heat is certainly very remarkable; and it seems to me very difficult of explanation except by adopting the hypothesis that heat is communicated by means of radiations.

As it might be supposed that a film of air, attached by a certain force of attraction to the surfaces of the metallic vessels, could have caused this apparent equality in their rate of cooling, I made the following experiments to elucidate this point.

One of the two brass vessels was covered, first with one, next with two, then with four, and finally with eight coatings of spirit varnish, and the experiment with the two vessels was repeated with each of these coatings. While the vessel, the surface of which was bare, cooled invariably through the given interval of 10 degrees in 45 minutes, the other vessel, which was varnished, cooled more or less rapidly according to the thickness of the coating of varnish with which its surface was covered, but always in a sensible degree more rapidly than the one whose surface was naked: —

| | Minutes. |
|---|---|
| With one coating of varnish it cooled in . . . . | $34\frac{1}{2}$ |
| With two coatings, in . . . . . . . . | 29 |
| With four coatings, in . . . . . . . . | $24\frac{1}{2}$ |
| And with eight coatings, in . . . . . . | 27 |

As the film of air which is supposed to have been attached to the surface of the vessel when this metallic surface was not covered with varnish ought to have been as completely driven off by *one* coating of varnish as by *two* or by a greater number, it seems very difficult to reconcile the results of these experiments with the supposition that a film of air attached to the surfaces of all the vessels, made as they were of different metals, was the cause of their cooling all equally slowly.

When I repeated the experiment with a vessel of glass, and with one of tinned iron of the same form and dimensions, I found that the glass vessel cooled much more rapidly in the air than the one made of tinned iron, although its walls were six times as thick as those of the latter. In water the vessel of tinned iron cooled most rapidly.

The results of all these experiments, and of a great number of others which it would take too long to detail here, convinced me that the ease with which a body is heated or cooled depends very much on the nature of the surface of that body, — these operations going on more slowly and with more difficulty as the surface of the body is more capable of reflecting the rays which fall upon it; I was therefore impatient to submit the theory of heat which I had adopted to the most searching of tests, by employing it to explain some of the grand and interesting phenomena of nature.

Close to us there occurs a most interesting phenomenon, and one which, assuredly, is calculated to excite our curiosity.

The people who inhabit hot countries are black, while those who dwell in cold climates are white.

What advantages do the negroes derive from their colour which makes them better fitted than the whites for supporting without inconvenience the excessive heats of their scorching climate?

In all climates a large amount of heat is necessarily

excited in the lungs by the act of breathing; and when man is placed in a situation where the air and all objects about him are almost as warm as his blood, the surface of his body ought to be of such a character as to be readily cooled; else the rays, very slightly cooling in their action, which reach him from the surrounding objects, would not suffice to free him from the heat generated continually in his lungs, and he would soon find himself oppressed and overcome by the accumulation of this heat.

In a cold country, where the cooling of the surface of a body by the cold objects which surround it is more than sufficient to counterbalance the heat continually produced by respiration, the body can be protected from this excessive cooling action by clothing; but we know of no sort of clothing fitted to promote sufficiently the cooling of the human body in a very hot climate.

What has Nature done to supply this want? She has given to the inhabitants of hot countries a black skin; this colour gives to the negro such facility for becoming cool that he feels perfectly comfortable in a situation where a white man would be overcome by the heat. But, in return, the negro shivers with cold in a climate which the white man finds perfectly agreeable.

Every one knows that a black surface reflects fewer rays of light than a white surface; and the results of all the experiments performed by myself and by others seem to show that those surfaces which are of such a character as to reflect light also reflect the calorific or frigorific rays which all bodies send continually from their surfaces; and if the temperature of a body is changed in consequence of the action of surrounding bodies through these radiations, it is seen clearly why the negro suffers less from the heat of the tropics, and more from the cold of the polar regions, than the man with a white skin.

But when the negro is exposed to the action of calorific rays — to those of the sun, for instance — must he not be heated more than a white man? It would be so, without doubt, if Nature had not foreseen the danger and provided means for warding off the evil.

When the negro is exposed to the rays of the sun, an oily matter appears immediately at the surface of his skin, and causes it to shine; the calorific rays which fall upon it are reflected to a great extent, and he finds himself but little heated.

The sun sets, or the negro enters his hut; the oil which covers the surface of his body retires under his skin, and he retains all the advantages which his colour affords in aiding him to become cool.

If a coating of oil on the skin serves to protect the body from the too violent action of calorific rays, it ought to serve also, without doubt, to protect it from the too violent action of frigorific rays in very cold countries, especially in winter, when the sun never rises. And, indeed, do not the Laplanders besmear themselves with oil?

But in the case of a question of so great interest, I wished to omit nothing which might throw light upon it.

The following experiment seemed to me to establish beyond doubt the principal facts.

Having covered two of my cylindrical vessels with an animal substance, namely, with gold-beater's skin, I painted one of them black with Indian ink, leaving the other of its natural white color. Having filled both of the vessels with hot water, I left them, at the same time, to cool in the air of a large quiet room.

The vessel covered with a black skin represented a negro; the one covered with a white skin represented a white man.

The negro cooled considerably more rapidly than the white man, requiring $23\frac{1}{2}$ minutes to cool through the usual interval of 10 degrees, while the white man required 28 minutes to cool through the same interval.

This interesting experiment was made at Munich, the 26th of March, 1803. The results of these experiments need no illustration; and I leave to physiologists and physicians to determine what advantages may be derived from them in taking measures for the preservation of the health of white men who are called upon to dwell in hot countries.

# AN INQUIRY

CONCERNING THE

## NATURE OF HEAT, AND THE MODE OF ITS COMMUNICATION.

HEAT is employed in such a vast variety of different processes, in the affairs of life, that every new discovery relative to it must necessarily be of real importance to mankind; for, by obtaining a more intimate knowledge of its nature and mode of action, we shall no doubt be enabled not only to excite it with greater economy, but also to confine it with greater facility, and direct its operations with more precision and effect.

Having many years ago found reason to conclude that a careful observation of the phenomena which attend the heating and cooling of bodies, or the communication of heat from one body to another, would afford the best chance of acquiring a farther insight into the nature of heat, my view, in all my researches on this subject, has been principally directed to that point; and the experiments of which I am now to give an account may be considered as a continuation of those I have already, at different times, had the honour of laying before the Royal Society, and of presenting to the public in my Essays.

In order that the attention of the Society may not be

interrupted unnecessarily by description of instruments in the midst of the accounts of interesting experiments, I shall begin by describing the apparatus which was provided for these researches; and, as a perfect knowledge of the instruments made use of is indispensably necessary in order to form distinct ideas of the experiments, I shall take the liberty to be very particular in these descriptions.

The thermometers, four in number, which were used in these experiments, were constructed under my own eye, and with the greatest possible care; and, after every trial I have been able to make with them, in order to ascertain their accuracy, they appear to be very perfect.

They are mercurial thermometers, graduated according to Fahrenheit; their bulbs are cylindrical, 4 inches long, and $\frac{4}{10}$ of an inch in diameter; and their tubes are from 15 to 16 inches long. The mercury with which they are filled is quite pure, and they are freed from air. Their scales were divided with the greatest care; and, by means of a nonius, they show eighth parts of a degree very distinctly; they are graduated from about 10 degrees below the freezing point to 5 or 6 degrees above the point of boiling water. Their bulbs are quite naked; their scales ending about 1 inch above the junction of the bulb with its tube. The freezing point is situated about 5 inches above the upper end of the bulb. The reason for placing it so high will be evident from the details of the experiments in which these instruments were used.

The instrument I contrived for ascertaining the warmth of clothing is extremely simple; it is merely a hollow cylindrical vessel made of thin sheet brass. It is closed at both ends, and has a narrow cylindrical neck, by which it is occasionally filled with hot water.

This vessel, being covered with a garment made to fit it, composed of any kind of cloth or stuff, or other warm covering, is supported in a vertical position on a wooden stand, which is placed on a table in a large quiet room; and one of the thermometers above described being placed in the axis of the vessel, the time employed in cooling the water, through the clothing with which the instrument is covered, is observed and noted down.

Now, as the time of cooling through any given interval of the scale of the thermometer (or from any given degree above the temperature of the air of the room to any other given lower degree, but still above the temperature of the air of the room) will be longer or shorter as the covering of the instrument is more or less adapted for confining heat, it is evident that the relative warmth of clothing of different kinds may be very accurately determined by experiments of this sort.

I provided four instruments of this kind, all very nearly of the same dimensions. Their cylindrical bodies are each 4 inches in diameter and 4 inches long; and their cylindrical necks are about $\frac{8}{10}$ of an inch in diameter, and 4 inches in length. This neck is placed in the centre of the circular flat top, or upper end, of the vertical cylindrical body; and opposite to it, in the centre of the flat bottom of the body, there is a hollow cylinder, $\frac{8}{10}$ of an inch in diameter and 3 inches long, projecting downwards, into which a vertical cylinder of wood is fitted, on the top of which the instrument is supported, in such a manner that the air has free access to every part of it. This cylinder of wood constitutes a part of the wooden stand above mentioned.

As the thermometer is placed in the axis of the cylindrical vessel, and as its bulb is just as long as the body of this vessel, it is evident that it must ever indicate the *mean temperature* of the water in the vessel, however different the temperature of that water may be at different depths.

The thermometer is firmly supported in its place by causing a part of the lower end of its scale to enter the neck of the cylindrical vessel, and to fit it with some degree of accuracy, but not so nicely as to be in danger of sticking fast in it.

The lower end of the bulb of the thermometer does not absolutely touch the bottom of the vessel, but it is very near touching it.

Figure 1 (Plate I.) will give a clear idea of this instrument placed on its wooden stand, which is so contrived that the instrument may be placed higher or lower at pleasure.

The foregoing description of this instrument is so particular that the figure will be easily understood without any further illustration. The cylindrical vessel is represented placed on the stand, with its thermometer in its place.

As, in some of the first experiments I made with this instrument, I found it difficult to apply the coverings which I used to the ends of the body of the instrument, I endeavoured, by covering up those ends with a permanent and very warm covering, to oblige most of the heat to pass off through the vertical sides of the instrument, to which it was easy to fit almost any kind of covering, and more especially coverings of various

Plate I.

Fig. 1.

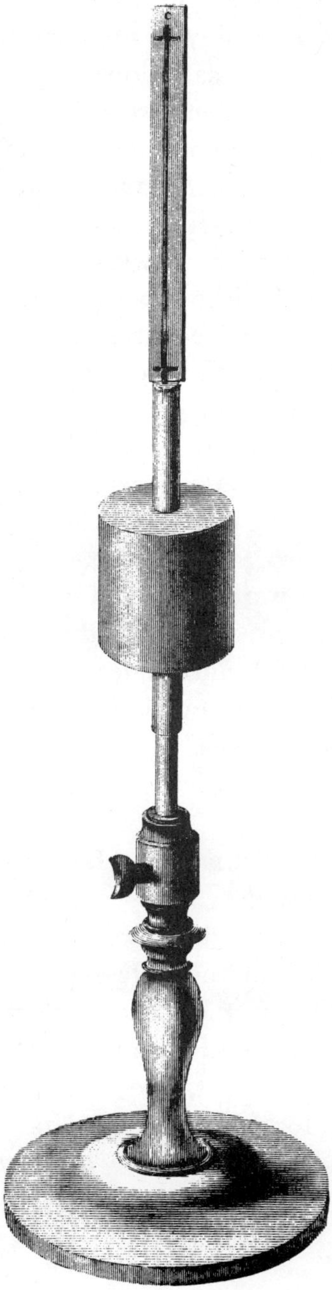

thicknesses of confined air, the relative warmth of which I was very desirous of ascertaining.

The means I employed for covering up the ends of the instrument were as follows. Having provided two thin cylindrical wooden boxes (like common pill-boxes, but much larger), something less in diameter than the body of the instrument, and $2\frac{1}{2}$ inches deep, I dried them as much as possible; and, after having varnished them within and without with spirit varnish, I covered them within and without with fine wove writing-paper, and then gave the paper three coats of the same varnish. I then perforated the bottoms of these boxes with round holes, just large enough to admit the neck of the instrument, and the cylindrical projection at its bottom; and then inverted them over the two ends of the instrument, filling the boxes at the same time with *eider-down.*

These boxes were fixed and confined in their places by means easy to be imagined; and, in order to confine the heat still more effectually, each of the boxes was covered on the outside with a cap of fur, as often as the instrument was used; as was also that part of the neck of the instrument which projected above the box.

Two of the instruments, which I shall distinguish by the numbers 1 and 2, were covered up at their ends in this manner; the other two instruments, No. 3 and No. 4, were left in the state represented by the Figure 1; that is to say, the ends of their cylindrical bodies were not covered with permanent coverings.

In each experiment, two similar instruments (No. 1 and No. 2, for instance, or No. 3 and No. 4) were used, the one *naked*, and the other *covered;* and, as the naked instrument always served as a standard, with which the results of the experiments made with the other were compared, it is evident that this arrangement rendered the general results of the experiments much more satisfactory and conclusive than they could possibly have been, had the experiments made on different days and with various kinds of covering been made singly, or unaccompanied by a fixed and invariable standard.

The experiments were made and registered in the following manner: The two instruments used in the experiment, placed on their wooden stands, being set down on the floor, were filled to within about 1½ inch of the tops of their cylindrical necks with boiling hot water; and, a thermometer being put into each of them, they were placed at the distance of three feet from each other, on a large table, in a corner of a large quiet room,* where they were suffered to cool undisturbed. Near them on the same table, and at the same height above the table, there was placed another thermometer (suspended in the air to the arm of a stand), by which the temperature of the air of the room was ascertained from time to time.

No person was permitted to pass through the room while an experiment was going on; and in order to prevent, as far as it was possible, all those currents of air in the room which were occasioned by partial heat, produced by the light which came in at the windows, the window-shutters were kept constantly shut; one of them only being opened for a moment, now and then, just to observe the thermometers, and note down the progress of the experiment.

* This room, which is adjoining to my laboratory, in my house at Munich, is 19 feet wide, 24 feet long, and 13 feet high.

The results of each experiment were entered on a separate sheet of paper; which paper was previously prepared for that use by being divided into separate vertical columns by lines drawn with a pen, and ruled in parallel horizontal lines with a lead-pencil.

"*Experiments on Heat, made at Munich,* 11*th March,* 1803. *The large cylindrical Vessels, No.* 1 *and No.* 2 (*made of thin sheet brass*), *were filled with hot Water, and exposed to cool in the Air of a large quiet Room. The Ends of both these Instruments were well covered with warm Clothing, Furs, &c. The vertical polished Sides of No.* 1 *were* naked. *The Sides of No.* 2 *were* covered *with one Thickness of fine white* Irish Linen, *which had been worn, strained over the metallic Surface.*"

| Time. | | Temperature | | Temperature of the Air. |
|---|---|---|---|---|
| h. | min. | of No. 1, *naked.* | of No. 2, *covered.* | |
| 10 | 10 | 126½° | 126° | 43¼° |
| .. | 30 | 109½ | 106½ | 43½ |
| .. | 45 | 105 | 100⅛ | 43¾ |
| 11 | .. | 101¼ | 94¾ | 44 |
| .. | 2½ | .. | 94 | .. |
| .. | 15 | 97½ | 90¼ | .. |
| .. | 30 | 94 | 86¼ | .. |
| .. | 39 | .. | 84 | .. |
| .. | 45 | 91¼ | 82½ | .. |
| 12 | .. | 88½ | 79⅜ | .. |
| .. | 15 | 85½ | 76 | .. |
| .. | 25 | 84 | .. | .. |
| .. | 30 | .. | 74½ | .. |
| .. | 45 | 80 | 70 | .. |
| 1 | .. | 78 | 68⅛ | .. |
| .. | 30 | 74¼ | 64¼ | .. |
| 2 | .. | 71⅛ | 61½ | 43¾ |
| .. | 30 | 68⅛ | 58¾ | 43½ |
| 3 | .. | 65¾ | 56¾ | .. |
| .. | 30 | 63½ | 54¾ | .. |

| Time. | | Temperature | | Temperature of the Air. |
|---|---|---|---|---|
| h. | min. | of No. 1, *naked.* | of No. 2, *covered.* | |
| 4 | .. | 61¾° | 53½° | 43½° |
| .. | 30 | 59½ | 52 | .. |
| 5 | 30 | 57 | 49¾ | 42½ |
| 6 | .. | 55½ | 49⅛ | .. |
| .. | 30 | 54¼ | 48¼ | .. |
| 7 | .. | 53½ | 47½ | 42 |
| 8 | .. | 51½ | 46½ | .. |
| 9 | .. | 50 | 45¾ | .. |
| 10 | .. | 49 | 45 | .. |
| 8 | 12th Mar. | 43 | 42 | 40 |
| | The instruments were now removed into a warm room. | | | |
| 8 | 2 | 43 | 42 | 62 |
| .. | 32 | 44¾ | 44¾ | 62½ |
| .. | 47 | 46 | 46½ | 63 |
| 9 | 24 | 48 | 49½ | .. |
| 10 | .. | 50 | 52 | .. |
| .. | 41 | 51½ | 53⅞ | .. |
| 12 | .. | 54 | 56½ | .. |
| 12 | 26 | 54½ | 57 | .. |
| An end was now put to the experiment. | | | | |

The above is an exact copy of one of these register-sheets, and contains the results of an actual and very interesting experiment, which lasted 26 hours.

Though it was easy to discover, by a single glance at the register, whether a covering which was put over one of the instruments prolonged the time of its cooling or not; yet, in order to compare the results of different experiments, and particularly of such as were made on different days, so as to determine with precision *how much* warmer one kind of covering was than another, it was necessary to fix on some particular interval in the scale of the thermometer, or number of degrees, commencing at some certain invariable number of degrees above the temperature of the air by which the instrument was surrounded, in order that the warmth of the covering, or its power of confining heat, might with certainty be estimated by the time employed in cooling through that interval.

By the results of a great number of experiments I found that the same instrument cooled through any given (small) number of degrees (10 degrees, for instance) in very nearly the same time, whatever was the temperature of the air of the room; provided always, that the point from which these 10 degrees commenced was at the same given number of degrees above the temperature of the air at the time being.

The interval I chose for comparing the results of my experiments is that which commences with the *fiftieth*, and ends with the *fortieth*, degree of Fahrenheit's thermometer *above the temperature of the air in which the instrument is exposed to cool.* When, for instance, the air was at 58°, the interval commenced at the 108th degree, and ended at the 98th. When the air was at $64\frac{1}{2}$°, it commenced at $114\frac{1}{2}$°, and ended at $104\frac{1}{2}$°.

That the same instrument, exposed to cool in the air, does in fact cool the same number of degrees in the same time, very nearly, when the given interval of the scale of the thermometer is reckoned from the same height, or given number of degrees above the temperature of the air at the time when the experiment is made, will appear from the following results of 11 different experiments, made on different days, and when the air in which the instrument was exposed to cool was at different degrees of temperature.

The large cylindrical vessel, No. 1, having its two ends well covered up with eider-down, furs, &c., its vertical sides being exposed *naked* to the air, in a large quiet room, was found to cool 10 degrees, *viz.* from the 50th to the 40th degree above the temperature of the air in which it was exposed, as follows: —

| Temperature of the air. | Degrees cooled. | Time employed in cooling. |
|---|---|---|
| ° | ° ° | |
| 44 | from 94 to 84 | 55 minutes. |
| $45\frac{1}{4}$ | " $95\frac{1}{4}$ to $85\frac{1}{4}$ | $55\frac{1}{2}$ " |
| 48 | " 98 to 88 | $55\frac{1}{4}$ " |
| $51\frac{1}{2}$ | " $101\frac{1}{2}$ to $91\frac{1}{2}$ | $55\frac{1}{2}$ " |
| 52 | " 102 to 92 | 55 " |
| 54 | " 104 to 94 | $54\frac{1}{4}$ " |
| 44 | " 94 to 84 | $55\frac{5}{6}$ " |
| $42\frac{1}{2}$ | " $92\frac{1}{2}$ to $82\frac{1}{2}$ | $55\frac{1}{3}$ " |
| 45 | " 95 to 85 | 56 " |
| 46 | " 96 to 86 | 55 " |
| 44 | " 94 to 84 | $55\frac{1}{3}$ " |

The fact which these experiments are here brought to prove has likewise been confirmed by other experiments, made with other instruments, and at times when the temperature of the air has been as high as 64°; but I will not take up the time of the Society by giving a particular account of them in this place.

As it sometimes happened, though very seldom, in the course of an experiment (which commonly lasted several hours) that I was called away, and was not present to observe the thermometer at the moment of the passage of the mercury through one or both of those points of its scale which formed the limits of the given interval chosen as the standard for a comparison of the results of the experiments with each other, it became a matter of considerable importance to find means for supplying these accidental defects, and ascertaining the points in question by interpolation.

In order to facilitate the means of doing this, I endeavoured to investigate the law of the cooling of hot bodies in a cold fluid medium; and I found reason to conclude,

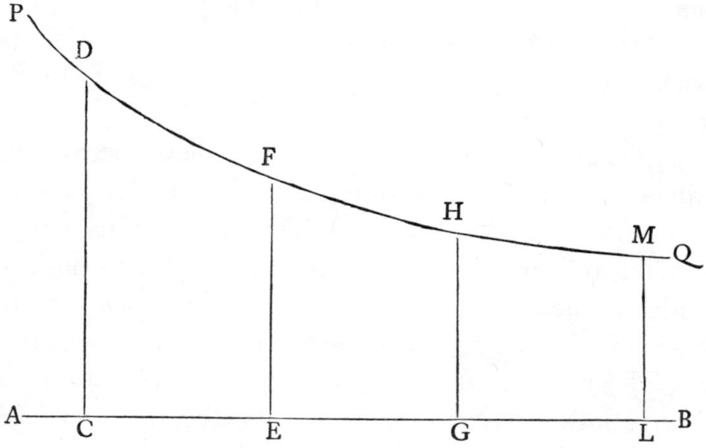


That if, on the right line AB, a perpendicular CD be taken, equal to the difference of the temperatures of the hot body and of the colder medium, expressed in degrees of the thermometer; and, after a certain given time, represented by CE, taken on the line AB at the

point E, another perpendicular EF be erected, and EF be taken equal to the difference of the temperatures after the time represented by CE has elapsed; and if the perpendiculars GH and LM be drawn, representing the difference of the temperatures after the times EG and GL have elapsed, a curved line PQ drawn through the points D, F, H, M, will be the logarithmic curve; or, if it vary from that curve, its variation, within the limits answering to a change of temperature amounting to a few degrees (especially if they be taken when the temperature of the hot body is about 40 or 50 degrees above that of the medium), will be so very small that no sensible error will result from a supposition that it is the logarithmic curve, in supplying, by computation, any intermediate observations which happen to have been neglected in making an experiment.

These computations are very easily made, with the assistance of a table of logarithms, in the following manner.

Supposing CD, CG, and GH, to have been determined by actual observation; and that it were required to ascertain, by computation, the absciss CE, corresponding to any given intermediate ordinate EF, or (which is the same thing) to determine at *what time* the cooling body was at any given intermediate temperature (= EF) between that (= CD) which it was found by observation to have at the point C, and that (= GH) which it was found to have after the time represented by the line GC had elapsed;

It is log. CD — log. GH is to CG as 1 to *m* (= modulus = the subtangent of the curve at the point D.)* And CE = *m* × log. CD — log. EF.

* The subtangent shows in what time the instrument would cool down to the tem-

If, for instance, in the experiment of the 11th March, (the details of which have just been given) the time when the instrument No. 2, in cooling, passed the important point of 94° had not been observed, this neglect might have been supplied, by computation, in the following manner.

It is CD = $94\frac{3}{4}°$, the nearest *observed* temperature higher than EF (= 94°), and GH = $90\frac{1}{4}$, the nearest observed temperature below that of 94°; and CG = 15 minutes, or 900 seconds = the time elapsed between the two observations.

$$\begin{aligned} \text{It is log. } 94\tfrac{3}{4} &= 1.9765792 \\ \text{And log. } 90\tfrac{1}{4} &= 1.9554472 \\ \text{Log. CD} - \text{log. GH} &= 0.0211320 \end{aligned}$$

And 0.0211320 is to 900 (= CG) as 1 to 42590 = *m*.

$$\begin{aligned} \text{And again, log. } 94\tfrac{3}{4} &= 1.9765792 \\ \text{Log. } 94 &= 1.9731279 \\ \text{Log. CD} - \text{log. EF} &= 0.0034513 \end{aligned}$$

42590 × 0.0034513 (= $m$ × log. CD — log. EF) = 147 seconds = 2 minutes and 27 seconds; which differs very little from $2\frac{1}{2}$ minutes, the observed time.

If, from the temperature observed at 11h. 30 min. = $86\frac{1}{4}°$, and the temperature observed at 11h. 45 min. = $82\frac{1}{2}°$, and the time which elapsed between these two

perature of the air in which it is placed, were its velocity of cooling at the point D to be continued *uniformly* from that point; and, as the subtangent of the logarithmic curve is *constant*, if PQ were the logarithmic curve, it would follow that the velocity with which a hot body cools in a fluid medium is everywhere such, that, were *that velocity* to be continued uniformly, the body would be cooled down to the temperature of the medium *in the same time*, whatever might be the excess of the temperature of the hot body above that of the medium, at the moment when its velocity of cooling became uniform.

observations (= 15 minutes), we were to determine by computation the time when the instrument was at the temperature of 84° (the lower point of the standard interval of 10 degrees answering to the temperature of the air, = 44°, in which the instrument was cooled), it will turn out, 8 minutes and 55 seconds after 11h. 30 min. The observed time was 11h. 39 min.; which differs from the computed time no more than 5 seconds.

If it were strictly true, as a very great philosopher and mathematician has advanced, that the velocity with which a hot body, exposed to cool in a cold fluid medium, parts with its heat is as the difference of the temperatures of the body and of the medium, it is most certain that the curve PQ could be no other than the logarithmic curve. Perhaps it may be so in fact, and that the variations from it which my experiments indicated were owing solely to the imperfection of the divisions of our thermometers. If it be so, it is not impossible to divide the scale of a thermometer in such a manner as to indicate with certainty *equal increments of heat*, as thermometers ought to do; but this is not the proper place to enlarge on this subject. I may perhaps return to it hereafter.

Passing over in silence a number of experiments I made in order to get thoroughly acquainted with my new instruments, and to assure myself that the results of similar experiments made with them were uniform and might be depended on, I shall now proceed to give an account of several experiments made with pointed views, the results of some of which were very interesting.

*Experiment No.* 1.—The large cylindrical vessel No. 1, with its ends covered with warm clothing, in the manner before described, and its vertical sides (which were polished, and very clean and bright) exposed naked to the air, was filled with water nearly boiling hot, and placed on its wooden stand, on a table, in a large quiet room to cool; the air of the room being at the temperature of 45° Fahrenheit.

Another cylindrical vessel, No. 2, in all respects like No. 1, and with its ends covered in the same manner, but with its vertical sides covered with a single covering of fine Irish linen (such as is sold in London for about 4 *s. per* yard), closely applied to the body of the instrument, was filled with hot water at the same time, and placed on the same table to cool.

This experiment lasted many hours; and, in that period, the temperature of the water in each of the instruments was carefully observed and noted down a great number of times.

The result of this experiment (the details of which have already been given) was very remarkable.

While the instrument No. 1, whose sides were *naked*, employed 55 minutes in cooling from the point of 94° to that of 84°, the instrument No. 2, whose sides were *covered with linen*, cooled through the same interval in 36½ minutes.

Hence it appears that clothing may, in some cases, expedite the passage of heat out of a hot body, instead of confining it in it.

Desirous of seeing whether the same covering would, or would not, expedite the passage of heat *into* the instrument, after having suffered both instruments to cool down to the temperature of about 42°, I removed them into a warm room, in which the air was at the temperature of 62°; and I found that the instrument

No. 2, which was clothed, acquired heat considerably faster than the other, No. 1, which was naked.*

The discovery of these extraordinary facts surprised me, and excited all my curiosity; and I immediately set about investigating their cause.

As it is well known that air adheres with considerable obstinacy to the surfaces of some solid bodies, I conceived it to be possible that the particles of air in immediate contact with the surface of the cylindrical vessel No. 1, might in fact be so attached to the metal as to adhere to it with some considerable force; and, if that were the case, as confined air is known to constitute a very warm covering, it appeared to me to be possible that the cooling of the vessel No. 1 might have been retarded by such an invisible covering of confined air; which covering, in the experiment with the vessel No. 2, had been displaced and in a great measure driven away by the colder covering of linen by which the body of the instrument was closely embraced.

I conceived that the linen must have accelerated the cooling of the instrument, either by facilitating the approach of a succession of fresh particles of cold air, or by increasing the effects of *radiation;* and, with a view to elucidate that important point, the following experiments were made.

*Experiment No.* 2. — Removing the linen with which the instrument No. 2 was clothed, I now covered the sides of that instrument with a thin transparent coating of glue; and, when it was quite dry and hard, I again filled the two instruments (No. 1 and No. 2)

* The details of this experiment (which was made on the 11th of March, 1803) may be seen on page 29.

with hot water, and observed the times of their cooling as before.

Result, or time of cooling 10 degrees, reckoned from the 50th to the 40th degree above the temperature of the air in which the instruments were exposed to cool: —

| | |
|---|---|
| Instrument No. 1, sides *naked*, . . . . . . | 55 min. |
| Instrument No. 2, sides *covered with one coating of glue*, | 43¼ " |

When we consider this experiment with attention, we shall find reason to conclude, that if it were by facilitating the approach and temporary contact of a succession of fresh particles of the cold air of the room to the surface of the glue (which was now in fact become the surface of the hot body), that the cooling of the instrument was accelerated, the metal being as completely covered, and the air, supposed to be attached and fixed to its surface, as completely excluded by one coating of the glue as it could be by two or more, two coatings could not possibly accelerate the cooling of the instrument more than one; but if, on the other hand, the cooling of the instrument in this experiment was accelerated, not by facilitating and accelerating the motions of the circumambient cold air, but by facilitating and increasing those *radiations* which are known to proceed from hot bodies, I conceived that two coatings of the glue might possibly accelerate the cooling of the vessel more than one. In order to put this conjecture to the test, I made the following decisive experiment.

*Experiment No. 3.* — I now gave the instrument No. 2 a second coating of glue; and, when it was thoroughly dry, I repeated the experiment last mentioned, with the above variation ; when I found the results to be as follows: —

| | Time of cooling the 10 degrees in question. |
|---|---|
| The instrument No. 1, *naked* metal, . . . . | $55\frac{1}{3}$ min. |
| The instrument No. 2, *covered* with two coatings of glue, | $37\frac{5}{6}$ " |

Finding that two transparent coatings of glue facilitated the cooling of this instrument even more than one coating, I washed off all the glue with warm water; then, making the instrument as clean and bright as possible, I covered its sides with a coating of very fine, transparent, and colourless spirit varnish; and, after this coating of varnish had become quite dry and hard, I repeated the experiment above mentioned ; and, finding that this covering, like that of glue, expedited the cooling of the instrument, I first added a *second* coating of the varnish, and repeated the experiment again, and then added two coatings more, making *four* in all. Finding that the cooling of the instrument was more and more rapid, as the thickness of the varnish was increased, I now added four coatings more, making *eight* coatings in the whole, giving time for each new coating to dry thoroughly before the next was applied; but I found, on repeating the experiment with this thick covering of varnish, that I had passed the limit of thickness which produced the greatest effect.

In order that the result of these experiments with coatings of different thicknesses of spirit varnish may be seen at one view, I shall here place them all together; and I shall place by the side of each the result of the standard experiment, which was made at the same time with the instrument No. 1, the sides which were *naked*.

| | Time employed in cooling through the given interval of 10 degrees. Instrument No. 1, *varnished.* | Instrument No. 2, *naked.* |
|---|---|---|
| *Experiment No.* 4. — 1 coating of varnish, | 42 min. | $55\frac{1}{2}$ min. |
| *Experiment No.* 5. — 2 coatings, . . . | $35\frac{3}{4}$ " | $55\frac{1}{4}$ " |
| *Experiment No.* 6. — 4 coatings, . . . | $30\frac{1}{4}$ " | $55\frac{1}{2}$ " |
| *Experiment No.* 7. — 8 coatings, . . . | $34\frac{1}{4}$ " | 55 " |

*Experiment No.* 8. — Desirous of finding out what effect *colour* would produce, I now painted the sides of the instrument No. 2 *black*, with lamp-black mixed up with size (this paint being laid upon the eighth coating of the varnish), and, repeating the experiment, its results were as follows: —

| | Time employed in cooling through the given interval. |
|---|---|
| The instrument No. 1, *naked*, . . . . . . | $55\frac{1}{4}$ min. |
| The instrument No. 2, covered with 8 coatings of varnish, and painted black, . . . . . | 34 " |

*Experiment No.* 9. — Finding that the painting of this thick coating of varnish *black* rendered the covering still colder, or accelerated the cooling of the instrument, I now washed off the black paint with warm water; then washing off all the varnish with hot spirit of wine, I painted the metallic sides of the instrument of a black colour with lamp-black and size; and when the paint was quite dry, I repeated the experiment so often mentioned, when the results were as follows: —

| | Time employed in cooling through the given interval. |
|---|---|
| The instrument No. 1, sides *naked*, . . . . . | $55\frac{1}{6}$ min. |
| The instrument No. 2, *painted black*, . . . . | 35 " |

*Experiment No.* 10. — In order to find out whether the *black* colour had any particular efficacy in expediting the cooling of the instrument, or whether another colouring substance would not produce the same effect, when mixed up with the same size, I now washed off the black paint and painted the sides of the instrument *white*, with whiting mixed up with size; and, on repeating the experiment, the results were as follows: —

| | Time of cooling through the given interval. |
|---|---|
| The instrument No. 1, *naked*, . . . . . . | $55\frac{1}{3}$ min. |
| The instrument No. 2, *painted white*, . . . . | 36 " |

As in both the two last experiments it was found necessary to paint the body of the instrument three or four times over, in order to cover the polished metal so completely as to prevent its shining through the paint; this, of course, occasioned the surface of the metal to be covered with a thick coating of size, which, no doubt, affected very sensibly the results of the experiment, and rendered it impossible to determine, in a satisfactory manner, what the effects really were, which were produced by the *different colours* used in the two experiments.

*Experiment No.* 11. — With a view to throw some more light on this interesting subject, having washed off the paint from the instrument No. 2, I now rendered its sides of a perfectly deep black colour, by holding it over the flame of a wax candle; and, repeating the usual experiment, the results were as follows: —

| | Time of cooling through the standard interval. |
|---|---|
| The instrument No. 1, *naked*, . . . . . . | $55\frac{7}{8}$ min. |
| The instrument No. 2, *blackened*, . . . . . | $36\frac{1}{8}$ " |

In order to ascertain the quantity of matter which composed this black covering, I weighed a small piece of clean and very fine linen; and, having wiped off with it all the black matter from the body of the instrument No. 2, in such a manner that the whole of it remained attached to the linen, I weighed it again, and by that means discovered that the whole of this black substance, which had so completely covered the sides of the instrument (a surface of polished brass = 50 superficial inches) that the metal did not shine through it in any part, weighed no more than $\frac{1}{18}$ of a grain Troy.

How this very thin covering, which, if the specific gravity of the black matter were only equal to that of water, would amount to no more than $\frac{1}{4509}$ of an inch in thickness, could expedite the cooling of the instrument, in the manner it was found to do, is what still remains to be shown; but, before I proceed any farther in these abstruse inquiries, I shall make a few observations relative to the results of the foregoing experiments.

Although we may with safety presume, that the velocities with which the heat escaped *through the sides of the instruments* * were nearly as the times inversely taken up in cooling through the given interval of 10 degrees; yet, as some heat must have made its way, in the course of the experiment, *through the ends of the instrument*, notwithstanding all the care that was taken to prevent it by covering them up with warm clothing, it is necessary, in order to be able to compare the results of the preceding experiments in a satisfactory manner, to

* I have found myself obliged in this, as in many other places, to make use of language which is far from being as correct as I could wish. I do not believe that heat ever *makes its escape* in the manner here indicated; but I could not venture to use uncommon expressions in pointing out the phenomena in question, however well adapted such expressions might be to describe the events which really take place. If it should be found that *caloric*, like *phlogiston*, is merely a creature of the imagination, and has no real existence (which has ever appeared to me to be extremely probable), in that case, it must be incorrect to speak of heat as *making its escape* out of one body, and *passing* into another; but how often are we obliged to use incorrect and figurative language, in speaking of natural phenomena!

find out how much of the heat made its escape through the covered ends of the instruments, during the time the instruments were cooling through the interval in question.

In order to determine that point, I now removed the covering from the ends of the instrument No. 1; and, when it was quite naked, I found, on making the experiment, that it cooled through the given interval in $45\frac{1}{2}$ minutes.

When its two ends and its cylindrical neck were covered up with warm clothing, I found, by taking the mean of the results of several experiments, that it required $55\frac{1}{2}$ minutes to cool through the same interval.

On measuring the instrument with care, I found its dimensions as follows: —

| | Inches. |
|---|---|
| Diameter of the body of the instrument, . . | = 4.03 |
| Length of the body, . . . . . . | = 3.96 |
| Diameter of the neck of the instrument, . . | = 0.8 |
| Length of the neck, . . . . . . | = 4. |

The superficies of the different parts of the instrument are therefore as follows: —

Superficies of the vertical sides of the body $(= 4.03 \times 3.14159 \times 3.96) = 50.136$ inches.

Superficies of the flat circular bottom of the instrument, $(= 4.03 \times 3.14159 \times \frac{4.03}{4}) = 12.755$ inches; deducting nothing for that part which is covered by the end of the tube, which serves as a support for the instrument.

Superficies of the flat circular top of the instrument (after deducting 0.502 of a superficial inch for the circular hole in its centre, made to receive the lower end of the cylindrical neck) = 12.253 inches.

Superficies of the cylindrical neck of the instrument (= 0.8 × 3.14159 × 4) = 10.051 inches.

Supposing, now, that the heat passes with equal velocity through the surface of all the different parts of the instrument, when the instrument is naked, we can determine the quantity of heat which escaped through the ends and neck of the instrument in the experiments in which those parts of the instrument were covered with warm clothing.

The whole of the metallic surface exposed to the air, in the experiments made with the instrument when it was quite naked, amounted to 85.195 superficial inches, namely: —

| | Inches. |
|---|---|
| Surface of its vertical sides, . . . . . | = 50.136 |
| Surface of its lower end, . . . . . | = 12.755 |
| Surface of its upper end, . . . . . . | = 12.253 |
| Surface of its neck, . . . . . . | = 10.051 |
| Total surface, . . . . . . . | = 85.195 |

When the instrument was exposed quite naked to the air, it was found to cool through the standard interval of 10 degrees in $45\frac{1}{2}$ minutes.

Assuming, now, any given number as the measure of the whole quantity of heat given off by the instrument during the period above mentioned, we can ascertain what part or proportion of that quantity passed off through the sides of the instrument; and what part of it must have made its escape through its ends, and through the sides of its neck.

As the quantities of heat given off are supposed to have been as the quantities of surface exposed to the air, if we suppose the whole quantity of heat lost by the instrument to be = 10,000 parts, the quantity which passed through the vertical sides of the instrument in $45\frac{1}{2}$ minutes, in the experiment, must have amounted to 5885 parts. For, the whole of the surface of the instrument, = 85.195 superficial inches, is to the whole of the heat given off, = 10,000, as the surface of the vertical sides of the instrument, = 50.136 superficial inches, to the quantity of heat which must have passed off through that surface in the given time, = 5885.

Now, as we may with safety conclude that the quantity of heat which passes off through a *given surface* must be as the times elapsed, all other circumstances being the same, we can determine how much of the heat given off by the instrument, in those experiments in which its ends were covered, passed through the sides of the instrument; and, consequently, how much of it must have made its way through its ends and neck, notwithstanding their being covered.

The instrument with its ends and neck covered up with eider-down, furs, &c., was found to cool through the standard interval of 10 degrees in $55\frac{1}{2}$ minutes. Now, as only 5885 parts of heat were found to pass through the naked vertical sides of the instrument in $45\frac{1}{2}$ minutes, no more than 7015 parts could have passed through the same surface in $55\frac{1}{2}$ minutes; consequently, the remainder of the heat lost by the instrument in the experiment in question, amounting to 2985 parts, must necessarily have made its way through the covered ends and neck of the instrument in the given period, $55\frac{1}{2}$ minutes.

Taking it for granted that these computations are well founded, we may now proceed to a more exact determination of the relative quantities of heat which made their way through the sides of the instrument

No. 2, when its sides were exposed naked to the air, and when they were covered with the different substances which appeared to facilitate the escape of the heat.

In the experiment No. 11, when the sides of the instrument were made quite black by holding it over the flame of a wax candle, the instrument cooled through the standard interval of 10 degrees in $36\frac{1}{8}$ minutes.

In that time a quantity of heat = 1942 parts must have passed off through the covered ends and neck of the instrument; for, if a quantity = 2985 parts could pass off that way in $55\frac{1}{2}$ minutes, the quantity above mentioned (= 1942 parts) must have escaped in $36\frac{1}{8}$ minutes.

This quantity, = 1942 parts, taken from the whole quantity, = 10,000 parts, lost by the instrument in cooling through the interval in question, leaves 8058 parts for the quantity which made its escape through the sides of the instrument in the experiment in question.

Now, if a quantity of heat = 7015 parts, requires $55\frac{1}{2}$ minutes to make its way through the naked sides of the instrument (as we have just seen), it would require $63\frac{3}{4}$ minutes for the quantity in question, = 8058 parts, to pass off through the same surface.

But, when that surface was blackened over the flame of a candle, that quantity of heat passed off through it in $36\frac{1}{8}$ minutes.

Hence it appears, that the velocity with which heat is given off from the naked surface of a heated metal exposed to cool in the air, is to the velocity with which it is given off by the same metal when its surface is blackened in the manner above described, as $36\frac{1}{8}$ to $63\frac{3}{4}$, or as 5654 to 10,000, very nearly; for the velocities are as the times of cooling, inversely.

Again, in the experiment No. 6, the sides of the instrument No. 2 being covered with four coatings of spirit varnish, the instrument was found to cool through the given interval of 10 degrees in $30\frac{1}{4}$ minutes.

In that time, a quantity of heat = 1627 parts, must have made its way through the covered ends of the instrument; and the remainder, = 8373 parts, must have made its way through its varnished sides.

This quantity, = 8373 parts, would have required $66\frac{1}{4}$ minutes to have made its way through the naked sides of the instrument; and, as it actually made its way through the varnished sides of the instrument in $30\frac{1}{4}$ minutes, it appears that the velocity with which the heat was given off from the naked metallic surface, was to the velocity with which it was given off from the same surface covered with four coatings of spirit varnish, as $66\frac{1}{4}$ to $30\frac{1}{4}$, or as 10,000 to 4566.

Without pursuing these computations any farther at present, and without stopping to make any remarks on the curious facts they present to us, I shall hasten to experiments, from the results of which we shall obtain more satisfactory information. But, before I proceed any farther, I must give an account of an instrument I contrived for measuring, or rather for *discovering*, those very small changes of temperature in bodies, which are occasioned by the radiations of other neighbouring bodies, which happen to be at a higher or at a lower temperature.

This instrument, which I shall take the liberty to call a *thermoscope*, is very simple in its construction. Like the hygrometer of Mr. Leslie (as he has chosen to call his instrument), it is composed of two glass balls, attached to the two ends of a bent glass tube; but the balls, instead of being near together, are placed at a considerable distance from each other; and the tube which connects them, instead of being bent in its middle, and its two extremities turned upwards, is quite straight in the middle, and its two extremities, to which its two balls are attached, are turned perpendicularly upwards, so as to form each a right angle with the middle part of the tube, which remains in a horizontal position.

At one of the elbows of this tube there is inserted a short tube of nearly the same diameter, by means of which a very small quantity of spirit of wine, tinged of a red colour, is introduced into the instrument; and, after this is done, the end of this short tube (which is only about an inch long) is sealed hermetically; and all communication is cut off between the air in the balls of the instrument and in its tube and the external air of the atmosphere.

A small *bubble* of the spirit of wine (if I may be allowed to use that expression) is now made to pass out of the short tube into the long connecting tube; and the operation is so managed that this bubble (which is about $\frac{3}{4}$ of an inch in length) remains stationary, at or near the middle of the horizontal part of the tube, *when the temperature* (*and consequently the elasticity*) *of the air in the two balls, at the two extremities of the tube, is precisely the same.*

By means of a scale of equal parts, attached to the horizontal part of the connecting tube, the position of the bubble can be ascertained, and its movements observed.

If now, the bubble being at rest in its proper place, one of the balls of the instrument be exposed to the calorific rays which proceed in all directions from a hot body, while the other ball is defended from those rays by a screen, the air in the ball so exposed to the action of these rays will be heated; and, its elasticity being increased by this additional heat, its pressure will no longer be counterbalanced by the elasticity of the colder air in the other ball, and the bubble will be forced to move out of its place and to take its station nearer to the colder ball.

By presenting two hot bodies at the same time to the two balls of the instrument, taking care that each ball shall be defended from the action of the hot body presented to the opposite ball, the distances of these hot bodies from their respective balls may be so regulated that their actions on those balls may be equal, however the temperatures of those hot bodies may differ, or however different may be the quantities or intensities of the calorific rays which they emit.

The instrument will show, with the greatest certainty, when the actions of these hot bodies on their respective balls are equal; for, until they become *unequal*, the bubble will remain immovable in its place.

And, when the actions of two hot bodies on the instrument are equal, the relative intensities of the rays they emit may be ascertained by the distances of the bodies from the balls of the instrument.

If their distances from their respective balls are equal, the intensities of the rays they emit must, of course, be equal.

If those distances are unequal, the intensities will probably be as the squares of the distances, inversely.

Plate II.

Fig. 2.

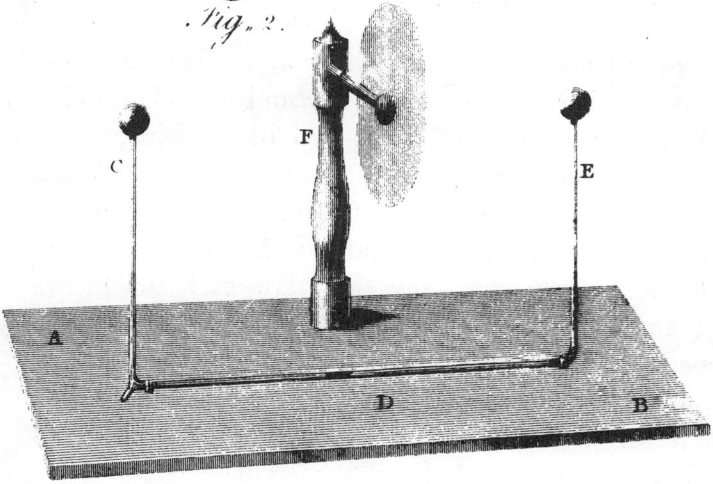


Fig. 3.

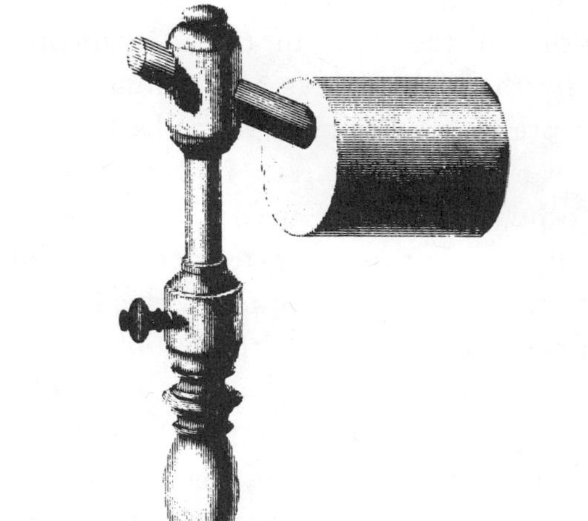

A distinct and satisfactory idea may be formed of the instrument I have been describing, from Fig. 2 (Plate II.).

AB is a board, 27 inches long, 9 inches wide, and 1 inch thick, which serves as a support for the bent tube CDE, at the two extremities of which the two balls are fixed. The two projecting ends of the tube, C and E, which are in a vertical position, are each 10 inches long; and the horizontal part D of the tube, which is fastened down on the board, is 17 inches in length.

The balls are each 1.625 inches in diameter. The diameter of the tube is such, that 1 inch of it in length would contain 15 grains Troy of mercury.

The pillar F, which, by means of a horizontal arm projecting from it, serves for supporting the circular vertical screen represented in the figure, is firmly fixed in the board AB.

This circular screen (which is made of pasteboard, covered on both sides with gilt paper) serves for preventing one of the balls of the instrument from being affected by the calorific rays proceeding from a hot body which is presented to the opposite ball.

Besides the circular screen represented in the figure, several other screens are used in making experiments; for the instrument is so extremely sensible, that the naked hand presented to one of the balls, at the distance of several inches, puts the bubble in motion; and it is affected very sensibly by the rays which proceed from the person who approaches it to make the experiments, unless care be taken, by the interposition of screens, to prevent those rays from falling on the balls. These screens can be best and most readily made by providing light wooden frames, about two feet square, and half an inch in thickness, and covering them on both sides, first with thick cartridge paper, and then with what is called gilt paper; the metallic substance (copper) with which one side of the paper is covered being on the outside.

To support a movable screen of this kind in a vertical position, it must of course be provided with a foot or stand. Those I use are fastened to one side of a pillar of wood by two screws, one of which passes through the centre of the screen where the cross-bars belonging to the frame of the screen meet, and the other through the middle of the piece of wood which forms the bottom of the screen. This pillar of wood, which is turned in a lathe, is 12½ inches high, and is firmly fixed, at its lower end, in a piece of wood 8 inches square and 1 inch thick, which serves as a stand or foot for supporting it.

As, in making experiments with this *thermoscope*, it is frequently necessary to remove the hot bodies that are presented to it farther from it or to bring them nearer to it, in order that this may be done easily and expeditiously by one person, and without its being necessary for him to remove his eye from the bubble (which he should constantly have in his view), I make use of a simple machine, which I have found to be very useful.

It is a long and shallow wooden box, open at both ends. It is 6 feet long, 12 inches wide, and 5 inches deep, measured on the outside; its vertical sides are made of 1½-inch deal; its bottom and top, of inch deal. A part only of the top or cover of this box is fixed down on the sides, and is immovable. The part of the cover which is fixed, and on which the thermoscope is placed, occupies the middle of the box, and is 13 inches in length. On the right and left of this fixed part, the top of the box is covered by a sliding board, 2 feet 3 inches long, which passes in deep grooves, made to receive it, in the sides of the box. A rack is fixed to the under side of each of these sliding boards; and there is a small cog wheel in the box, the axis of which passes through the sides of the box, and is furnished with a winch in the front of the box. By turning round these wheels by means of their winches (both of which can be managed by the same person, at the same time), the sliders may be moved backwards and forwards at pleasure.

In order to ascertain with facility and dispatch the distances of the hot bodies from their respective balls, the top of the front side of the wooden box is divided into inches on each side of the fixed part of the cover of the box; and there is a *nonius* belonging to each of the sliders, which is placed in such a manner as to indicate, at all times, the exact distance of the hot body from its corresponding ball.

The level of the upper surface of that part of the cover which is fixed is about $\frac{1}{8}$ of an inch higher than the level of the upper surface of the sliders, in order that, when a thermoscope longer than this fixed part is placed on it, the sliders may pass freely under its two projecting ends without deranging it.

It is evident, from this description, that by placing the thermoscope on the fixed part of the cover of the box, with its two balls in a line parallel to the axis of the box, and by placing the two hot bodies presented to the two balls of the instrument (elevated to a proper height) on stands set down on the sliders, an observer, by taking the two winches in his hands, keeping his eye fixed on the bubble, may, with the greatest facility, so regulate the distances of the hot bodies from their respective balls that the bubble shall remain immovable in its place.

In order to be able to ascertain precisely the temperatures of the hot bodies presented to this instrument, and in order that their surfaces might be equal, two equal cylindrical vessels, of thin sheet brass, with oblique cylindrical necks, were provided, of the form represented in Figure 3 (Plate II.).

This cylindrical vessel, which is placed in a horizontal position in order that its flat bottom may be presented *in a vertical position* to one of the balls of the thermoscope, is so fixed to a wooden stand, of a peculiar construction, that it may be raised or lowered at pleasure. This is necessary, in order that its axis may be in the continuation of a line passing through the centres of the two balls of the thermoscope.

This cylindrical vessel is 3 inches in diameter and 4 inches in length, and its oblique cylindrical neck is 0.86 of an inch in diameter and 3.8 inches in length.

The neck of this vessel is inserted *obliquely* into its cylindrical body, in order that the water with which it is occasionally filled may not run out of it, when the body of the vessel is laid down in a horizontal position, in the manner represented in the above-mentioned figure.

A thermometer, with a cylindrical bulb 4 inches in length, being inserted into the body of this vessel, through its neck, shows the temperature of the contained water.

Care is necessary, in constructing a thermoscope, to choose a tube of a proper diameter; if its bore be too small, it will be found very difficult to keep the spirit of wine in one mass; and if it be too large, the little horizontal column it forms (which I have called a bubble) will be ill defined at its two ends, which will render it difficult to ascertain its precise situation. After a number of trials I have found that a tube, the bore of which is of such a size that 1 inch of it in length contains about 15 or 18 grains Troy of mercury, answers best. For a tube of that size the balls may be about $1\frac{1}{2}$ inch in diameter; and they should both be painted black with Indian ink, which renders the instrument more sensible.

I have an instrument of this kind, the tube of which is quite filled with spirit of wine, excepting only the space occupied by a small bubble of air, which is introduced into the middle of the horizontal part of the tube; but it does not answer so well as those which contain only a very small quantity of that liquid, sufficient to form a small bubble.

But, without enlarging any farther, at present, on the construction of these instruments, I now proceed to give an account of the experiments for which they were contrived.

Having found abundant reason to conclude, from the results of the experiments of which an account has already been given, that all the heat which a hot body loses when it is exposed in the air to cool is not given off to the air which comes into contact with it, but that a large proportion of it escapes in rays, which do not heat the transparent air through which they pass, but, like light, generate heat only when and where they are stopped and absorbed, — I suspected that in every case when, in the foregoing experiments, the cooling of my instruments was expedited by coverings applied to their metallic surfaces, those coverings must, by some means or other, have facilitated and accelerated the emission of calorific rays from the hot surface.

Those suspicions implied, it is true, the supposition that different substances, heated to the same temperature, emit unequal quantities of calorific rays ; but I saw no reason why this might not be the case in fact; and I hastened to make the following experiments, which put the matter beyond all doubt.

*Experiment No.* 12.—Two equal cylindrical vessels, made of sheet brass, and polished very bright, each 3 inches in diameter and 4 inches long, suspended by their oblique necks in a horizontal position (being placed on their wooden stands), were filled with water at the temperature of 180°; and their circular flat bottoms were presented in a vertical position to the two balls of the thermoscope, at the distance of 2 inches.

When the two hot bodies were presented, at the same moment, to the two balls of the instrument, or, what was still better, when two screens were placed before the two balls, at the distance of about an inch, and, after the hot bodies were placed, these screens were both removed at the same instant, the small column of spirit of wine, which I have called a *bubble*, remained immovable in its place, in the middle of the horizontal part of the tube of the instrument.

If one of the hot bodies was now brought nearer the ball to which it was presented (the other hot body remaining in its place), the bubble immediately began to move from the hot body which was advanced forward, towards the opposite ball to which the other hot body was presented.

If, instead of advancing one of the hot bodies nearer

the ball to which it was presented, it was drawn backward to a greater distance from it, the action of its calorific rays on the ball was diminished by this increase of distance; and, being overcome by the action of the rays from the hot body presented to the opposite ball (at a smaller distance), the bubble was forced out of its place, and obliged to move towards the ball which had been drawn backward.

When one of the hot bodies only was presented to one of the balls, the bubble was immediately put in motion, and by bringing the hot body nearer to the ball, it might be driven quite out of the tube into the opposite ball; this, however, should never be done, because it totally deranges the instrument, as it is easy to perceive it must do.

Having, by these trials, ascertained the sensibility and the accuracy of my instrument, I now proceeded to make the following decisive experiment.

*Experiment No.* 13. — Having blackened the flat circular bottom of one of the cylindrical vessels by holding it over the flame of a wax candle, I now filled both vessels again with water at the temperature of 180° F., and presented them, as before, to the two opposite balls of the instrument at equal distances.

The bubble was instantly driven out of its place by the superior action of the blackened surface, and did not return to its former station till after the vessel which was blackened had been removed to more than 8 inches from the ball to which it was presented; the other vessel, which had not been blackened, remaining in its former situation, at the distance of 2 inches from its ball.

The result of this experiment appeared to me to

throw a new light on the subject which had so long engaged my attention, and to present a wide and very interesting field for farther investigation.

I could now account, in a manner somewhat more satisfactory, for those appearances in the foregoing experiments which were so difficult to explain, — for the acceleration of the passage of the heat out of my instruments, which resulted from covering them with linen, varnish, &c.; and I immediately set about making a variety of new experiments, from which I conceived I should acquire a farther insight into those invisible mechanical operations which take place when bodies are heated and cooled.

Finding so great a difference in the quantities of calorific rays which are thrown off by the polished surface of a metal when exposed *naked* to the cold air and when *blackened*, I now proceeded to make experiments to ascertain whether or not all those substances with which the sides of my cylindrical vessels had been covered, and which had been found to expedite the cooling of those instruments, would also facilitate the emission of calorific rays from the surfaces of the instruments I presented to the balls of my thermoscope; and I found this to be the case in fact.

As the results of all these experiments proved, in the most decisive manner, that all the substances which, when applied to the metallic surfaces of my large cylindrical vessels, had expedited their cooling, facilitated and expedited the emission of calorific rays, I could no longer entertain any doubts respecting the agency of *radiation* in the heating and cooling of bodies. Many important points, however, still remained to be investigated before distinct and satisfactory ideas could be formed respecting the nature of those rays and the mode of their action.

I had hitherto made use of but one metal (brass) in my experiments; and that was not a simple, but a compound metal. The first subject of inquiry which presented itself, in the prosecution of these researches, was to find out whether or not similar experiments made with other metals would give similar results.

*Experiment No.* 14. — Procuring from a gold-beater a quantity of leaf gold and leaf silver about three times as thick as that which is commonly used by gilders, I covered the surfaces of the two large cylindrical vessels, No. 1 and No. 2, with a single coating of oil varnish; and, when it was sufficiently dry for my purpose, I gilt the instrument No. 1 with the gold leaf, and covered the other, No. 2, with silver leaf. When the varnish was perfectly dry and hard, I wiped the instruments with cotton, to remove the superfluous particles of the gold and silver, and then repeated the experiment, so often mentioned, of filling the instruments with boiling-hot water, and exposing them to cool in the air of a large quiet room.

The time of cooling through the given interval of 10 degrees was just the same as it was before, when the natural surface of these brass vessels was exposed *naked* to the air. I repeated the experiment several times, but could not find that the difference in the metals made any difference in the times of cooling.

*Experiment No.* 15. — Not satisfied to rest the determination of so important a point on a trial with three metals only, — brass, gold, and silver, — I now provided myself with two new instruments, — the one made of lead, and the other covered with tinned sheet-iron, improperly, in England, called tin.

As the *conducting power* of lead, with respect to heat, is much greater than that of any other metal, I conceived that, if the *radiation* of a body were any way connected with its *conducting power*, the cooling of the water contained in the leaden vessel would necessarily be either more or less rapid than in a vessel constructed of any other metal.

The result of this experiment, as also the results of several others similar to it, showed that heat is given off with the same facility, or with the same celerity, from the surfaces of all the metals.

Is not this owing to their being all equally wanting in *transparency?* And does not this afford us a strong presumption that heat is in all cases excited and communicated by means of radiations, or *undulations*, as I should rather choose to call them?

I am sensible, however, that there is another and most important question to be decided before these points can be determined; and that is, whether bodies are cooled in consequence of the rays they emit or by those they receive.

The celebrated experiment of Professor Pictet, which has often been repeated, appears to me to have put the fact beyond all doubt, that rays, or emanations, which, like light, may be concentrated by concave mirrors, proceed from cold bodies; and that these rays, when so concentrated, are capable of affecting, in a manner perfectly sensible, a delicate air thermometer.

One of the objects I had principally in view, in contriving the before-described instrument, which I have called a thermoscope, was to investigate the nature and properties of those emanations, and to find out, if possible, whether they are not of the same nature as those calorific rays which have long been known to proceed from hot bodies.

My first attempts, in these investigations, were to ascertain the existence of those emanations universally, and to discover what visible effects they might be made to produce independently of concentration by means of concave mirrors.

*Experiment No.* 16. — My two horizontal cylindrical vessels of sheet brass (of the same form and dimensions), having been made very clean and bright, were fixed to their stands ; and, being elevated to a proper height to be presented to the balls of the thermoscope, were set down near that instrument (which was placed on a table in a large quiet room), where they were suffered to remain several hours, in order that the whole of this apparatus might acquire precisely the same temperature.

Daylight was excluded by closing the window-shutters ; and, in order that the thermoscope might not be deranged by the calorific rays proceeding from the person of the observer on his entering the room to complete the intended experiments, screens were previously placed before the instrument in such a manner that its balls were completely defended from those rays.

Things having been thus prepared, I entered the room as gently as possible, in order not to put the air of the room in motion, and, approaching the thermoscope, presented first one and then the other cylindrical vessel to one of the balls of the instrument ; but it was not in the least degree affected by them, the bubble of spirit of wine remaining immovably in the same place.

*Experiment No.* 17. — Having assured myself, by these

previous trials, that the instrument was not sensibly affected by a bright metallic surface being presented to it, provided the temperature of the metal and that of the instrument were the same, I now withdrew one of the cylindrical vessels, and, taking it into another room, I filled it with pounded ice and water.

Entering the room again, I now presented the flat vertical bottom of this horizontal cylindrical vessel, filled with ice and water, to one of the balls of the thermoscope at the distance of four inches.

The bubble of spirit of wine began instantly to move with a slow, regular motion towards the cold body; and, having advanced in the tube about an inch, it remained stationary.

On bringing the cold body nearer the ball to which it was presented, the bubble was again put in motion, and advanced still farther towards the cold body.

*Experiment No.* 18. — Although the result of the foregoing experiment appeared to me to afford the most indisputable proof of the *radiation* of cold bodies, and that the rays which proceed from them have a power of *generating cold* in warmer bodies which are exposed to their influence, yet in a matter so extremely curious, and of such high importance to the science of heat, I was not willing to rest my inquiries on the result of a single experiment.

In order to vary the substance, or species of matter, presented cold to the instrument, and at the same time to remove all suspicion respecting the possibility of the effects observed being produced by currents of cold air occasioned in the room by the presence of the cold body, I now repeated the experiment with the following variations.

The thermoscope was laid down on one side, so that the two ends of its tube, to which its balls were attached, instead of being vertical, were now in a horizontal position; and the cold body, instead of being presented to the ball of the instrument on one side of it, and on the same horizontal level with it, was now placed *directly under it*, and at the distance of 6 inches.

This cold body, instead of being a metallic substance, was a solid cake of ice, circular, flat, and about 3 inches thick, and 8 inches in diameter. It was placed in a shallow earthen dish, about 9 inches in diameter below, 12 inches in diameter above, at its brim, and 4 inches deep. The cake of ice being laid down on the bottom of the dish, the top of the dish was covered by a circular piece of thick paper, 14 inches in diameter, which had a circular hole in its centre, just 6 inches in diameter.

This earthen dish, containing the ice, and thus covered, was placed perpendicularly under one of the balls of the thermoscope, at such a distance that the centre of the upper surface of the flat cake of ice was 6 inches below the ball.

The result of this experiment was just what might have been expected: the ice was no sooner placed under the ball of the instrument than the bubble of spirit of wine began to move towards that side where the cold body was placed; and it did not remain stationary till after it had advanced more than an inch in the tube.

*Experiment No.* 19.—Desirous of discovering whether the surface of a liquid emits frigorific or calorific rays, as solid bodies have been found to do, I now removed the cake of ice from the earthen dish, and replaced it with an equal mass of ice-cold water.

The result of this experiment was, to all appearance, just the same as that of the last. The bubble moved towards the cold body, and took its station in the same place where it had remained stationary before. I found reason, however, to conclude, after meditating on the subject, that although the last experiment proves, in a most decisive manner, that radiations actually proceed from the surface of *water*, yet the proof of the radiation from the surface of ice, afforded by the preceding experiment, is not equally conclusive; for, as the temperature of the air of the room in which these experiments were made was many degrees above the freezing point, it is possible, and even probable, that the surface of the ice was actually covered with a very thin, and consequently invisible, coating of water during the whole of the time the experiment lasted.

Finding reason to conclude that frigorific rays are always emitted by cold bodies, and that these emanations are very analogous to the calorific rays which hot bodies emit, I was impatient to discover whether all cold bodies, at the same temperature, emit the same quantity of rays, or whether (as I had found to be the case with respect to the calorific rays emitted by hot bodies) some substances emit more of them and some less.

With a view to the ascertaining of this important point, I made the following experiments.

*Experiment No.* 20. — Having found that a metallic surface, rendered quite black by holding it over the flame of a wax candle, emits a much larger quantity of calorific rays when hot, than the same metal, at the same temperature, throws off when naked, I was very curious to find out whether blackening the surface of a cold metal would or would not increase, in like manner, the quantity of frigorific rays emitted by it.

Having blackened, in the manner already described, the flat bottom, or rather end, of one of my horizontal cylindrical brass vessels with an oblique neck, I filled it with a mixture of ice and common salt; and, filling another vessel of the same kind, the bottom of which was not blackened, with the same cold mixture, I presented them both, at the same instant, and at the same distance, to the two opposite balls of my thermoscope.

The result of this experiment was perfectly conclusive: the bubble of spirit of wine began immediately to move towards the ball to which the *blackened* cold body was presented; indicating thereby that that ball was more cooled by the frigorific rays which proceeded from the blackened surface than the opposite ball was cooled by the rays which proceeded from an equal surface of naked metal, at the same temperature.

As this experiment appeared to me to be of great importance, I repeated it several times, and always with the same results; the motion of the bubble, which constituted the index of the instrument, constantly showing that the frigorific rays from the blackened surface were more powerful in generating cold than those which proceeded from the naked metal.

The bubble, it is true, did not move so far out of its place as it had done in the experiments in which hot bodies were presented to the balls; but this was not to be expected, for though I had taken pains, by mixing salt with the ice, to produce as great a degree of cold as I conveniently could, yet still the difference between the temperature of the balls and that of the bodies presented to them was much greater when the hot bodies were used than when the experiments were made with the cold bodies; and it is evident, that the distance to which the bubble is driven out of its place must necessarily be greater or less in proportion as that difference is greater or less.

In those experiments in which the horizontal cylindrical vessels were filled with hot water, and then presented to the balls of the instrument, the temperature of the circular flat surfaces was that of 180°, while the temperature of the air of the room in which those experiments were made, and consequently that of the balls, was about 60°; the difference amounts to no less than 120 degrees of Fahrenheit's scale; but, in these experiments with cold, the difference of the temperatures at the moment when the cold bodies were first presented to the instrument did not probably amount to more than 40, or at the most 50 degrees; and in a very few seconds it must have been reduced to less than 30 degrees, in consequence of the freezing of the water precipitated by the air of the atmosphere on the surface of the vessel containing the cold mixture.

This precipitation of water by the surrounding air was so copious that the brilliancy of the polish of the metallic surface was almost instantly obscured by it, and the vessels were very soon covered with a thick coat of ice. These accidents, which were not to be prevented, affected in a very sensible manner the results of the experiment. The bubble, instead of remaining stationary for some time after it had reached the point of its greatest elongation, as it had done in the experiments with hot bodies, had no sooner reached that point than it began to return back towards the place from which it had set out; and, as often as I wiped off the ice from the surface of the flat end of the vessel which was not blackened, and presented it clean and bright to the ball of the instrument, the bubble began again to move towards the opposite side, — which, by the bye, shows that ice emits a greater quantity of frigorific rays than a bright metallic surface, at the same temperature.

Having frequently observed, on presenting my hand to one of the balls of the thermoscope, that the instrument was greatly affected by the calorific rays which proceeded from it, apparently much more so than it would have been by a much hotter body of the same quantity of surface, but of a different kind of substance, placed at the same distance, I was extremely curious to find out whether *animal substances* do not emit calorific (and consequently frigorific) rays much more copiously than other substances, and whether living animal bodies do not emit them in greater abundance than dead animal matter.

The first experiment I made, with a view to the investigation of this particular point, was as simple as its result was striking and conclusive.

*Experiment No.* 21. — Having procured a piece of gold-beater's skin (which, as is well known, is one of the membranes that line the larger intestines in cattle, and is exceedingly thin), I moistened it with water; and, applying it, while moist, to the flat circular end of one of my horizontal cylindrical vessels, it remained firmly attached to the surface of the metal when it became dry. I now filled this vessel, and another, of equal dimensions, the end of which was not covered, with hot water (at the temperature of 180°), and presented them both, at the same moment, to the two balls of the thermoscope, and at the same distance.

The bubble of spirit of wine was immediately driven out of its place to a great distance; and did not return to its former station till after the vessel whose end was covered with gold-beater's skin had been removed to a distance from the ball to which it was presented which was *five times* greater than the distance at which the other vessel was placed from the opposite ball.

I was induced to conclude, from the result of this interesting experiment, that an animal substance emits 25 *times* more calorific rays than a polished metallic surface of the same dimensions, both substances being at the same temperature.

*Experiment No.* 22 — Having emptied both the vessels used in the last experiment, and refilled them with pounded ice and water, I now presented them again to the thermoscope, at equal distances from their respective balls.

The result of this experiment confirmed the conclusion I had been induced to draw from a former experiment of the same kind (No. 13), the motion of the bubble towards the vessel whose surface was covered with gold-beater's skin showing that the rays which proceeded from that animal substance were considerably more efficacious in producing cold than those which proceeded from the naked metal.

The radiation of cold bodies appearing to me to have been proved beyond all doubt by the preceding experiments, I now set about to investigate a very important point which still remained to be determined: I endeavoured to find out whether the intensity of the action of the frigorific rays which proceed from cold bodies, or their power of affecting the temperatures of other warmer bodies, *at equal intervals of temperature*, is, or is not, equal to the intensity of the action of the calorific rays which proceed from hot bodies. To ascertain this point, I made the following very simple and decisive experiment.

*Experiment No.* 23.—Having placed the thermoscope on a table, in the middle of a large quiet room, at the temperature of 72° F., I presented to one of its balls, at the distance of 3 inches, the flat circular end of one of the horizontal cylindrical vessels (A) above described, with an oblique cylindrical neck, this vessel being filled with pounded ice and water; and, at the same moment, an assistant presented to the opposite side of the same ball of the thermoscope, at the same distance (3 inches), the flat end of the other similar and equal cylindrical vessel (B), filled with warm water at the temperature of 112° F., the opposite ball of the thermoscope being hid and defended, by means of screens, from the actions of the bodies presented to the other ball, as also from the calorific rays which proceeded from the bodies of the persons present.

From this description it appears, that while one of the balls of the thermoscope was so defended by screens that it could not be sensibly affected by the radiations of the neighbouring bodies, the other ball was exposed to the simultaneous action of two equal bodies, at equal distances (two vertical metallic disks, 3 inches in diameter, placed on opposite sides of the ball, at the distance of 3 inches); one of these bodies being at the temperature of 32° F., or 40 degrees below that of the ball, while the other was at 112° F., or 40 degrees above the temperature of the ball.

I knew, from the results of former experiments, that this ball would, at the same time, be heated by the calorific rays from the hot body and cooled by the frigorific rays from the cold body; and I concluded that if its mean temperature should remain unchanged under the influence of these two opposite actions, that event would be a decisive proof of the equality of the intensities of those actions.

The result of the experiment showed that the intensities of those opposite actions were in fact equal; the bubble of spirit of wine, which, by its motion, would have indicated the smallest change of temperature in the ball of the thermoscope to which the hot and the cold bodies were presented, remained at rest.

On removing the cold body a little farther from the ball, — to the distance of 3½ inches, for instance, — the hot body remaining in its former station, at the distance of 3 inches, the bubble began immediately to move towards the opposite ball of the thermoscope, indicating an increase of heat in the ball exposed to the actions of the hot and the cold bodies; but, when the hot body was removed to a greater distance, the cold body remaining in its place, the bubble indicated an increase of cold.

The celerity with which the ball of the thermoscope acquired heat or cold might be estimated by the velocity with which the bubble of spirit of wine advanced or retired in its tube; but, on the most careful and attentive observation, I could not perceive that it moved faster when the ball was acquiring heat than when it was acquiring cold, provided that the hot and the cold bodies from which the calorific and frigorific rays proceeded were at the same relative distances.

From these experiments, which I lately repeated at Geneva, in the presence of Professor Pictet, Mons. de Saussure, M. Senebier, and several other persons, we may venture to conclude, that, *at equal intervals of temperature,* the rays which generate cold are just as real, and just as intense, as those which generate heat; or, that their actions are equally powerful in changing the temperatures of neighbouring bodies.

On a superficial view of this subject, it might appear extraordinary that so important a fact as that of the frigorific radiations of cold bodies should have been so long unnoticed, while the calorific radiations of hot bodies have been so well known; but, if we consider the matter with attention, our surprise will cease. Those radiations by means of which the temperatures of neighbouring bodies are gradually changed and equalized are not sensible to our feeling unless the intervals of temperature be very considerable; and the constitution of things is such, that, while we are often exposed to the influence of bodies heated several thousand degrees (as measured by the thermometer) above the mean temperature of the surface of the skin, it is very seldom that we have opportunities of experiencing the effects of the radiations of bodies much colder than ourselves, and we have no means of producing degrees of cold which bear any proportion to the intense heats excited by means of fire.

From the result of the experiment of which an account has just been given, it is evident that we should be just as much affected by the calorific rays emitted by a cannon bullet at the temperature of 160 degrees of Fahrenheit's scale (= 64 degrees above that of the blood) as by the frigorific rays of an equal bullet, ice cold, placed at the same distance; and that a bullet at the temperature of freezing mercury could not affect us much more sensibly, by its frigorific rays, than an equal bullet at the temperature of boiling water would do by its calorific rays; — but at these comparatively small intervals of temperature, the radiations of bodies are hardly sensible, and could never have been perceived, much less compared and estimated, without the assistance of instruments much more delicate than our organs of feeling. Hence we see how it happened that the frigorific radiations of cold bodies remained so long unknown. They were suspected by Bacon; but their existence was first ascertained by an experiment made at Florence towards the end of the seventeenth century. And it is not a little curious, that the learned academicians who made that experiment, and who made it with a direct view to determine the fact in question, were so completely blinded by their prejudices respecting the nature of heat that they did not believe the report of their own eyes; but, regarding the reflection and concentration of cold (which they considered as a negative quality) as *impossible*, they concluded that the indication of such reflection and concentration which they observed must necessarily have arisen from some error committed in making the experiment.

Happily for the progress of science, the matter was again taken up, about twenty years ago, by Professor Pictet; and the interesting fact, which the Florentine academicians would not discover, was put beyond all doubt. But still, this ingenious and enlightened philosopher did not consider the appearances of a reflection of cold, which he observed in his experiments, as being *real;* nor was he led by them to admit the existence of frigorific emanations from cold bodies, analogous to those calorific emanations from hot bodies which he calls radiant heat. He everywhere speaks of the reflection of cold (by metallic mirrors) as being merely *apparent;* and it is on that supposition that the explanation he has given of the phenomena is founded.

On a supposition that the *caloric* of modern chemists has any real existence, and that heat, or an increase of temperature in any body, is caused by an *accumulation* of that substance in such body, the reflection of cold would indeed be impossible; and the supposition that such an event had taken place would be absurd, and could not be admitted, however striking and convincing the appearances might be which indicated that event. But, to return from this digression: —

Having found that the intensity of the calorific rays emitted by a hot body, at any given temperature, depends much on the surface of such body, — that a polished metallic surface, for instance, throws off much fewer rays than the same surface, at the same temperature, would emit if painted, or blackened in the smoke of a lamp or candle, — I was desirous of finding out whether the frigorific rays from cold bodies are affected in the same manner, by the same means, and in the same degree.

It was to ascertain that point that the experiment No. 20 was made; and although the result of that experiment afforded abundant reason to conclude that those substances which, when hot, throw off calorific rays in the greatest abundance, actually throw off great quantities of frigorific rays when they are cold, yet, as the relative quantities of these rays could not be exactly determined by that experiment, in order to ascertain so important a fact I had recourse to the following simple contrivance.

*Experiment No.* 24.—Having found, by the result of the last experiment (No. 23), that the calorific emanations of a circular disk of polished brass, 3 inches in diameter, at the temperature of 112° F., were just counterbalanced by the frigorific emanations of an equal disk of the same polished metal, at the temperature of 32° F., placed opposite to it, so that one of the balls of the thermoscope placed between these two disks, at equal distances, was just as much heated by the one as it was cooled by the other, I now blackened the two disks, by holding them over the flame of a wax candle, and repeated the experiment with them so blackened.

I knew, from the results of former experiments, that the intensity of the calorific radiations from the hot disk would be very much increased, in consequence of its surface being blackened; and I was certain that, if the intensity of the frigorific radiations of the cold disk should not be increased in *exactly the same degree*, the ball of the thermoscope, exposed to the simultaneous actions of these two disks, could not possibly remain at the same constant temperature, that of 72°.

The result of the experiment was very decisive; the bubble of spirit of wine remained at rest,—which proved that the intensities of the rays emitted by the two disks still continued to be equal at the surface of the ball of the thermoscope, which, at equal distances, was exposed to their simultaneous action.

Hence we may conclude, that those circumstances which are favourable to the copious emission of calorific rays from the surfaces of hot bodies are equally favourable to a copious emission of frigorific rays from similar bodies when they are cold.

But it is time to consider these emanations in a new point of view. What difference can there be between calorific rays and frigorific rays? Are not the same rays either calorific or frigorific according as the body at whose surface they arrive is hotter or colder than that from which they proceed?

Let us suppose three equal bodies, A, B, and C, (the globular bulbs of three mercurial thermometers, for instance,) to be placed at equal distances (3 inches) in the same horizontal line; and let A be at the temperature of freezing water, B at the temperature of 72° F., and C at that of 102° F. The rays emitted by B will be *calorific* in regard to the colder body A, but in respect to the hotter body C they will be frigorific; and, from the results of the two last experiments, we have abundant reason to conclude that they will be just as efficacious in heating the former as in cooling the latter.

Before I proceed to give an account of the experiments which were made with a view to determine the relative quantities of rays emitted from the surfaces of various substances, from living animals, dead animal matter, &c. (which I must reserve for a future communication), I shall lay before the Society the results of several experiments, of various kinds, which were made with a view to the farther investigation of the radiations of hot and of cold bodies, and of the effects produced by them.

*Experiment No.* 25. — Having found, from the results of the experiments No. 21 and No. 22, that great quantities of rays are thrown off from the surface of the animal substance used in those experiments (gold-beater's skin), I now covered the whole of the external surface of one of my large cylindrical passage thermometers (No. 4) with that substance; and, filling it with boiling-hot water, exposed it to cool gradually in the air of a large quiet room, in the manner often described in former parts of this paper; another similar *naked* standard instrument (No. 3) being filled with hot water at the same time, and exposed to cool in the same situation.

The temperature of the air of the room being $51\frac{1}{2}°$, the instruments were found to cool through the standard interval of 10 degrees, namely, from $101\frac{1}{2}$ to $91\frac{1}{2}$, in the following times: —

| | |
|---|---|
| No. 4, *covered* with gold-beater's skin, . | in $27\frac{3}{4}$ minutes. |
| No. 3, which was *naked*, . . . | in 45 " |

*Experiment No.* 26. — Being desirous of finding out whether or not the covering of animal matter, which had so remarkably facilitated the cooling of the instrument No. 4, would be equally efficacious in facilitating the passage of heat *into* the instrument, I suffered both instruments to remain in the cold room all night; and, entering the next morning, at half an hour past seven o'clock, I found the temperature of the water in the *naked* instrument, No. 3, to be $50\frac{1}{8}°$; that in the instrument No. 4, which was covered with gold-beater's skin, was $49\frac{1}{4}°$; while the air of the room was at 48°.

At 7 h. 30 m. A. M. I removed both instruments into a warm room, and observed the times of their acquiring heat to be as expressed in the following table.

| Times when the observations were made. | Observed Temperature. No. 3, *naked.* | Observed Temperature. No. 4, *covered.* | Temperature of the air of the room. |
|---|---|---|---|
| At 7 h. 30 m. . . | $50\frac{1}{8}°$ | $49\frac{1}{4}°$ | 64° |
| 7 45 . | $51\frac{1}{2}$ | $51\frac{1}{2}$ | $64\frac{1}{2}$ |
| 8 .. . . | $52\frac{1}{2}$ | $53\frac{1}{8}$ | 65 |
| 8 15 . | $53\frac{3}{4}$ | $54\frac{7}{8}$ | .. |

| Times when the observations were made. | Observed Temperature. No. 3, *naked.* | No. 4, *covered.* | Temperature of the air of the room. |
|---|---|---|---|
| At 8 h. 30m. | $54\frac{3}{8}$ | 56 | .. |
| 8 45 | $55\frac{1}{2}$ | $57\frac{1}{8}$ | .. |
| 9 .. | $56\frac{1}{4}$ | $58\frac{1}{2}$ | .. |
| 9 30 | $57\frac{1}{2}$ | 60 | .. |
| 10 .. | $58\frac{1}{4}$ | $61\frac{1}{4}$ | .. |
| 10 30 | $59\frac{1}{2}$ | $62\frac{1}{8}$ | .. |
| 11 .. | $60\frac{1}{2}$ | 63 | .. |
| 11 30 | 61 | $63\frac{1}{2}$ | $64\frac{1}{2}$ |

The results of this experiment, and of several others similar to it, showed, in a manner which appeared to me to be perfectly conclusive, that those substances which part with heat with the greatest facility, or celerity, are those which also acquire it most readily, or with the greatest celerity.

If we might suppose that the temperatures of bodies are changed, not by the rays they *emit*, but by those they *receive* from other neighbouring bodies, this fact might easily be explained; but, without stopping to form any hypothesis for the explanation of these appearances, I shall proceed in my account of the various attempts I have made to elucidate, by new experiments, those parts of this interesting subject which still appeared to be enveloped in obscurity.

As the cooling of hot bodies is so much accelerated by covering their surfaces with such substances as emit calorific rays in great abundance, or with such as are much affected by the frigorific rays of the colder bodies by which they are surrounded, it seems to be highly probable that a comparatively small part of the heat which a body so cooled actually loses is acquired by the air; a much greater proportion of it passing off through that *transparent* fluid, under the form of calorific rays, without affecting its temperature.

If this supposition should turn out to be well founded, the knowledge of the fact would enable us to explain several interesting phenomena, and particularly that most curious process by means of which living animals preserve an equal temperature, notwithstanding the vast quantities of heat that are continually generated in the lungs, and notwithstanding the great variations which take place in the temperature of the air in which they live.

It is evident, that the greater the power is which an animal possesses of *throwing off* heat from the surface of his body, independently of that which the surrounding air takes off, the less will his temperature be affected by the occasional changes of temperature which take place in the air, and the less will he be oppressed by the intense heats of hot climates.

It is well known that *negroes* and people of colour support the heats of tropical climates much better than white people. Is it not probable that their *colour* may enable them to throw off calorific rays with great facility, and in great abundance; and that it is to this circumstance they owe the advantage they possess over white people in supporting heat? And, even should it be true, that bodies are cooled, not in consequence of the rays they emit, but by the action of those frigorific rays they receive from other colder bodies (which I much suspect to be the case), yet, as it has been found by experiment that those bodies which emit calorific rays in the greatest abundance are also most affected by the frigorific rays of colder bodies, it is evident that in a very hot country, where the air and all other surrounding bodies are but very little colder than the surface of the skin, those who by their colour are prepared and disposed to be cooled with the greatest facility will be the least likely to be oppressed by the accumulation of the heat generated in them by respiration, or of that excited by the sun's rays.

With a view to throw some light on this interesting subject, I made the following experiments.

*Experiment No.* 27. — Having covered the flat ends of both my horizontal cylindrical vessels with gold-beater's skin, I painted one of these coverings (of this animal substance) black, with Indian ink; and then, filling both vessels with boiling-hot water, I presented them, at equal distances, to the two opposite balls of the thermoscope.

The bubble of spirit of wine was immediately driven out of its place by the superior efficacy of the calorific rays which proceeded from the blackened animal substance.

On repeating this experiment a great number of times, and when the water in the vessels was at different degrees of temperature (the temperature being the same in the two vessels in each experiment), the results uniformly indicated that calorific rays were thrown off from the *black* surface in greater abundance than from the equal surface which was not blackened.

Although the results of these experiments appeared to me to be so perfectly conclusive as to establish the fact in question beyond all possibility of doubt, yet, in so interesting an inquiry, I was desirous, by varying my experiments, to bring, if possible, a variety of proofs to support the important conclusions which result from it.

*Experiment No.* 28. — Having covered the two large cylindrical vessels, No. 3 and No. 4, with gold-beater's skin, I painted one of them black, with Indian ink; and, filling them both with boiling-hot water, I exposed them to cool, in the manner already often described, in the air of a quiet room.

No. 4, which was *blackened*, cooled through the standard interval of 10 degrees in $23\frac{1}{2}$ minutes; while the other, No. 3, which was not blackened, took up 28 minutes in cooling through the same interval.

In a former experiment (No. 25), the instrument No. 4, covered with gold-beater's skin, but not blackened, had taken up $27\frac{3}{4}$ minutes in cooling through the given interval, as we have before seen.

The results of these experiments do not stand in need of illustration; and I shall leave to physicans and physiologists to determine what advantages may be derived from a knowledge of the facts they establish, in taking measures for the preservation of the health of Europeans who quit their native climate to inhabit hot countries.

All I will venture to say on the subject is, that were I called to inhabit a very hot country, nothing should prevent me from making the experiment of blackening my skin, or at least of wearing a black shirt, in the shade, and especially at night; in order to find out if, by those means, I could not contrive to make myself more comfortable.

Several of the savage tribes which inhabit very cold countries besmear their skins with oil, which gives them a shining appearance. The rays of light are reflected copiously from the surface of their bodies. May not the frigorific rays, which arrive at the surface of their skin, be also reflected by the highly polished surface of the oil with which it is covered?

If that should be the case, instead of despising these poor creatures for their attachment to a useless and loathsome habit, we should be disposed to admire their ingenuity, or rather to admire and adore the goodness of their invisible Guardian and Instructor, who teaches them to like, and to practise, what he knows to be useful to them.

The Hottentots besmear themselves, and cover their bodies, in a manner still more disgusting. They think themselves *fine*, when they are besmeared and dressed out according to the loathsome custom of their country. But who knows whether they may not in fact be *more comfortable*, and better able to support the excessive heats to which they are exposed? From several experiments which I made, with a view to elucidate that point, (of which an account will be given to this Society at some future period,) I have been induced to conclude that the Hottentots derive advantages from that practice exactly similar to those which negroes derive from their black colour.

It cannot surely be supposed that I could ever think of recommending seriously to polished nations the filthy practices of these savages. That is very far indeed from being my intention, for I have ever considered cleanliness as being so indispensably necessary to comfort and happiness that we can have no real enjoyment without it; but still I think that a knowledge of the physical advantages which those savages derive from such practices may enable us to acquire the same advantages by employing more elegant means. A knowledge of the manner in which heat and cold are excited would enable us to take measures for these important purposes with perfect certainty; in the mean time, we may derive much useful information by a careful examination of the phenomena which occasionally fall under our observation.

If it be true that the black colour of a negro, by rendering him more sensible to the few frigorific rays which are to be found in a very hot country, enables him to support the great heats of tropical climates without inconvenience, it might be asked how it happens that he is able to support, naked, the direct rays of a burning sun.

Those who have seen negroes exposed naked to the sun's rays, in hot countries, must have observed that their skins, *in that situation*, are always very shining. An oil exudes from their skin, which gives it that shining appearance; and the polished surface of that oil reflects the sun's calorific rays.

If the heat be very intense, sweat makes its appearance at the surface of the skin. This watery fluid not only reflects very powerfully the calorific rays from the sun which fall on its polished surface, but also, by its evaporation, generates cold.

When the sun is gone down, the sweat disappears; the oil at the surface of the skin retires inwards; and the skin is left in a state very favourable to the admission of those feeble frigorific rays which arrive from the neighbouring objects.

But I shall refrain from pursuing these speculations any farther at present.

I shall now proceed to give an account of several experiments, of various kinds, which were made with a view to a farther investigation of the radiations of cold bodies.

Having found, by several of the foregoing experiments, that the radiations of cold bodies affected my thermoscope very sensibly, even when placed at a considerable distance from it, and in situations where currents of cold air could not be suspected to exist, I was desirous of finding out whether the cooling of a hot body would or would not be *sensibly* accelerated by those rays. To determine that point, I made the following experiment.

*Experiment No.* 29. — Having provided two conical vessels, made of thin sheet brass, each 4 inches in diameter at the base, and 4 inches high, ending above in a cylindrical neck, 0.88 of an inch in diameter, I enclosed each of them in a cylinder of thin pasteboard, covered with gilt paper, and then covered them up with rabbit-skins, which had the hair on them, in such a manner that no part of these vessels, except their flat bottoms, was exposed naked to the air. I then covered their bottoms with gold-beater's skin, painted black with Indian ink, in order to render them as sensible as possible to calorific and frigorific rays.

This being done, I suspended these two vessels in an erect position, or with their bottoms downwards, to the two opposite horizontal arms of a wooden stand, provided for the experiment; and I placed under each of them a pewter platter, blackened on the inside by holding it over a lighted wax candle.

Each of these platters was 12 inches in diameter, and they were supported on the top of two shallow earthen dishes, each of which was $11\frac{1}{2}$ inches in diameter at its brim; these earthen dishes being supported on circular wooden stands 10 inches in diameter.

A circular piece of thick drawing-paper, $12\frac{1}{2}$ inches in diameter, with a circular hole in its centre, just 6 inches in diameter, was placed on each of the platters, and served as a perforated cover to it.

The stands on which the platters were supported were of such a height that the upper surface of the flat bottom of each of the platters was elevated just 40 inches above the level of the floor of the room; and the horizontal arms of the wooden stand which supported the conical vessels were of such a height that the flat bottoms of these vessels (which were placed perpendicularly over the centres of the platters) were just 4 inches above the flat horizontal surface of the bottoms of the platters.

One of the platters was at the temperature of the air of the room (63° F.), but the other was kept constantly ice-cold, during the whole of the time the experiment lasted, by means of pounded ice and water, which was put into the earthen dish, over which, or rather in which, this platter was placed.

Each of the platters was just 1 inch deep, measured from the level of the top of its brim to the level of the upper surface of the flat part of its bottom; this flat part was about 8 inches in diameter.

The two conical vessels were now filled with boiling-hot water, and the times of their cooling were carefully observed.

From the above description of the apparatus used in this experiment, it is evident that the vessel which was suspended over the ice could not be reached by any streams of cold air that might be occasioned by that ice, or by the cooled sides of the vessel which contained it; for the air which, coming into contact with the sides of that vessel, was cooled by it, becoming specifically heavier than it was before, naturally descended, and spread itself out on the floor of the room; and the perforated circular sheet of paper, which was laid down horizontally on the platter, effectually prevented any of the air so cooled from being thrown upwards against the bottom of the conical vessel (placed immediately over the platter), by any occasional undulation of the air in the room.

To preserve the air of the room in a state of perfect quietness, not only the doors and windows, but even the window-shutters of the room were kept shut; so much light only being admitted occasionally as was necessary to observe the thermometers which were placed in the conical vessels.

In order to guard still more effectually the bottoms of the vessels which were cooling from the effects of occasional undulations in the air of the room, over each of these vessels there was drawn a cylindrical covering of very fine thin post paper, the lower open end of which projected just half an inch below the horizontal level of the flat bottom of the vessel. These cylindrical coverings of post paper were made to fit as exactly as possible the cylinders of pasteboard by which the sides of the conical vessels were covered and defended from the air; and the warm coverings of fur (rabbit-skins) were put over all.

To confine the heat still more effectually, a quantity of eider-down had been introduced between the outside of each conical vessel and its cylindrical neck, and the inside of the hollow cylinder of pasteboard in the axis of which it was fixed and confined.

The result of this experiment was very conclusive. The conical vessel which was suspended over the *ice-cold* pewter platter cooled through the standard interval of 10 degrees (namely, from the point of 50 degrees to that of 40 degrees above the temperature of the air of the room) in 33 minutes and 42 seconds; whereas the other vessel, which was not over ice, required 39 minutes and 15 seconds to cool through the same interval.

*Experiment No.* 30.— On repeating this experiment the next day, the air of the room still remaining at 63°, the times of cooling through the given interval were as follows: —

| | Min. | Sec. |
|---|---|---|
| The vessel suspended over the ice-cold platter, in . | 33 | 15 |
| The other vessel, in . . . . . . . | 39 | 30 |

From the results of these experiments (which were made with the greatest possible care) it appears that the radiations of cold bodies act on warmer bodies *at a distance*, and gradually diminish their temperatures.

It will likewise be evident, when we consider the matter with attention, that the cooling of the vessel which was suspended over the ice-cold platter was in fact considerably more accelerated by the frigorific radiations from that cold surface than it appears to have been when we estimate the effects produced simply by the difference of the times taken up in the cooling of the two vessels, without having regard to any other circumstance.

These times are, no doubt, inversely as the velocities of cooling; but, as all the heat lost by the vessels during the time of their cooling did not pass off through their flat bottoms, and as the rays from the cold surface fell on the *bottom only* of the vessel which was suspended over it, without at all affecting its covered sides, the velocity with which the heat made its way through the covered sides of the vessels was the same in both; consequently, more heat must have passed that way, and of course less through the bottom of the vessel, when the time of cooling was the longest, that is to say, in the vessel which was not p aced over ice.

As the cooling of these vessels is a complicated process, I will endeavour to elucidate the subject still farther.

As the two conical vessels were of the same form and dimensions, and contained equal quantities of hot water, the quantities of heat they parted with, in being cooled the same number of degrees, must of course have been equal.

Expressing that quantity by the algebraic symbol $a$, and putting $x =$ the quantity of heat which passed off through the covered sides of the vessel which was suspended over ice during the time it was cooling through the given interval of 10 degrees, and $y =$ the quantity which passed off through the covered sides of the other vessel during the time that vessel was coo ing through the same interval, the quantity of heat which passed off through the bottom of the vessel which was placed over ice during the time it was cooling through the given interval must have been $= a - x$, and that which passed off through the bottom of the other vessel during the time of its cooling through the same interval $= a - y$.

But, as the velocities of the heat through the covered sides of both vessels must have been equal, the quantities of heat which passed off *that way* must have been as the times of cooling.

The times of cooling in the last-mentioned experiment (No. 30) were as follows: —

| | Min. Sec. | Seconds. |
|---|---|---|
| Of the vessel suspended over ice, . . . | 33 15 = | 1995 |
| Of the other vessel, . . . . . | 39 30 = | 2370 |

$x$ is therefore to $y$, as 1995 to 2370; consequently, $x = \frac{1995\,y}{2370} = 0.84177\,y$;

And, substituting for $x$ its value $= 0.84177\,y$, the quantities of heat which passed off through the bottoms of the two vessels, in the experiment in question (No. 30), must have been $= a - 0.84177\,y$ for the vessel which was suspended over ice, and $= a - y$ fort he other vessel.

And, as $y$ is greater than $0.84177\,y$, consequently $a - 0.84177\,y$ is greater than $a - y$, or the quantity of heat which passed off *through the bottom* of the vessel which was cooled the most rapidly was greater than that which passed off *through the bottom* of the other vessel; and hence we perceive that the effect produced by the frigorific rays from the cold surface, in the experiments in question, was *greater* than it appeared to be at first sight, when it was estimated by the times of cooling.

To determine exactly *how much* the cooling was accelerated by the presence of the cold body, it is necessary to find out how much heat actually passed off through the bottoms of the two vessels, in the experiments in question. This we will endeavour to do by comparing the results of those experiments with the results of some other experiments of a similar nature.

In the experiment No. 28, a cylindrical vessel of thin sheet brass, 4 inches in diameter, and 4 inches in height, covered with gold-beater's skin painted black with Indian ink, being filled with hot water and exposed to cool in the air of a large quiet room, cooled from the point of 50 degrees to that of 40 degrees above the temperature of the air of the room in $23\frac{1}{2}$ minutes.

The quantity of surface by which this vessel was exposed to the cold air was = 74.5581 superficial inches, exclusive of its neck, which was well covered up with fur.

The quantity of surface which was exposed to the air, in the foregoing experiments with the conical vessels, or the area of the bottom of each of the vessels, was (4 × 3.14159) = 12.4263 superficial inches.

As the diameters and heights of the conical and cylindrical vessels were equal, the contents of the former must have been to the contents of the latter as 1 to 3; and the quantities of heat which they lost in cooling were as their contents.

If now the cylindrical vessel lost a quantity of heat = 3 in $23\frac{1}{2}$ minutes, it would have disposed of a quantity = 1 (equal to that which the conical vessel lost) in one third part of that time, or in 7 minutes and 50 seconds.

But the quantity of surface exposed to the air in the experiment with the cylindrical vessel was to that so exposed in the experiment with the conical vessel as 74.5581 to 12.4263, or as 6 to 1.

Now, as the time in which any given quantity of heat can pass out of any closed vessel into or through any cold fluid medium by which the vessel is surrounded must be inversely as the surface of the vessel, other things being equal, if a quantity of heat = 1 could pass out of the cylindrical vessel in 7 minutes and 50 seconds, it would require 6 times as long, or 47 minutes, to pass out of the conical vessel *through its flat bottom*, supposing no heat whatever to escape through the covered sides of that vessel.

If now the whole of the heat which the conical vessel

actually lost would have required 47 minutes to have passed through the bottom of that vessel, it is evident that the quantity which actually passed through that surface, in the experiment in question (No. 30), could not have been to the whole quantity actually lost in a greater proportion than that of the times, or as $39\frac{1}{2}$ to 47.

Assuming any given number — as 10,000, for instance — to represent the whole of the heat lost in the experiment, we can now determine what part or proportion of it passed off through the bottom of the conical vessel, and consequently how much of it must have made its way through its covered sides.

If the whole quantity, = 10,000, would have required 47 minutes to have passed through the bottom of the vessel, the quantity which actually passed through that surface in $39\frac{1}{2}$ minutes could not possibly have amounted to more than 8404, $= a - y$.

For it is 47 minutes to 10,000, as $39\frac{1}{2}$ minutes to 8404. The remainder of the heat, = 10,000 — 8404 = 1396 parts, $(= y)$ must have made its way through the covered sides of the vessel.

And, if a quantity of heat = 1396 required $39\frac{1}{2}$ minutes to make its way through the covered sides of one of the conical vessels, the quantity which made its way through the covered sides of the other in $33\frac{1}{4}$ minutes could not have amounted to more than 1175 parts; and the remainder of that which was actually disposed of in the experiment = 10,000 — 1175 = 8825 $(= a - x,)$ must have passed off through the bottom of the instrument.

Hence it appears, that the quantity of heat which actually passed off through the bottom of the conical vessel which was placed over ice, in $33\frac{1}{4}$ minutes, was to that which passed off in $39\frac{1}{2}$ minutes through the bottom of the other vessel as 8825 to 8404; and consequently, that the velocity with which the heat passed through the bottom of the vessel which was exposed to the frigorific rays from the surface of the cold platter was to the velocity with which it passed through the bottom of the other vessel in the compound ratio of 8825 to 8404, and of $39\frac{1}{2}$ to $33\frac{1}{4}$; or as 10,000 to 8025, which is as 5 to 4, very nearly.

From these experiments and computations it appears that the cooling of the hot body which was placed over the ice-cold platter was sensibly, and very considerably, accelerated by the vicinity of that cold body,—may we not venture to say, by the frigorific rays which proceeded from it?

I made several other experiments similar to those just described, and with similar results; but I shall not take up the time of the Society by giving a detailed account of them. I may, perhaps, at a future time find occasion to mention some of them more particularly.

In the two last-mentioned experiments, as the conical vessels were suspended in an erect position, and had a circular band or hoop of fine post paper, by which the lower end of each of them was surrounded, and which projected downwards half an inch below the horizontal level of the bottom of the vessel, and as the air which came into immediate contact with the bottom of the vessel, and received heat from it (though it became specifically lighter than it was before), could not make its escape *upwards* into the atmosphere, being confined and prevented from moving upwards by the thin projecting hoop of paper, there is no doubt but that the time of cooling was prolonged by this arrangement; for, there being much reason to believe that the propagation of heat downwards, in air, from one particle of that fluid to another, is either quite impossible or so extremely slow as to be imperceptible, as a succession of fresh particles of cold air was prevented from coming into contact with the bottoms of the vessels, but very little heat could have been given off *immediately* to the air in those experiments.

In order to be able to form some probable conjecture respecting the quantity so given off in cases where the succession of fresh particles of air is free and uninterrupted, I made the following experiment.

*Experiment No.* 31. — The two conical vessels used in the last experiment (which I shall now distinguish by calling the one No. 5 and the other No. 6) being left suspended in the air to the two horizontal arms of their wooden stand, at the height of 44 inches above the floor of the room (the pewter platters, the earthen dishes, and the stands on which they were placed being removed), both the vessels were again filled with boiling hot water, and exposed to cool in the air.

The vessel No. 5 remained in a vertical position, or with its flat bottom in a horizontal position, as before; but the vessel No. 6 was now reclined, so that its axis, and consequently the plane of its flat bottom, made an angle with the plane of the horizon of 45 degrees. In this position of the vessel No. 6, it is evident that the air, heated by coming into contact with its bottom, had full liberty to escape *upwards*, and to make way for other particles of colder air to come into contact with the hot surface and be heated, rarefied, and forced upwards in their turns; and under these circumstances it might reasonably be expected that as much heat as possible would be communicated *immediately* to the air by the hot body, and that the heat so communicated would of course accelerate the cooling of that vessel.

It was in fact cooled in a shorter time than the other, No. 5, which was suspended in a vertical position; but the difference of the times of cooling was very small; which indicates, if I am not mistaken, that a comparatively small quantity of the heat a hot body loses when it is cooled in air is communicated to that fluid, much the greater part of it being sent off through the air, to a distance, in calorific rays.

The vessel No. 5 was found to cool through the standard interval of 10 degrees in $38\frac{1}{2}$ minutes; and No. 6, which was in a reclined position, in $37\frac{1}{4}$ minutes.

It will no doubt be remarked that the vessel No. 5 cooled somewhat faster in this experiment than it had done in the two preceding experiments (No. 29 and No. 30), when it stood over a pewter platter which (at the beginning of the experiment at least) was at the same temperature as the air of the room.

The calorific rays from the bottom of the vessel heating the platter in some small degree, and still more, perhaps, the upper surface of the perforated sheet of paper which covered it, the frigorific rays from these bodies were, on that account, somewhat less powerful in lowering the temperature of the neighbouring hot body; and the time of its cooling was consequently a little prolonged.

In one of the preceding experiments it cooled through the given interval in $39\frac{1}{2}$ minutes, and in the other in

$39\frac{1}{4}$ minutes; but in this experiment it took up only $38\frac{1}{2}$ minutes in cooling through it, as we have just seen.

Supposing now (what appears to me to be not improbable) that all, or very nearly all, the heat lost by the instrument No. 5 passed off in rays *through* the air, we can ascertain what part of the heat lost by the instrument No. 6 was communicated *to the air* which came into contact with its surface.

Putting the total quantity of heat lost by each of the instruments in cooling through the given interval = 10,000, as we have just seen that a quantity of heat = 1396 passes through the covered sides of each of these instruments in $39\frac{1}{2}$ minutes, the quantities so lost in this experiment must have been as follows: By the instrument No. 5, in $38\frac{1}{2}$ minutes, = 1081; by No. 6, in $37\frac{1}{4}$ minutes, = 1046; and, deducting these quantities so lost (through the covered sides of the instruments) from the total quantity lost by each (= 10,000), we shall find out how much heat passed off *through the bottom* of each of the instruments.

| | | | |
|---|---|---|---|
| For the instrument | No. 5 | it is . | 10,000 — 1081 = 9919 |
| And for " | No. 6 | . | 10,000 — 1046 = 9954 |

If now the whole of the heat lost through the bottom of the instrument No. 5 passed off *through* the air in rays, as there is no reason to suppose that a less quantity passed off in the same time, *in the same way*, through the bottom of the instrument No. 6, it appears that this last-mentioned instrument must have lost *by radiation*, or in rays which passed *through* the air, a quantity of heat = 9597.

For it is $38\frac{1}{2}$ minutes to 9919 as $37\frac{1}{4}$ minutes to 9597.

And if of the total quantity of heat which passed off through the bottom of the conical instrument No. 6, = 9954, a quantity = 9597 passed off *through* the air in calorific rays, the remainder only (9954 — 9597), which amounts to no more than 357 parts, could have been communicated to the air.

Hence it would appear that when a hot body is cooled in air $\frac{1}{27}$ part only of the heat which it loses is acquired by the air; for 357 is to 9597 as 1 to 27, very nearly. But I shall refrain from enlarging farther on this subject at present.

One of the objects which I had in view in the last experiment was to find out whether the cooling of a hot body in air is or is not sensibly accelerated or retarded by the greater or lesser distance at which the body is placed from other neighbouring solid bodies, when these neighbouring bodies are at the same temperature as the air; and, as a comparison of the result of this experiment with the results of the two preceding experiments so strongly indicated that the cooling of the conical vessel in the preceding experiments had in fact been *retarded* by the vicinity of the pewter platter over which it was suspended, I was now induced to repeat these experiments with some variations.

These investigations appeared to me to be of the more importance, as I conceived that the results of them might lead to a discovery of one of the causes of the warmth of clothing.

*Experiment No.* 32. — I now placed the pewter platters once more in their former stations, perpendicularly under the bottoms of the two conical vessels, but at the distance of 3 inches only; that which was under the vessel No. 5 being at the temperature of the air of the room

(62°), while that placed under the vessel No. 6 was kept ice-cold, by means of pounded ice and water, which was put into the earthen dish on the brim of which it was supported.

The times of the cooling of the vessels, through the standard interval of 10 degrees, were as follows : —

| | | |
|---|---|---|
| No. 5 | . . . . . . | in $40\frac{1}{4}$ minutes. |
| No. 6, which was over ice, | . | in $33\frac{1}{4}$ " |

*Experiment No.* 33. — I repeated this experiment once more, but varied it by bringing the pewter platters still nearer to the bottoms of the conical vessels. The flat horizontal part of each of the platters was now only 2 inches below the flat surface of the bottom of the conical vessel which was suspended over it. Both the platters still remained covered by their flat circular perforated covers of paper; but it should be remembered that the circular hole in the centre of each of these covers was no less than 6 inches in diameter, and consequently that a large portion of the flat part of the bottom of the platter was in full view (if I may use that expression) of the bottom of the vessel which was suspended over it.

The times of cooling in this experiment were as follows : —

| | | |
|---|---|---|
| No. 5 cooled through the given interval | | in $42\frac{3}{4}$ minutes. |
| No. 6, which was over ice, | . . | in $32\frac{1}{2}$ " |

The results of these experiments show (what indeed might have been expected, especially on the supposition that the heating and cooling of bodies is effected by means of radiations) that, although the cooling of the hot body suspended over a surface kept constantly cold by artificial means was accelerated by being brought nearer to that cold surface, yet, in a case where the cold surface was less intensely cold, and where its temperature could be sensibly raised by the calorific rays from the hot body, the cooling of the hot body was retarded by a nearer approach of that cold surface.

From the results of these experiments we may safely conclude that, if the hot body, instead of being a conical vessel covered up on all sides except its flat bottom, had been a globe, and if this hot globe had been suspended in the centre of another larger thin hollow sphere (this last being, at the beginning of the experiment, at the same temperature as the air and walls of the room), the vicinity of the surface of this hollow globe to the surface of the hot body would have retarded the cooling of the hot body in the same manner as the cooling of the conical vessel No. 5 was retarded in the foregoing experiments; and if, instead of inclosing the hot body in the centre of a single hollow sphere of any given thickness, it were placed in the common centre of a number of much thinner concentric spheres, of different diameters, the time of cooling would be still more retarded.

By tracing the various operations which would take place in the cooling of the hot body in this imaginary experiment, we shall become acquainted with the nature of those which actually take place when the cooling of a hot body is prolonged by means of warm clothing.

From the results of several of the foregoing experiments we may conclude that, supposing the thin concentric hollow spheres in which the hot body is confined to be made of metal, the cooling will be slower if the surfaces of these spheres are polished than if they are unpolished or blackened; and hence we might very naturally be led to suspect (what is probably true in fact) that the *warmth* of any kind of substance used as clothing, or its power of preventing our bodies from being cooled by the influence (frigorific radiations) of surrounding colder bodies, depends very much *on the polish of its surface.*

If, with the assistance of a microscope, we examine those substances which supply us with the warmest coverings, — such, for instance, as furs, feathers, silk, &c.,— we shall find their surfaces not only smooth, but also very highly polished; we shall also find that, other circumstances being equal, those substances are the warmest which are the finest, or which are composed of the greatest number of fine polished detached threads or fibres.

The fine white shining fur of a Russian hare is much warmer than coarse hair; and fine silk, as spun by the silkworm is warmer than the same silk twisted together into coarse threads; as I found by actual experiments, an account of which has already been laid before this Society and published in the Philosophical Transactions.[7]

I formerly considered the warmth of natural and artificial clothing as depending *principally* on the obstacle it opposes to the motions of the cold air by which the hot body is surrounded; but, by a patient and careful examination of the subject, I have been convinced that the efficacy of radiation is much greater than I had supposed it to be.

From the result of the experiment No. 31, we might be led to conclude that a very small part only of the heat which a hot body appears to lose when it is cooled in air is in fact communicated to that fluid, a much greater portion of it being communicated to other surrounding bodies at a distance; and, in one of my former experiments, a hot body was cooled, though it was placed in a Torricellian vacuum.

These researches appear to me to be the more interesting, as I have long been of opinion that it must be by experiments of this kind (showing in what manner the temperature of bodies are affected reciprocally at different degrees of temperature and at different distances) that the hypothesis of radiation must be established or proved to be unfounded.

When I speak of heat as being communicated to air *immediately* by a hot body which is cooled in it, I mean only that it is not first communicated to other neighbouring bodies, and then given *by them* to the particles of air with which they happen to be in contact. In this last-mentioned way much of the heat, no doubt, which a hot body loses when cooled in air is ultimately communicated to that fluid.

I am far from supposing that the particles of air which, coming into contact with a hot body, are heated in consequence of that near approximation receive heat in any other *manner* than that in which other bodies at a greater distance receive it. If in the one case it be generated or excited by the agency of calorific rays or undulations caused by the hot body, it must, I am persuaded, be excited in the same manner in the other.

The reason why the particle of air which is in immediate contact with a hot body is heated, while other particles near it are not affected by the calorific rays from the hot body which are continually passing by them through the air, is, I conceive, because the particle heated is at *the surface of the fluid* (air), where these rays are either reflected, refracted, or absorbed; but when a ray has once passed the surface of a transparent fluid, it proceeds straight forward, without being farther affected by it, *and consequently without affecting it*, till it comes to the confines of the medium, or to the surface of some other body.

If this hypothesis of the communication, or rather *generation*, of heat and of cold by radiation be true, it will enable us to explain, in a satisfactory manner, what has been called the *non-conducting power* of transparent fluids with respect to heat; for, if heat be really communicated or excited in the manner above described, it is quite evident that *a perfectly transparent fluid* can receive heat only at its surface, and, consequently, that heat cannot be propagated in such a fluid by communication from one particle of the fluid to another.

By a *transparent* fluid I mean such an one as admits the calorific and frigorific rays emitted by hot and by cold bodies to pass freely through it without obstructing their passage or diminishing their intensities.

Whether any of the fluids with which we are acquainted be *perfectly* transparent in this sense of the word or not, I will not pretend to say; but there is reason to think that pure water and air and most other fluids which are transparent to light, possess a high degree of transparency in regard to calorific and frigorific rays, or that they give a very free passage to them when they have once passed their surfaces.

An even or polished surface has been found to facilitate very much the reflection of the rays of light. May it not, in all cases, have an equal tendency to facilitate the reflection of calorific and frigorific rays?

In the experiments with the large cylindrical vessels, where they were exposed *naked* to cool in the air, their surfaces were polished, and they were a long time in cooling. But, when the surface of the vessel was blackened or covered with other substances, the vessel was found to cool much more rapidly.

A large proportion of the frigorific rays from the surrounding colder bodies were, in the former case, reflected at the polished surface of the metallic vessel; but, in the latter case, more of them were absorbed.

When a large drop of water rolls about without being evaporated upon the flat surface of a piece of red-hot iron, the surface of the drop is *polished;* and, the calorific rays being mostly reflected, the water is very little heated, notwithstanding the extreme intensity of the heat of the iron and its nearness to the water.

If the iron be *less hot*, the water penetrates the pores of the oxide which covers the metal, the drop ceases to have a polished surface, acquires heat very rapidly, and is soon evaporated.

If a drop of water be placed on the clean and polished surface of a metal not so easily oxidable as iron, it will retain its spherical form and polished surface under a lower degree of temperature than on iron; and consequently will be less heated, and less rapidly evaporated by a moderate heat.

If a large drop of water be put carefully into a clean silver spoon, previously heated very hot (that is to say, so hot as to give a loud hissing noise when touched with the wetted finger, but much below the heat of red-hot metal), the drop will support, or rather *resist*, this heat for a considerable time; but after the spoon has been suffered to cool down nearly to the temperature of boiling water a drop of water put into it will be evaporated instantaneously.

It appears, from the results of these experiments, to be probable that under high temperatures air is attracted by metals so much more strongly than water that even the weight of a drop of water is not sufficient to force away the stratum of air which covers and adheres to the surface of a metal on which the drop reposes; but at lower temperatures this does not seem to be the case.

The following experiment, which I made several months ago with a view to investigate the cause of the slow evaporation of drops of water placed on hot metals, will, I think, throw much light on this subject.

*Experiment No.* 34. — Taking a clean polished silver spoon, I blackened the inside of it by holding it over the flame of a wax candle; then, putting a large drop of water into it, I found, as I expected, that the drop took a spherical form, and rolled about in the spoon without wetting its blackened surface.

I now held the spoon over the flame of a candle, and attempted to make the water boil; but I found it to be absolutely impossible. The handle of the spoon became so very hot that I could not hold it in my hand without being burnt, though it was wrapped up in three or four thicknesses of linen; but still the drop of water did not appear to be at all affected by this intense heat. If the bowl of the spoon were touched with the finger, a hissing noise announced that it was extremely hot; but still the water remained perfectly quiet in the spoon without being evaporated.

Having in vain attempted to make this drop of water boil, and not being able to hold the spoon over the flame of the candle any longer on account of the heat of its handle, I now poured the drop into the palm of my hand. I found it to be warm, but by no means scalding hot.

By holding the spoon with a pair of tongs over the

flame of the candle for a longer time, I found that a drop of water in the spoon gradually *changed its form*, became less, and was at length evaporated; from being spherical and lucid, it gradually took an oblong form, and its surface became obscure; and when it was evaporated it left a kind of skin behind it, which was evidently composed of the particles of black matter which had by degrees attached themselves to its surface, and which probably had contributed not a little to its being at last heated and evaporated.

The change in the form of the drop of water, and more especially the gradual loss of its lucid appearance, made me suspect that it had turned round during the experiment. If it really did so its motion must either have been extremely rapid or very slow, for, though I examined it with great attention, I could not perceive that it had any rotatory motion.

I will take the liberty to mention another little experiment which I have often made to amuse myself and others, though it may perhaps be thought too trifling to deserve the attention of the Royal Society.

*Experiment No. 35.*—If a large drop of water be formed at the end of a small splinter of light wood (deal, for instance), and this drop be thrust quickly into the centre of the flame of a newly snuffed candle, which burns bright and clear, the drop of water will remain for a considerable time in the centre of the flame and surrounded by it on every side, without being made to boil, or otherwise apparently affected by the heat; and if it be taken out of the flame and put upon the hand, it will not be found to be scalding hot.

If it be held for some time in the flame, it will be gradually diminished by evaporation; but there is much reason to think that the heat which it acquires is not communicated to it by the flame, but by the wood to which it adheres, which is soon heated by the flame, and even set on fire.

I cannot refrain from just observing that it appears to me to be extremely difficult to reconcile the results of any of the foregoing experiments with the hypothesis of modern chemists respecting the *materiality of heat*.

Deeply sensible of the insufficiency of the powers of the human mind to unfold the mysteries of nature and discover the agents she employs and their mode of action in her secret and invisible operations, and being, moreover, fully aware of the danger of forming an attachment to a false theory, and of the folly of wasting time in idle speculation, I have ever, in my philosophical researches, been much more anxious to discover new facts, and to show how the discoveries of others may be made useful to mankind, than to invent plausible theories, which much oftener tend to misguide than to lead us in the path of truth and science.

There are, however, situations in which an experimental inquirer sometimes finds himself, where it is almost impossible for him to abstain from forming or adopting some general theory for the purpose of explaining the phenomena which fall under his observation, and directing him in his future researches.

Finding myself in that situation at this time, I beg the attention and, above all, the *indulgence* of the Society while I endeavour to explain the conjectures I have formed respecting the nature of heat and the mode of its communication.

*Hot* and *cold*, like *fast* and *slow*, are mere relative terms; and, as there is no relation or proportion between motion and a state of rest, so there can be no relation between any degree of heat and absolute cold, or a total privation of heat; hence it is evident that all attempts to determine the place of *absolute cold*, on the scale of a thermometer, must be nugatory.

It seems probable that *motion* is an essential quality of matter, and that rest is nowhere to be found in the universe.

We well know that all those bodies which fall under the cognizance of our senses are in motion; and there are many appearances which seem to indicate that the constituent particles of all bodies are also impressed with continual motions among themselves, and that it is these motions (which are capable of augmentation and diminution) that constitute the *heat* or temperature of sensible bodies.

The only effects of which we have any idea resulting from the action of one body on another are a change of velocity or a change of direction, or both. We perceive, it is true, that certain bodies have a power of affecting certain other bodies *at a distance;* but this is no proof that the effects produced are essentially different from those which result from collision; for, if an elastic body be interposed between the two bodies, their actions on each other may be communicated through such intermediate elastic body, which, when the action is at an end, and the effects resulting from it on the two bodies have taken place, will be in the same state precisely in which it was before the action began.

If a bell or any other solid body, *perfectly elastic*, placed in a perfectly elastic fluid, and surrounded by other perfectly elastic solid bodies, were struck and made to vibrate, its vibrations would by degrees be communicated, by means of the undulations or pulsations they would occasion in the elastic fluid medium, to the other surrounding solid and elastic bodies. If these surrounding bodies should happen to be already vibrating, and with the same velocity as that with which the bell is made to vibrate by the blow, the undulations in the elastic fluid occasioned by the bell would neither increase nor diminish the velocity or frequency of the vibrations of the surrounding bodies; neither would the undulations caused by the vibrations of these bodies tend to accelerate or to retard the vibrations of the bell. But if the vibrations of the bell were more frequent than those of the surrounding bodies, the undulations it would occasion in the elastic fluid would tend to accelerate the vibrations of the surrounding bodies; on the other hand, the undulations occasioned by the slower vibrations of the surrounding bodies would retard the vibrations of the bell, and the bell and the surrounding bodies would continue to affect each other until, by the vibrations of the latter being gradually increased and those of the former diminished, in consequence of their actions on each other, they would all be reduced to the same *tone*.

Supposing now that heat be nothing more than the motions of the constituent particles of bodies among themselves (an hypothesis of ancient date, and which always appeared to me to be very probable), if for the bell we substitute a hot body, the cooling of it will be attended by a series of actions and reactions exactly similar to those just described.

The rapid undulations occasioned in the surrounding ethereal fluid, by the swift vibrations of the hot body, will act as calorific rays on the neighbouring colder solid bodies, and the slower undulations, occasioned by the vibrations of those colder bodies, will act as frigorific rays on the hot body ; and these reciprocal actions will continue, but with decreasing intensity, till the hot body and those colder bodies which surround it shall, in consequence of these actions, have acquired the same temperature, or until their vibrations have become isochronous.

According to this hypothesis, *cold* can with no more propriety be considered as the absence *of heat* than a low or grave sound can be considered as the absence of a higher or more acute note ; and the admission of rays which generate cold involves no absurdity and creates no confusion of ideas.

On a superficial view of the subject, it may perhaps appear difficult to reconcile solidity, hardness, and elasticity with those never-ceasing motions which we have supposed to exist among the constituent particles of all bodies ; but a patient investigation of the matter will show that the admission of that supposed fact, instead of rendering it more difficult to form distinct and satisfactory ideas of the causes on which those qualities of bodies depend, will rather facilitate those abstruse researches.

Judging from all the operations of nature, of the causes of which we are able to form any distinct ideas, we are certainly led to conclude that the force of dead matter (and perhaps of living matter also), or its power of affecting, that is to say, of *moving* other matter, or of *resisting its impulse*, depends on its motion.

If, therefore, solid (or fluid) bodies have any powers whatever, either of impulse or of resistance, it appears to me to be more reasonable to ascribe them to the living forces residing in them — to the never-ceasing motions of their constituent particles — than to suppose them to be derived from their want of power, and their total indifference to motion and to rest.

No reasonable objection against this hypothesis (of the incessant motions of the constituent particles of all bodies), founded on a supposition that there is not room sufficient for these motions, can be advanced; for we have abundant reason to conclude that if there be in fact any indivisible solid particles of matter (which, however, is very problematical), these particles must be so extremely small, compared to the spaces they occupy, that there must be ample room for all kinds of motions among them.

And whatever the nature or directions of these internal motions may be among the constituent particles of a solid body, as long as these constituent particles, in their motions, do not break loose from the systems to which they belong (and to which they are attached by gravitation), and run wild in the vast void by which each system is bounded (which, as long as the known laws of nature exist, is no doubt impossible), the form or external appearance of the solid cannot be sensibly changed by them.

But if the motions of the constituent particles of any solid body be either increased or diminished, in consequence of the actions or radiations of other distant bodies, this event could not happen without producing some visible change in the solid body.

If the motions of its constituent particles were *diminished* by these radiations, it seems reasonable to conclude that their elongations would become less, and consequently that the volume of the body would be contracted; but if the motions of these particles were increased, we might conclude, *a priori*, that the volume of the body would be expanded.

We have not sufficient data to enable us to form distinct ideas of the nature of the change which takes place when a solid body is melted; but as fusion is occasioned by heat, that is to say, by an augmentation (from without) of that action which occasions expansion, if expansion be occasioned by an increase of the motions of the constituent particles of the body, it is, no doubt, a certain additional increase of those motions which causes the form of the body to be changed, and from a solid to become a fluid substance.

As long as the constituent particles of a solid body which are at the surface of that body do not, in their motions, *pass by each other*, the body must necessarily retain its form or shape, however rapid those motions or vibrations may be; but as soon as the motion of these particles is so augmented that they can no longer be restrained or retained within these limits, the regular distribution of the particles which they acquired in crystallization is gradually destroyed, and the particles so detached from the solid mass form new and independent systems, and become a liquid substance.

Whatever may be the figures of the orbits which the particles of a liquid describe, the mean distances of those particles from each other remain nearly the same as when they constituted a solid, as appears by the small change of specific gravity which takes place when a solid is melted and becomes a liquid; and, on a supposition that their motions are regulated by the same laws which regulate the solar system, it is evident that the additional motion they must necessarily acquire, in order to their taking the fluid form, cannot be lost, but must continue to reside in the liquid, and must again make its appearance when the liquid changes its form and becomes a solid.

It is well known that a certain quantity of *heat* is requisite to melt a solid, which quantity disappears or remains *latent* in the liquid produced in that process; and that the same quantity of heat reappears when this liquid is congealed and becomes a solid body.

But before I proceed any farther in these abstruse speculations, I shall endeavour to investigate some of the consequences which would necessarily result from the radiations of hot and of cold bodies, supposing those radiations to exist, and their motions and actions to be regulated by certain assumed laws.

And first, it is evident that the intensity of the rays emitted by a luminous point, in a perfectly transparent medium, is everywhere as the squares of the distance from that point inversely; for the intensity of those rays must be as their condensation; and their condensation being diminished in proportion as the space they occupy is increased, if we suppose all the rays which proceed in all directions from any point to set out at the same instant and to move with the same velocity in right lines, these simultaneous rays (or undulations) will in their progress form a sphere, which sphere will increase continually in size as the rays advance; and as all the rays must be found at the surface of this sphere, their intensity or condensation must necessarily be as the surface of the sphere inversely, or as the squares of the distance inversely from the centre of the sphere, or, which is the same thing, from the luminous point from which these rays proceed; the surfaces of spheres being to each other as the squares of their radii.

Supposing now (what, indeed, appears to be incontrovertible) that the intensity of the rays which hot and cold bodies emit, in a medium perfectly transparent, follows the same law, we can determine what effects must be produced by the largeness or smallness of the confined space (of a room, for instance) in which a hot body is placed to cool.

To simplify this investigation, we will suppose this confined space to be a hollow sphere of ice 9 feet in diameter, at the temperature of freezing water; and the hot body to be a solid sphere of metal 2 inches in diameter, at the temperature of boiling water, placed in the centre of it; and we will suppose, farther, that this hollow sphere is void of air, and that the cooling of the hot body is effected solely by the frigorific rays from the ice.

The question to be determined is, in what manner the cooling of the hot body would be affected by increasing the diameter of this hollow sphere of ice.

Let us suppose its diameter to be increased to 18 feet. Its internal surface will then be to the surface of a sphere 9 feet in diameter as the square of 18 to the square of 9, that is to say, as 324 to 81, or as 4 to 1. And as the quantity of frigorific rays emitted are, *cæteris paribus*, as the surface from which they proceed, the quantity of rays emitted by the internal surface of the larger sphere will be to the quantity emitted by the internal surface of the smaller as 4 to 1.

But the intensities of these rays at the common centre of these spheres (where the hot body is placed) being as the squares of the distances from the radiating points inversely, the intensity of the rays from the internal surface of the smaller sphere must be to the intensity of the rays from the internal surface of the larger sphere as 4 to 1, at the common centre of those spheres.

Now, as the time of the cooling of the hot body will depend on the *quantity* of frigorific rays which arrive at its surface, and on the *intensity* of their action, and as the intensity of the rays from the internal surface of the sphere at its centre is diminished in the same proportion as the surface of the sphere is augmented when its diameter is increased, it follows that a hot body placed in the centre of a hollow sphere at any given constant temperature below that of the hot body, will be cooled in the same time, or with the same celerity, whatever may be the size of the sphere.

If this conclusion be well founded (and I see no reason to suspect that it is not so), it will follow, from the principles assumed, that the hot body will be cooled in the same time, in whatever part of the hollow sphere it be situated. And as the cooling of the body is not affected, that is to say, accelerated or retarded, either by the greater or smaller size of the enclosed space in which it is confined, or by its situation in that confined space, so it cannot be in any manner affected either by the form of that hollow space or by the presence of a greater or less number of other solid bodies; provided always, that all these surrounding bodies be at the same constant temperature.

If, however, any of these surrounding bodies, the temperature of which is liable to be sensibly changed during the experiment by the calorific rays emitted by the hot body, be placed *very near* that body, the cooling of that hot body will be retarded, the rays from this neighbouring body, *so heated*, being less frigorific than those from other bodies at a greater distance, which it intercepts.

The results of all my experiments on the cooling of bodies tended uniformly to confirm the above conclusions.

Admitting that the cooling of a hot body is effected solely by the rays which proceed from colder bodies, and that these rays, like those of light, are reflected, refracted, and concentrated, according to certain known laws, by the polished surfaces of mirrors and lenses, it might perhaps be imagined that the cooling of a hot body might be accelerated or retarded by giving it some peculiar form; or by placing near it, and in certain positions with respect to it, two or more highly polished reflecting mirrors.

As these conjectures, if well founded, might lead to experiments from the results of which the truth or falsehood of the hypothesis in question might be demonstrated, it is of much importance that this matter should be thoroughly investigated. I shall therefore beg the indulgence of the Society while I endeavour to examine it with that careful attention which it appears to me to deserve.

When different solid substances, heated to the same degree of temperature, are exposed in the air to cool, those among them which appear to the touch to be the hottest are not those which cool the fastest, or which send off calorific rays through the air in the greatest abundance.

As polished metals reflect a great part of the rays from other bodies which arrive at their surfaces, and as they are neither heated nor cooled by the rays so reflected, their temperatures are slowly changed by the actions of the surrounding bodies at a different temperature.

When a hot polished metallic body is exposed in the air to cool, surrounded by other bodies at the same temperature as that of the cold air, most of the rays from the surrounding bodies are reflected at the polished surface of the hot body; it is evident, then, that two sorts of rays must proceed from the surface of that body, namely, those calorific rays which it emits, and those other rays (which with regard to the surrounding bodies are neither calorific nor frigorific) which it reflects.

On a cursory view of the subject, one might be led to imagine that, as the rays which proceed from the hot metallic body are of two kinds, the energy of the calorific rays, which properly belong to the hot body, might be diminished by those other reflected rays by which they are accompanied, and with which they may be said to be mixed; but a more careful examination of the matter will show that this cannot be the case, that is to say, as long as all the surrounding bodies continue to be at the same temperature. If the temperature of the surrounding bodies be different, such of them will be affected by the reflected rays as happen to be of a temperature different from that from which the ray originated; but still the effects produced by the rays emitted by the hot body will be the same, or their power of effecting changes in the temperatures of other (hotter or colder) bodies will remain undiminished and unchanged.

The reason why their effects are not more powerful

than they are found to be, is not because they are mixed with other reflected rays, but because they are few, the greater part of the rays which the hot body actually emits being reflected and turned back upon itself by the reflecting surface by which it is immediately surrounded.

The reflecting surface at which the rays of light which impinge against the polished surface of any solid or fluid body are turned back and reflected is actually situated *without the body*, and even at some distance from it; this has been proved by the most decisive experiments; and there are so many striking analogies between the rays of light and those invisible rays which all bodies at all temperatures appear to emit, that we can hardly doubt of their motions being regulated by the same laws.

Perhaps there may be no other difference between them than exists between those vibrations in the air which are audible and those which make no sensible impression on our organs of hearing.

If the ear were so constructed that we could hear all the motions which take place in the air, we should, no doubt, be stunned by the noise; and if our eyes were so constructed as to see all the rays which are emitted continually, by day and by night, by the bodies which surround us, we should be dazzled and confounded by that insupportable flood of light poured in upon us on every side.

Taking it for granted that these invisible radiations exist, we will endeavour to trace the effects which must necessarily be produced by them, and see if these investigations will not lead us to a discovery of the causes of some appearances which have hitherto been enveloped in much obscurity.

Suppose two concave reflecting mirrors, of highly polished metal, each 18 inches in diameter, and 18 inches focal distance, to be placed opposite to each other at the distance of 10 feet, in a large quiet room, in which the air and the walls of the room remain constantly at the same temperature (that of freezing water, for instance), without any variation.

If we suppose the floor, ceiling, walls of the room, and doors and windows, to be lined with a covering of ice, at the temperature of freezing water, we can then, without any difficulty, conceive that the temperature of the room may remain the same, notwithstanding the presence of hotter bodies, which are brought into it for the purpose of making experiments.

Let us now suppose one of the mirrors to be at the temperature of freezing, and the other at that of boiling water; and let us see what effects they would produce on each other by their radiations.

And first, with respect to the hot mirror, it is evident that it will be cooled, not only by the frigorific rays which proceed from the cold metal of which the opposite mirror is constructed, but also by such of the frigorific rays from the sides of the room as, impinging against the polished reflecting surface of the cold mirror, and being reflected by that surface, happen to fall on the surface of the hot mirror without being reflected by it.

But, as the quantity of rays which the cold mirror *reflects* is greater in proportion as the reflecting surface is more perfect, while the quantity of rays emitted by this cold mirror is less in proportion as its reflecting surface is more perfect, it is extremely probable that the *total* quantity of frigorific rays (emitted and reflected) which, coming from the surface of the cold mirror, impinge against the surface of the hot mirror, will be the same, whatever may be the degree of polish, or reflecting power, of the cold mirror. And, if this be the case, we may conclude that the presence of this mirror will have no effect whatever on the hot mirror; or that it will no more expedite its cooling than any other body, of any other form, would do, at the same distance and occupying the same space.

It might perhaps be imagined that the *form* of the cold mirror might concentrate the rays it emits and reflects, and, by such concentration, produce a greater effect on the opposite mirror than if its surface were flat, or of any other form; but a more attentive examination of the matter will show that no such concentration actually takes place: for, with regard to those rays which are *emitted* by this cold body, as they proceed from each point of its surface *in all directions*, it is perfectly evident that these are not concentrated; and with respect to those which are *reflected*, it is equally certain that they are not concentrated, because, in order to their being concentrated, they must arrive at the surface of the mirror in parallel lines, and in the direction of the axis of the mirror, which, under the given circumstances, is evidently impossible.

Hence we see that the presence of the cold mirror will not tend, in the smallest degree, either to accelerate or to retard the cooling of the hot mirror; that is to say, provided its temperature be not raised by the calorific rays from the hot mirror.

If its temperature be raised by those rays, it will tend to retard the cooling of the hot mirror; but, even in this case, it will not retard it more than any other polished metallic body would do, of any other form, having the same area or quantity of surface opposed to the hot mirror, and being placed at the same distance from it.

By a similar train of reasoning it may be shown that the *form* of the hot body (that of a concave mirror) will contribute nothing to the effect it will produce on the cold mirror, in heating it by the calorific rays it emits; and that it will itself be cooled neither faster nor slower on account of its peculiar form.

Let us now suppose both mirrors to be at the temperature, precisely, of the room (that of freezing water), and that a bullet, or other small body of a spherical form, at the temperature of boiling water, be placed in the focus of one of the mirrors, which mirror we shall call A.

As the rays emitted by this hot body are sent off in right lines, in all directions, in the same manner as light is emitted by luminous bodies, all those rays which fall on the concave polished surface of the mirror A will be reflected (as is well known) in lines nearly parallel to the axis of the mirror; they will consequently fall on the concave polished surface of the opposite mirror B, and, being there again reflected, they will be *concentrated* at the focus of the second mirror.

If now a sensible thermometer, at the temperature of the room, be placed in this focus, it will immediately begin to rise, in consequence of the heat generated in it by the action of these calorific rays, so accumulated in that place.

If, instead of being placed in the focus of this second mirror, the thermometer be placed at a very small distance from that focus, on one side of it, the instrument, however sensible it may be, will not be apparently affected by the rays from the hot body.

This experiment, which is of ancient date, has often been made, and always with the same results.

Let us now suppose the hot body to be removed from the focus of the mirror A, and that a colder body be substituted in place of it. And, in the first place, we will suppose the temperature of this colder body to be that of freezing water, or just equal to that which reigns in the room.

As the rays which bodies at the same temperature send off from one to the other have no tendency to increase or to diminish the temperature of those bodies, the concentration of rays in the focus of the mirror B, proceeding from the ice-cold body placed in the focus of the mirror A, can have no effect on a thermometer, at the same temperature, which is exposed to their action.

If heat be a vibratory motion of the constituent particles of bodies, and if the rays which sensible bodies send off in all directions be undulations in an ethereal elastic fluid by which they are surrounded, occasioned by those motions; as the pulsations in this fluid must be isochronous with the vibrations by which they are occasioned, these pulsations or undulations can neither accelerate nor retard the vibrations of other bodies at the surfaces of which they arrive, provided the vibrations of the constituent particles of such bodies are, at that time, isochronous with the vibrations of the constituent particles of the body from which these undulations proceed. But to return to our experiment.

Suppose now that, instead of this ice-cold body, another much colder — at the temperature of freezing mercury, for instance — be placed in the focus of the mirror A, and that a thermometer at the temperature of freezing water be placed in the focus of the mirror B; what might be expected to be the result of this experiment? — That the thermometer would fall, in consequence of its being cooled by the accumulation of frigorific rays proceeding from this very cold body.

Now this is what actually happened in the celebrated experiment of my ingenious friend, Professor Pictet, of Geneva.

Several attempts have been made to explain the result of that experiment, on the supposition that caloric has a real or material existence, and that radiant heat is that substance, emitted and sent off in right lines in all directions from the surfaces of hot bodies. But none of these explanations appear to me to be satisfactory. One of the most plausible of them is that which is founded on a supposition that caloric is emitted continually, under the form of radiant heat, by all bodies, at all temperatures, but in greater abundance by hot bodies than by such as are colder; and that a body, at the same time that it sends off radiant caloric in all directions to the bodies by which it is surrounded, receives it in return, in greater or less quantities, from all those bodies; that in all cases where a body, in any given time, receives more radiant caloric than it gives off, an accumulation of caloric in the body takes place, in consequence of which accumulation it becomes hotter, but when it gives off more caloric in any given time than it receives, its quantity of caloric is gradually diminished and it becomes colder; and that a constant temperature results from the quantities of caloric emitted and received continually being equal. But besides the difficulty of explaining how, or by what mechanism, it can be possible for the same body to receive and retain, and reject and drive away, the same kind of substance, at one and the same time (an operation not only incomprehensible, but apparently impossible, and to which there is nothing to be found analogous, to render it probable), many other reasons might be brought to show that this hypothesis of the supposed continual interchanges of caloric between neighbouring bodies is very improbable; and, among the rest, there is one which appears to me to be quite conclusive.

As the point in dispute seems to be of great importance to the science of heat, I shall endeavour to examine it with all possible attention; and, in order to put the hypothesis in question to the test, we will see if it will accord with the results of some of the foregoing experiments, which, in order to their being more easily comprehended and examined, I shall elucidate by figures.

Let the two opposite ends of the cylinders A and B (Plate III. Fig. 4) represent the two vertical metallic disks of equal dimensions, which were presented at the same time to the ball of the thermoscope C, in the experiment No. 23.

In that experiment the disk A being at the temperature of 32° F. (that of freezing water), and the disk B at 112° F., while the ball of the thermoscope C and all other surrounding bodies were at 72°, it was found that the temperature of the thermoscope was not changed by the simultaneous actions of these two bodies, the one hot and the other cold.

In order to account for this result on the hypothesis before mentioned, we must begin by supposing that the

Plate III.

Fig. 4.

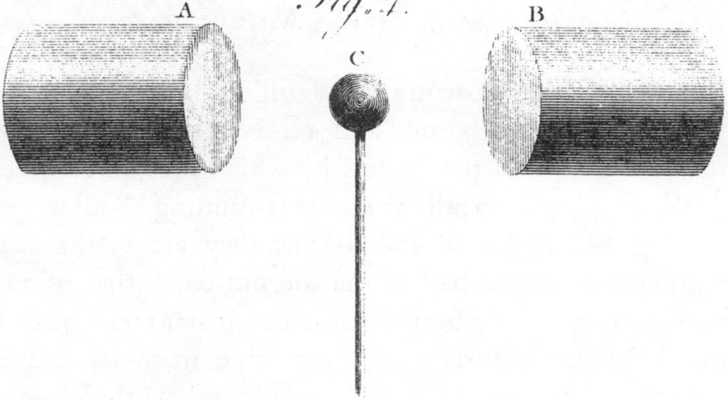


Fig. 5.

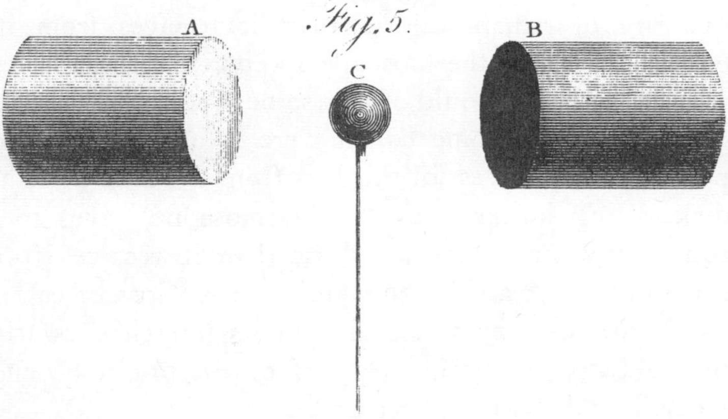


Fig. 6.

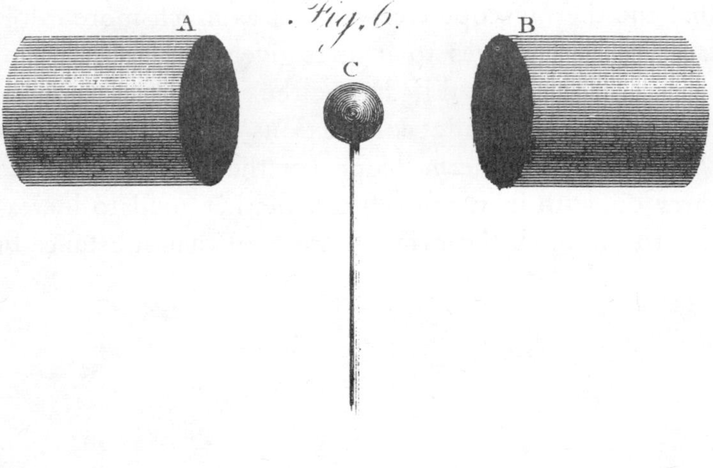

ball of the thermoscope gives off radiant caloric continually in all directions, and receives it in return from the surfaces of all the bodies by which it is surrounded.

With regard to all these surrounding bodies (excepting the disks A and B), as they are at the same temperature as the ball of the thermoscope (that of 72°), they will give continually to that instrument just as much radiant caloric as they receive from it, and no change of temperature will result from these equal interchanges.

But in respect to the disk A, as that is colder than the ball of the thermoscope, it returns to it a smaller quantity of radiant caloric than it receives from it; consequently the thermoscope receives continually less than it gives: it would of course be gradually exhausted of caloric and become colder were it not for the compensation it receives for this loss from the disk B. This disk, being hotter than the thermoscope, gives to it continually more radiant caloric than it receives from it; and were it not for the simultaneous loss of caloric which the instrument sustains in its interchanges with the cold disk A, its quantity of caloric would be augmented, and it would become hotter.

Now, as the temperature of the ball of the thermoscope is an arithmetical mean between that of the disk A and that of the disk B, it is reasonable to suppose that the thermoscope receives just as much more caloric from B than it gives to it as it gives to A more than it receives from it; and if that be the case in fact, it is evident that the simultaneous actions of the two disks on the ball of the thermoscope (or the traffic which they carry on with it in caloric) can neither tend to increase nor to diminish the original stock of that substance belonging to that instrument; consequently the instrument will neither be heated nor cooled by these interchanges, but will continue invariably at the same constant temperature.

This explanation is plausible, but, before the hypothesis on which it is founded can be admitted, we must see if it will agree with the results of other experiments, — for the greatest care ought always to be used in the admission of hypotheses in physical researches, and in no case can it be more indispensably necessary than where an hypothesis has evidently been contrived for the sole purpose of explaining a single experiment, or elucidating a new fact.

When the surface of the metallic disk B was blackened by holding it over the flame of a candle, the intensity of its radiation at the given temperature (that of 112°) was found to be very considerably increased; and when (being so blackened) it was again presented to the ball of the thermoscope at the same distance as in the last-mentioned experiment, and the cold disk A (at the temperature of 32°) was placed opposite to it at an equal distance, as represented in Fig. 5, the thermoscope, instead of continuing to retain its original temperature (that of 72°), was now gradually heated.

There is nothing, it is true, in that event, which appears difficult to explain on the assumed principles; for, if the quantity of radiant caloric emitted by the disk B be increased by blackening its surface, the quantity received from it by the ball of the thermoscope must be increased also, and that additional quantity must, of course, tend to raise the temperature of the instrument. But here is an experiment which cannot be explained on those principles.

The surface of the cold disk A having been blackened as well as that of the hot disk B, when both disks (blackened) were again presented at equal distances to the ball of the thermoscope, as represented in Fig. 6, it was found that the original temperature of the thermoscope remained unchanged.

The result of this most interesting experiment proves that the ball of the thermoscope was just as much cooled by the influence of the cold blackened disk as it was heated by the hot blackened disk.

Now, as it was found by experiment that the intensity of the radiation of the disk B was *increased* by the blackening of the surface of that disk, we must conclude that the intensity of the radiation of the disk A was likewise *increased* by the use of the same means; but if those radiations be *caloric*, emitted by those bodies (which the hypothesis in question supposes), how did it happen that the ball of the thermoscope, instead of being *more heated* by the additional quantity of caloric which it received in consequence of the blackening of the disk A, was actually *more cooled?*

It may perhaps be said by the advocates for the hypothesis in question, that the blackening of the surface of the disk A caused a greater quantity of caloric to be sent off to it by the ball of the thermoscope. Without insisting on an explanation of the mode of action of the cause which is supposed to produce this effect (which I might certainly do, as the supposition is perfectly gratuitous), I will content myself with just observing that as the surface of the opposite disk *was also blackened*, this supposed augmentation of the quantity of caloric emitted by the ball of the thermoscope, *occasioned by the blackening of the surfaces of the bodies presented to it*, can be of no use in explaining the phenomena in question.

The results of the two last mentioned experiments appear to me to be very important; and I do not see how they can be reconciled with the opinions of modern chemists respecting the nature of heat.

In order to simplify our speculations on this abstruse subject, we have hitherto supposed that *difference of temperature* depends solely on the difference *of the times* of the vibrations of the component particles of bodies. It is possible, however, and even probable, that it depends principally on the *velocities* of those particles; for it is easy to perceive that, the more rapid the motions of those particles are, the greater their elongations must be in their vibrations, and the more, of course, will the volume of the body they compose be expanded.

It is well known that the pulsations occasioned in an elastic fluid by the vibrations of an elastic solid body proceed from that body in all directions, and that these pulsations are everywhere (that is to say, at all distances from the body) isochronous with the vibrations of the solid body; it is known, also, that the mean velocity of any individual particle of the fluid is less in proportion as the distance of the particle is greater from the centre from which these pulsations proceed.

In the case of the pulsations occasioned in the air by the vibrations of sonorous bodies, those pulsations are everywhere isochronous with the vibrations of the sonorous body, and the time, or *frequency*, of these pulsations, determines the *note;* but it is the *velocity* of the particles of the air, or the breadth of the wave, on which the *force* or *strength* of the sound depends; and this velocity becoming less as the distance from the sonorous body increases, the sound is weakened in the same proportion.

There are several circumstances which might lead us to suspect that *colour* depends on the *frequency* of those pulsations which have been supposed to constitute light; and that the *heat* produced by them is in proportion to their *force*.

If this supposition should be well founded, a knowledge of that important fact might perhaps enable us to explain several very interesting phenomena, — the combustion of inflammable bodies, for instance, and the great intensity of the heat which is produced by the *concentration* of calorific rays.

There are several well-known experiments with burning-glasses which show that the intensity of the heat generated by the concentration of the solar rays is not simply as the *condensation* of those rays, but in a higher proportion; and that it depends much on their *direction*, being greater as the angle is greater at which they meet at the focus of the lens.

That fact is certainly very remarkable. It has often been the subject of my meditations, and it has contributed not a little to the opinion I have been induced to adopt respecting the nature of light and of heat. I never could reconcile it with the supposition that heat is caused by the *accumulation* of anything *emitted* by the sun, or by any other body which sends off calorific radiations.

Reserving for a future communication an account of the sequel of my inquiries respecting the subject which I have undertaken to investigate, I shall conclude this long paper with some observations concerning the *practical uses* that may be derived from a knowledge of the facts which have been established by the results of the foregoing experiments.

In all cases where it is designed to *preserve the heat* of any substance which is confined in a metallic vessel, it will greatly contribute to that end if the external surface of the vessel be very clean and bright; but if the object be to *cool* anything quickly in a metallic vessel, the external surface of the vessel should be painted, or covered with some of those substances which have been found to emit calorific rays in great abundance.

Polished tea-urns may be kept boiling hot with a much less expense of spirit of wine (burnt in a lamp under them) than such as are varnished; and the cleaner and brighter the dishes and covers for dishes are made, which are used for bringing victuals on the table, and for keeping it hot, the more effectually will they answer that purpose.

Saucepans and other kitchen utensils which are very clean and bright on the outside may be kept hot with a smaller fire than such as are black and dirty; but the bottom of a saucepan or boiler should be blackened, in order that its contents may be made to boil quickly, and with a small expense of fuel.

When kitchen utensils are used over a fire of sea-coal or of wood, there will be no necessity for blackening their bottoms, for they will soon be made black by the smoke; but when they are used over a clear fire made with charcoal, it will be advisable to blacken them, — which may be done in a few moments by holding them over a wood or coal fire, or over the flame of a lamp or candle.

Proposals have often been made for constructing the

broad and shallow vessels (flats), in which brewers cool their wort, of metal, on a supposition that the process of cooling would go on faster in a metallic vessel than in a wooden vessel; but this would not be found to be the case in fact, a metallic surface being ill calculated for expediting the emission of calorific rays.

The great thickness of the timber of which brewers' flats are commonly made is a circumstance very favourable to a speedy cooling of the wort; for, when the flats are empty, this mass of wet wood is much cooled, not only by the cold air which passes over it, but also and more especially by evaporation; and when the flat is again filled with hot wort a great part of the heat of that liquid is absorbed by the cold wood.

In all cases where metallic tubes filled with steam are used for warming rooms or for heating drying-rooms, the external surface of those tubes should be painted or covered with some substance which facilitates the emission of calorific rays. A covering of thin paper will answer that purpose very well, especially if it be black, and if it be closely and firmly attached to the surface of the metal with glue.

Tubes which are designed for *conveying* hot steam from one place to another should either be well covered up with *warm* covering or should be kept clean and bright. It would, I am persuaded, be worth while, in many cases, to gild them, or at least to cover them with what is called gilt paper, or with tin foil, or some other metallic substance which does not easily tarnish in the air.

The cylinders and principal steam-tubes of steam-engines might be covered first with some warm clothing, and then with thin sheet brass kept clean and bright. The expense of this covering would, I am confident, be amply repaid by the saving of heat and fuel which would result from it.

If garden walls painted black acquire heat faster when exposed to the sun's direct rays than when they are not so painted, they will likewise cool faster during the night; and gardeners must be best able to determine whether these rapid changes of temperature are, or are not, favourable to fruit-trees.

Black clothes are well known to be very warm in the sun; but they are far from being so in the shade, and especially in cold weather. No coloured clothing is so cold as black when the temperature of the air is below that of the surface of the skin, and when the body is not exposed to the action of calorific rays from other substances.

It has been shown that the warmth of clothing depends much on the *polish* of the surface of the substance of which it is made; and hence we may conclude that, in choosing the colour of our winter garments, those dyes should be avoided which tend most to destroy that polish; and, as a white surface reflects more light than an equal surface, equally polished, of any other colour, there is much reason to think that white garments are warmer than any other in cold weather. They are universally considered as the coolest that can be worn in very hot weather, and especially when a person is exposed to the direct rays of the sun; and if they are well calculated to reflect calorific rays in summer, they must be equally well calculated to reflect those frigorific rays by which we are cooled and annoyed in winter.

I have found, by direct and decisive experiments (of which an account will hereafter be given to this Society), that garments of fur are much warmer in cold weather when worn with the fur or hair outwards than when it is turned inwards. Is not this a proof that we are kept warm by our clothing, not so much by confining our heat as by keeping off those frigorific rays which tend to cool us?

The fine fur of beasts, being a highly polished substance, is well calculated to reflect those rays which fall on it; and if the body were kept warm by the rays which proceed from it being reflected back upon it, there is reason to think that a fur garment would be warmest when worn with the hair inwards; but if it be by reflecting and turning away the frigorific rays from external (colder) bodies that we are kept warm by our clothes in cold weather, we might naturally expect that a pelisse would be warmest when worn with the hair outwards, as I have found it to be in fact.

The point here in question is by no means a matter of small importance; for until the principles of the warmth of clothing be understood, we shall not be able to take our measures with certainty, and with the least possible trouble and expense, for defending ourselves against the inclemencies of the seasons, and making ourselves comfortable in all climates.

The fur of several delicate animals becomes white in winter in cold countries, and that of the bears which inhabit the polar regions is white in all seasons. These last are exposed alternately, in the open air, to the most intense cold and to the continual action of the sun's direct rays during several months. If it should be true that heat and cold are excited in the manner above described, and that white is the colour most favourable to the reflection of calorific and frigorific rays, it must be acknowledged, even by the most determined sceptic, that these animals have been exceedingly fortunate in obtaining clothing so well adapted to their local circumstances.

The excessive cold which is known to reign, in all seasons, on the tops of very high mountains and in the higher regions of the atmosphere, and the frosts at night which so frequently take place on the surface of the plains below in very clear and still weather in spring and autumn, seem to indicate that frigorific rays arrive continually at the surface of the earth from every part of the heavens.

May it not be by the action of these rays that our planet is cooled continually, and enabled to preserve the same mean temperature for ages, notwithstanding the immense quantities of heat that are generated at its surface, by the continual action of the solar rays?

If this conjecture should be well founded, we should be led to conclude that the inhabitants of certain hot countries who sleep at night on the tops of their houses, in order to be more cool and comfortable, do wisely in choosing that situation to pass their hours of rest.

# RESEARCH ON HEAT:

## SECOND MEMOIR

In a memoir which I had the pleasure of presenting to the class recently, I gave an account of several experiments performed with the thermoscope which appear to prove that all bodies emit rays.

I beg the members of the class to be good enough to grant me their attention for a few moments; I shall attempt to acquaint them with the results of the continuation of my research on this interesting subject.

I shall begin by noting that, since the rays which bodies are continually giving off from their surfaces do not manifestly affect any of our organs, except in a few special cases, it is hardly surprising that their existence was so long unknown. The things which escape our senses are most difficult to seize; but it is sometimes possible to surround them with our nets and to clasp them so strongly that they are obliged to reveal themselves and unveil the mystery of their invisible operations.

If it is true that the particles composing sensible bodies are continually agitated by very rapid vibratory movements and that, as a result of these movements, bodies at all temperatures give off, from every point of their surfaces, rays or waves analogous to the waves given off into the air by sound-producing bodies, and if bodies of different temperatures act upon one another at a distance by means of these rays, simultaneously effecting reciprocal changes in temperature and gradually reducing them to a common temperature, one must view the cooling of an isolated hot body as a result of the action of less warm bodies surrounding it; and since rays from hot bodies and, consequently, from cold bodies are reflected by polished surfaces, and since reflected rays produce little or no effect upon the surface of bodies whence they are reflected, one can conclude *a priori* that polished bodies must change temperature more slowly than unpolished ones.

Following are the results of a series of experiments which I performed with a view to illustrating this important fact.

Two metal vases, of identical form and capacity (both yellow copper foil cylinders 4 inches in diameter and 4 inches in height) — the outer surface of one was smooth and polished while that of the other was blackened with candle flame — were filled with boiling water and exposed together in winter to the undisturbed air of a large room. The blackened (and consequently unpolished) vase cooled approximately twice as rapidly as the polished vase.

The blackened vase was cleaned and covered with a simple, fitted envelope of thin cloth; upon repeating the experiment with the two vases, the one which was polished required 45 and a half minutes to drop 10 degrees on the Fahrenheit thermometer scale, that is to say, from 50 degrees above the temperature of the room to 40 degrees above this same temperature, while the other (cloth-covered) vase cooled through the same interval in 26 minutes.

The cloth-covered vase having been "undressed," its surface was covered with one, then two, and finally four coats of ethyl alcohol varnish; with each of these various coatings, the experiment with the two vases was repeated. While the polished vase continued to traverse the temperature interval in question (10 degrees) in 45 and a half minutes, the other cooled more or less rapidly, depending on the thickness of the varnish coating upon its surface, but always perceptibly more quickly than the polished vase.

| | |
|---|---|
| With one coat of varnish, it lost 10 degrees in | 31 minutes |
| With two coats, in | 25 and a half minutes |
| With four coats, in | 20 and three-fourths minutes |
| With eight coats, in | 24 minutes |

When the varnish had been removed, the surface of this same vase was painted successively in white and black with water colors and then covered with goldbeater's skin, first white and then painted black with Indian ink; all these various coverings and coatings more or less accelerated the cooling of the vase.

Upon covering the surface of one of the vases with gold-foil and then with silver-foil — both attached with ordinary gilding size (for wood) — I found that, despite the presence of the varnish used to hold the gold- and silver-foil to its surface, the vase thus covered over cooled in precisely the same time as the other vase of which the naturally smooth and polished surface had not been altered.

Animal hair, which provides the warmest fur, is very glossy and hence capable of reflecting to a considerable degree. In fact, I have found, by direct experiments, that a hot body wrapped in a simple fur cover retains its heat much longer when the hair is outside than when it is inside next to the hot body.

If isolated hot bodies are cooled from without by the action of rays given off by cold bodies surrounding them, it is natural to conclude not only that cold bodies must be heated in like manner by rays from nearby hot bodies, but also that they must heat less quickly when their surfaces are polished then when they are unpolished; and this important fact was indeed demonstrated by a large number of experiments which I performed for the purpose of rendering it more intelligible.

After cooling my water-filled vases by exposure, in winter, to the cold air of a large room, I transported them to a room heated by a stove where I constantly found that unpolished surfaces, which facilitate the cooling of a warm body, also, and even equally, facilitate heating of the same body when it is cold.

If the results of all these experiments do not furnish convincing proof that the communication of heat and cold is analogous to the communication of sound between sound-producing bodies, it seems to me that they confer such conjectures with a high degree of probability.

There follow some other facts which appear to confirm this opinion. Since a drop of water which one allows to fall on a red-hot ion retains its spherical form and, consequently, its reflecting surface, it heats and evaporates very slowly; it rolls about upon the surface of the metal and appears little affected by its extreme heat, the great proximity notwithstanding.

When a metal is very hot, it would appear that the air of the atmosphere clings to its surface with such force that water cannot easily displace it; but when the metal is somewhat less hot, the weight of a drop of water is sufficient to drive away this layer of air, and the surface of the metal then becomes wet, and the form and reflecting capacity of the drop are destroyed, and this small quantity of water evaporates in an instant, with a sizzling sound.

In hopes of shedding some light on this remarkable fact (of a drop of water rolling about intact on the surface of a red-hot metal), I devised the following experiment. Using a candle flame, I blackened the inner bowl of a silver spoon, and then, having put a large drop of water upon it, I attempted to make this small quantity of water boil over the candle flame. Instead of spreading out upon the black substance which lined the inner surface of the spoon, the drop retained its spherical form and, consequently, its reflecting capacity, and it was absolutely impossible to make it boil.

When the handle of the spoon (which was swathed in cloth to permit holding it in the hand without being burned) had become so hot, even at its extremity, that touching it with moistened fingertips produced a sizzling sound, the drop of water in the spoon, in spite of its proximity to the point of the candle flame, was so slightly heated that one could drop it upon one's hand and not be troubled by its heat.

This small experiment, which is very easy to repeat, gives rise to some very profound reflections.

Here is another experiment of the same sort which requires even less equipment and which is nonetheless very striking. If one suspends a large drop of water from the end of a small piece of wood (from a match, for example) and then carefully places it in the middle of a candle flame, without touching the wick, one can keep it in this situation, surrounded by the flame on all sides, for a considerable period of time without its evaporating or even becoming greatly heated; the piece of wood will immediately take fire and will gradually transmit heat to the drop; but one will observe, by the slowness of evaporation, that the flame acts but slightly upon the drop itself.

Since the reflecting surface of a polished body is not the true surface of the body but rather a surface or barrier situated a certain distance — a very small distance, no doubt — away from that body, it would appear that this surface must reflect not only rays coming from without but also those given off in all directions from the true surface of the body itself. If this conjecture is well founded, since a large portion of the rays given off by a polished body must be reflected off its reflecting surface and driven back whence they came, it is only a relatively small number of them which, by forcing their way through the barrier, do succeed in traveling any notable distance. For this reason, the degree to which (hot and cold) polished bodies affect and change the temperature of neighboring bodies must be less, all initial temperatures being equal, than the degree of influence of unpolished bodies; and this important fact was fully supported by those experiments described in my first memoir, which I presented to the class at its last meeting.

In dealing with heat, one must always remain on one's guard against those prejudices which are born of the misleading impressions of our senses. It is extremely difficult to free ourselves entirely from their dominion, but, with skill and perseverance, one can achieve all things.

To convince ourselves that the difference between hot and cold is but a matter of more or less or of more quickly–less quickly, one need but inquire what our opinions on this subject would have been if, rather than remaining habitually at a temperature of about 30 degrees (Réaumur scale) above that at which ice melts, our bodies were instead formed in such fashion as to remain constantly at a temperature lower than that of any body now known to us. In such a case, we would certainly have had no notion of cold and our languages would have possessed no word to express it.

The only real difference between a hot body and another less hot but otherwise similar body seems to me to be perfectly analogous to that maintained between a sound-producing body which gives a high-pitched sound and another sound-producing body which gives a low-pitched sound; if sound-producing bodies were so organized as to be able to render with equal facility all the various notes of the gamut and all the intermediate sounds, and if these bodies mutually affected one another at a distance by means of the waves that they gave off, in such manner that all of them were gradually made, by this reciprocal action, to give out a *common intermediate note*, the analogy between the communication of heat and the communication of sound would then be complete.

If, admitting the hypothesis herein set forth, one wished to retain the word *fire* — a term consecrated by the remotest antiquity — it would be necessary to attribute it to that eminently rarefied and elastic fluid in which *heat* and *light* are propagated; thereupon, this element would regain its high prerogatives and occupy its vast domain once more.

# HISTORICAL REVIEW

OF THE

## VARIOUS EXPERIMENTS OF THE AUTHOR ON THE SUBJECT OF HEAT.

A WRITER who directs the attention of the public to a work upon a subject as important as it is difficult of investigation must assuredly be allowed at the very outset to state modestly the reasons which entitle him to a hearing. It is also equally true that a natural philosopher can with justice lay claim to the confidence and approbation of the learned only so far as his claims are based upon his own labours, upon toilsome and accurate observations, as well as upon experiments planned and executed with all possible care.

To engage in experiments on heat was always one of my most agreeable employments. This subject had already begun to excite my attention when, in my seventeenth year, I read Boerhave's admirable Treatise on Fire. Subsequently, indeed, I was often prevented by other matters from devoting my attention to it, but whenever I could snatch a moment I returned to it anew, and always with increased interest. Even now this object of my speculations is so present to my mind, however busy I may be with other affairs, that everything taking place before my eyes, having the slightest bearing upon it, immediately excites my curiosity and attracts my attention.

This habit of many years' standing, by force of which I seize with the greatest eagerness, and endeavour to investigate, each and every phenomenon related even in the slightest manner to heat and its operations which comes to my knowledge, has suggested to me almost all the experiments that I have performed with reference to this subject.

In the year 1778 I was engaged in investigating the force of gunpowder and the velocity of bullets discharged from fire-arms. For this purpose I discharged many times a musket-barrel which was loaded in various ways, and which rested on two iron rods, perfectly free (that is, without any stock), in a horizontal position, about four feet from the ground.* This gave me occasion to make a very striking observation.

Since these experiments were intended principally to determine, from the recoil of the barrel, the velocities with which the bullets were discharged, it was first necessary to ascertain how much the weight of the powder which caused the discharge of the bullets had to do with this recoil. In order to solve this problem, I made several successive experiments, — some with a charge of powder without any bullet, and some with two, three, or even four bullets, one upon another.

According to my usual practice, I seized the piece with my left hand immediately after each discharge, in order to hold it firmly until I had wiped it out with some tow fastened to the rammer. I was therefore not a little astonished to notice, on this occasion, that

* A detailed description of these investigations may be found in the seventy-first volume of the Philosophical Transactions, and in the first volume of my Philosophical Papers, which was published at London, in the year 1802, by Cadell and Davies.[11]

the barrel was always hotter when the charge had consisted of powder alone than when loaded with one or more bullets.

I had, up to this time, no suspicion but that the piece, on being discharged, became warm as an immediate consequence of the heat caused by the burning of the gunpowder; now, however, I was convinced by the result of the above-mentioned experiment, that this supposition was entirely without foundation.

For if we should hold that the gun in question was actually heated by the inflammation of the powder, since the flame would issue from the piece much more rapidly when the charge consisted of powder alone than when the same charge had to force out one or more bullets, it would follow that a much higher degree of temperature would be reached in the latter case than in the former. But since the above-mentioned experiment shows the contrary, it follows that the heating of the piece in question is not due to the combustion of the powder, but to the vibrations caused by the concussion within the barrel, and to the operation, as rapid as it is brief, of the elastic fluid generated by this combustion.

No one is ignorant of the fact that a heavy blow is much more effective in producing heat in a solid body than a lighter one; and if the hypothesis be well founded that heat is nothing more than a continuous, more or less rapid, vibratory motion among the particles of solid bodies, this phenomenon is easily explained.

Nothing is more certain than that the shock taking place within the barrel, in the case of the above-mentioned experiment, by the combustion of the powder, was more vibrating or heavier when the charge was fired without a bullet than when the elastic fluid generated by the combustion was obliged, in order to get room for action, to push slowly before it one or more balls, which were anything but light. On careful consideration it seems to me that this circumstance is more than sufficient to explain in a satisfactory manner the results of the experiments in question, although I am perfectly free to confess that I never could reconcile myself to the hypothesis which has been developed with regard to *caloric*.

The above-mentioned occurrence made so deep an impression upon me, that I could hardly wait long enough to procure the necessary instruments before undertaking a number of successive experiments upon heat, in order to arrive at some conclusion with regard to its character, as well as to the manner of its operation.

I proposed, first of all, to undertake various experiments on what has since been called the *specific heat* of bodies. For this purpose, I procured from Mr. Fraser, New Bond Street, London (now physical and mathematical instrument maker to the King of England), a considerable number of solid balls of precisely the same diameter, namely, one inch. Some of these balls were of gold, some of silver; in short, they all were of one metal or another, or of some solid substance easily turned in a lathe. Each of these balls was suspended by a thin silken cord, and I proposed to heat the balls in certain liquids up to a given temperature, and then to plunge them into a known quantity of water which had been cooled in the same proportion. I drew this inference, — that the degree of temperature which the balls communicated to the known amount of water, as shown by the thermometer, would be more than sufficient to calculate therefrom the proportional amount of heat necessary to bring to the same temperature the balls and an equal quantity of water.

I had already begun upon these experiments, but before I could finish them the war made it necessary for me to go to America. These researches were therefore interrupted for several years; and when, after the peace of 1783, I returned to England, I learned that Wilkin, in Sweden, had already carried out exactly what I had proposed to myself. Since I had not the slightest occasion to doubt the accuracy of the experiments performed by this philosopher, I laid aside, as useless, the apparatus which I had designed for my own investigations.

In the following year I left England and went to Bavaria, where I was received into the service of the late Elector. I brought with me several instruments belonging to the above-mentioned apparatus, which are still to be seen in the museum of the military school in Munich.

For more than twenty years I have never in any of my writings mentioned either my project and the preparations made for carrying out experiments on this point, or the experiments I really made and which agree with those of Wilkin, simply because I hate, and always have hated, the character of a man who appropriates the discoveries of another. I speak of them now rather to convince the public that I have long thought about this subject, than from any motive which might perhaps have its origin in personal vanity.

My relations at the court at Munich, and that, too, with a prince who was much interested in the promotion of knowledge, afforded me during a period of four years abundance of leisure to pursue, almost without interruption, my physical investigations, and I employed this leisure in making a considerable number of experiments on heat.

In the years 1785 and 1786 I was occupied in researches as to the manner in which heat passes through various substances and communicates itself still farther. A detailed description of these experiments is to be found in the two papers which I inserted in the Philosophical Transactions of the Royal Society of London.[2] The first is in the seventy-sixth, the other in the eighty-third, volume of this work. For the latter I received the gold medal which this Society is accustomed to confer annually.*

In the summer of 1785 I discovered that heat could be transmitted through, or excited in, a Torricellian vacuum.

Since this discovery has contributed not a little towards strengthening me in the opinion which I have since adopted with regard to the real character of heat, I do not consider it at all superfluous to give here, with all its details, an account of the experiment by which this fact was established beyond doubt. This experiment was conducted as follows.

After a skilful workman in Mannheim, Artaria by name, had succeeded in fixing firmly the globular bulb of a mercurial thermometer, half an inch in diameter, in the centre of another glass bulb an inch and a half in diameter, the space between the outer surface of the thermometer bulb and the inner surface of the outside ball, or the *globe*, was filled with mercury by means of a barometer tube which was soldered to a small hollow tube or point projecting outwards from the globe. This projection extended downwards when the thermometer fastened to the globe was in its natural upright position.

As soon as the vacant space inside of the globe and around the thermometer bulb, as well as the barometer tube (thirty-six inches in length), was filled with mercury, the end of the tube was dipped into a vessel of mercury; the tube was then inverted and brought into a perpendicular position, so that the globe in which the thermometer was fastened was at the top.

Since the instrument was converted in this way into a true barometer, the mercury in the globe and in the upper part of the barometer tube fell until the upper surface of the mercury in the tube was twenty-eight inches above the surface of the mercury in the vessel, where it remained at rest, being kept at this height by the pressure of the outside air. A lighted wax-candle was now held at the upper part of the tube where it entered the globe, and where the diameter of the tube had previously been contracted, and the flame was directed, by means of a blow-pipe, against that part of the tube which it was desired to melt together.

As the glass was softened by the heat, the pressure of the outside air immediately forced the walls of the tube together; the whole operation was successful.

The barometer tube was then detached, and the bulb of the thermometer was now surrounded on all sides by a vacuum, as may be seen from the figure on the opposite page. The thermometer was filled with mercury,

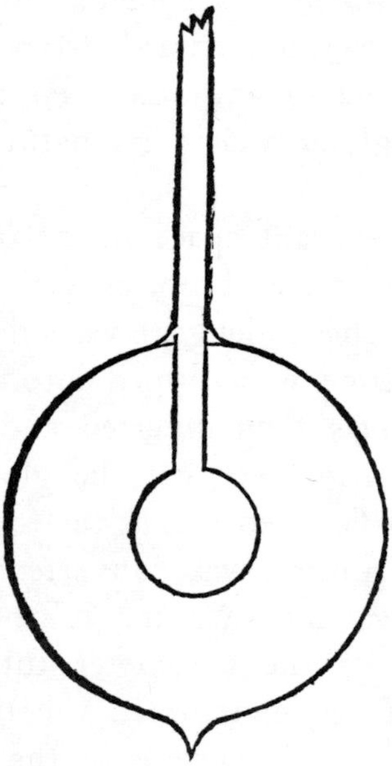

and provided with a scale, and I could then scarcely master my impatience and wait for the time when I should satisfy myself whether heat would be able to pass through this vacuum.

I now put the apparatus into a vessel filled with water at 18° Reaumur, and left it there until I was sure (from the scale of the instrument) that the bulb filled with mercury, which was in the centre of the vacuum, had reached this temperature of 18 degrees. I then took the instrument out of this vessel, and held it for some minutes in another full of hot water, which was kept constantly boiling by a lamp placed under it.

Since the mercury in the tube of the thermometer began to rise, although slowly, there remained no longer any doubt that the heat of the boiling water really passed through the vacuum into the bulb of the thermometer.

The mercury in the thermometer rose in the following manner: After the instrument had remained in the boiling water 1 min. 30 sec. the mercury had risen from 18° to 27°. After the lapse of 4 minutes, it had risen to $44\frac{9}{10}$°, and at the end of 5 minutes to $48\frac{1}{5}$°.

In order to estimate more accurately the relative rapidity with which heat passed through a vacuum and through air, I broke off the end of the small pointed tube which projected from the under side of the globe so that the air could freely enter the globe; I then melted the tube together a second time, by means of a candle; cooled my apparatus in water, and plunged it, as soon as it had acquired the temperature of this water, that is 18°, again into boiling water. The mercury rose much more rapidly than in the preceding experiment.

The manner in which the temperature gradually increased in both experiments is shown in the following table.

When the spherical reservoir of the mercurial thermometer, which was fastened in the centre of a glass globe an inch and a half in diameter, was plunged into boiling water, the times of ascent were as follows: —

| | In a Torricellian vacuum. (Exp. No. 1.) | | Surrounded by air. (Exp. No. 2.) | |
|---|---|---|---|---|
| | Time elapsed. | Heat acquired. | Time elapsed. | Heat acquired. |
| Upon being plunged into boiling water | | 18° | | 18° |
| | m. s. | ° | m. s. | ° |
| After remaining in it | 1 30 | 27 | 0 45 | 27 |
| | 4 0 | $44\frac{9}{10}$ | 2 10 | $44\frac{9}{10}$ |
| | 5 6 | $48\frac{1}{5}$ | 5 0 | $60\frac{9}{10}$ |

From the results of these experiments it is evident that the heat increases nearly twice as fast when the bulb is surrounded by air as when it is in a vacuum.

I afterwards performed other experiments of the same

kind without discovering the least difference from those mentioned above. It would take too much time and space to describe them all here. They are to be found, however, in my memoir in the Philosophical Transactions and in my eighth Essay.[3]

I had subsequently several instruments of the same sort made, in order to repeat and vary my experiments. Sometimes I observed the time which they took in cooling, sometimes that necessary for the heat to penetrate them. Sometimes I performed the experiment in the open air, sometimes in water. All these experiments gave the same result, namely, that the thermometer bulb in a vacuum became warm or cold as the case might be, the only difference being that it always took nearly twice as long to effect this change of temperature as was required when the bulb was surrounded by air.

The passage of heat through a vacuum was a fact of such importance in the investigation of the nature of heat, that I wished to confirm it by experiments which would not allow a shadow of doubt.

That part of the thermometer tube which was inserted in the glass globe was in contact with this globe. Hence the thought might suggest itself that a part of the heat received or given out by the thermometer bulb, which was surrounded by the vacuum, was communicated by means of the tube of the thermometer, since a portion of this tube was surrounded by air or water in which the heating or cooling was effected. In order to be fully satisfied as far as this circumstance was concerned, it occurred to me to repeat the experiment with a thermometer suspended by a very fine silken thread in the middle of a glass body of such size that the thermometer with its tube was entirely contained in it.

This glass body was then voided of air by means of mercury.

The results of the experiments performed with these instruments differed little or not at all from those made with the apparatus previously described, therefore the fact of the transmission of heat through the Torricellian vacuum was established beyond any doubt.

These results are sufficiently known to the learned world; now the question arises as to how these results can be reconciled with the theory which at the present day has been adopted in regard to *caloric*. I must confess freely, that, however much I might desire it, I never could reconcile myself to it, because I cannot by any means imagine how heat can be communicated in two ways entirely different from each other.

Philosophers have made little or no mention of the results of these investigations: I do not assume to explain their silence; if I myself mentioned them as little as they, it is easy to imagine the cause of my silence. It will at least be admitted that I have pointed out plainly enough the doubts which the results of my experiments could give rise to.

I afterwards undertook many other experiments to determine accurately the various degrees of rapidity with which heat passes into mercury when surrounded by common or atmospheric air, by air saturated with moisture, by carbonic acid gas, and by air brought to various degrees of density.

In the year 1787 I made a series of experiments which are described in the Philosophical Transactions for 1792;[7] my principal object was to investigate the conducting power with regard to heat possessed by various substances, especially by those which we are accustomed to use for clothing. The instrument which I used in these experiments, and which I called a *passage-thermometer*, differs but slightly from that described above. I fixed the bulb of a mercurial thermometer half an inch in diameter within a glass globe an inch and a half in diameter, with a long cylindrical neck; I then filled the space between the outer surface of the thermometer bulb and the inner surface of the glass globe with a certain quantity of the substance whose conducting power was to be determined, and allowed the instrument to cool in a mixture of pounded ice and water. As soon as the thermometer showed me that its bulb (which was in the middle of the glass globe) had acquired and retained constantly the temperature of the cooling mixture (that is, 0° of Reaumur's scale), I took the apparatus out of this cold mixture, plunged it into boiling water, observed the times required for the heat to pass into the bulb of the thermometer through the surrounding substance, and inserted them in a table, noting every ten degrees as accurately as possible.

Since the water into which I plunged my apparatus was kept constantly boiling, it is evident that the outside of the instrument, that is, the outer surface of the globe, was always of the same temperature; hence the more or less rapid heating of the thermometer bulb within the globe indicated the resistance which the covering of the bulb offered to the passage of the heat from the inner surface of the globe to the bulb of the thermometer.

In this way I made several experiments; but as I was inconvenienced by the steam rising from the boiling water, and so experienced difficulty in noting the rising and falling of the mercury, I changed my method of operation, and no longer observed the time necessary for the instrument to grow warm, but that necessary for it to grow cold.

When, therefore, my apparatus, plunged in boiling water, had acquired such a temperature that the mercury had reached 77° of Reaumur's scale, I took it out of the boiling water and held it in the air, over the large vessel filled with pounded ice and water, ready to plunge it into this cooling mixture the very moment that the mercury had fallen to 75°.

As soon as the mercury had reached this division of the scale, I plunged my apparatus immediately into the cooling mixture, and holding at the same time at my ear a watch which beat half-seconds (which I carefully counted), I waited for the moment when the mercury had fallen to 70°. I then noted and recorded the time elapsed, and in the same way observed the time when the mercury had fallen to 60°, and thus proceeded, noting every ten degrees, until the apparatus had cooled to the temperature of 10°.

Sometimes the apparatus cooled to such an extent that the mercury in the thermometer stood at 0°; this, however, took up much time, and was attended with no particular advantage, as the determination of the times taken up in cooling from 70° to 10° was quite sufficient for calculating the conducting power of every sort of covering; on this account I generally ended the experiment when the mercury had just passed the 10° mark on the scale.

During the time of cooling the apparatus in ice and water, I moved it about in the mixture very slowly and constantly from one place to another; moreover, I always mixed the water with such a quantity of ice that the temperature of this mixture remained constant.

Since in such experiments the thermometer bulb in the middle of the glass globe was entirely surrounded as well by the air contained in the globe as by the substances of which the covering consisted, I made a few experiments to determine the time necessary for the bulb of the thermometer to become cold again when the globe contained nothing but air. I thus learned that when the apparatus previously warmed in boiling water was plunged into the mixture of cold water and pounded ice, it required 576 seconds to cool from 70° to 10° Reaumur.

The following table contains the results of several experiments undertaken with a view to determine the relative warmth of various substances such as are commonly used for clothing.

I only remark, in addition, that I always determined the amount of the substance by weight (16 grains standard weight), and endeavoured to distribute it as equally as possible in the globe, and in such a manner that the bulb of the thermometer was surrounded by it.

| Gradual Loss of Heat. | Substances used for Covering. | | | | | | | |
|---|---|---|---|---|---|---|---|---|
| | Air. | Raw Silk. | Sheep's-wool. | Cotton-wool. | Fine Lint. | Beaver's Fur. | Hare's Fur. | Eider-down. |
| | Time elapsed | Time elapsed. | Time elapsed | Time elapsed. | Time elapsed. | Time elapsed. | Time elapsed. | Time elapsed. |
| From 70° to 60° | 38′ | 94″ | 79″ | 83″ | 80″ | 99″ | 97″ | 98″ |
| 60 50 | 46 | 110 | 95 | 95 | 93 | 116 | 117 | 116 |
| 50 40 | 59 | 133 | 118 | 117 | 115 | 153 | 144 | 146 |
| 40 30 | 80 | 185 | 162 | 152 | 150 | 185 | 193 | 192 |
| 30 20 | 122 | 273 | 238 | 221 | 218 | 265 | 270 | 268 |
| 20 10 | 231 | 489 | 426 | 378 | 376 | 478 | 494 | 485 |
| From 70° to 10° | 576 | 1284 | 1118 | 1046 | 1032 | 1296 | 1315 | 1305 |

In order to determine what influence the *density* of a covering or clothing of a given thickness exerted on the warmth of this covering or on its power to confine heat, I made three consecutive experiments with different quantities of one and the same substance, namely, with eider-down. For the first experiment I took 16 grains of this substance, for the second 32 grains, and for the third 64 grains. In all cases I used the same apparatus, so that the thickness of the covering always remained the same.

The results of these three experiments are contained in the following table.

| Loss of Heat. | The covering of Eider-down consisted of the following quantities of the substance. | | |
|---|---|---|---|
| | 16 grains. | 32 grains. | 64 grains. |
| | Time elapsed. | Time elapsed. | Time elapsed. |
| From 70° to 60° | 97″ | 111″ | 112″ |
| 60 50 | 117 | 128 | 130 |
| 50 40 | 145 | 157 | 165 |
| 40 30 | 192 | 207 | 224 |
| 30 20 | 267 | 304 | 326 |
| 20 10 | 486 | 565 | 658 |
| From 70° to 10° | 1304 | 1472 | 1615 |

Having convinced myself by these experiments that the *density* of any covering or clothing exercises a very considerable influence on its power to confine heat, its *thickness* remaining the same, I now sought to discover what effect the internal structure or constitution of the covering has on this power, its mean density and its thickness remaining the same.

By the expression *internal structure* I mean the state of division, whether fine or coarse, of the substance of which the covering consists, in the space which it occupies. This substance may be very fine and of delicate texture, and may be equally distributed through the whole space occupied by it, — as raw silk, for example; or it may be coarser and have larger interstices, — as, for example, a covering consisting of bits of stout sewing-thread, or one consisting of ravellings of cloth.

If heat really passed *through* the substances of which the covering is made, and if the efficiency of such a covering in restraining the same depended solely on the greater or less difficulty which the heat meets in passing through the solid parts of the covering, in that case the warmth of a covering would be, *cæteris paribus*, the same as that of the raw materials employed in its construction. It is evident, however, from the foregoing experiments, as well as from those to be detailed hereafter, that heat is not propagated in any such manner.

In one of my previous experiments I had endeavoured to determine the warmth of 16 grains of raw silk, which I had distributed equally in a certain space about the bulb of a thermometer. I now repeated this experiment twice, but with this difference: the first time I surrounded the bulb of the thermometer with 16 grains of a sort of lint made from a piece of white taffety; the second time with 16 grains of white sewing-silk, cut into small pieces, two inches long. The results of these experiments are recorded in the following table. My apparatus was warmed in boiling water, and then cooled in a mixture of water and pounded ice.

| Loss of Heat. | Substances of which the Covering consisted. | | |
|---|---|---|---|
| | Raw Silk, 16 grains. | Ravellings of Taffety, 16 grains. | Silk Threads, 16 grains. |
| | Time elapsed. | Time elapsed. | Time elapsed. |
| From 70° to 60° | 94″ | 90″ | 67″ |
| 60 50 | 110 | 106 | 79 |
| 50 40 | 133 | 128 | 99 |
| 40 30 | 185 | 172 | 135 |
| 30 20 | 273 | 246 | 195 |
| 20 10 | 489 | 427 | 342 |
| From 70° to 10° | 1284 | 1169 | 917 |

Having convinced myself by these experiments that the fineness of the particles or fibres of the substance used as a covering contributes very much to the warmth of the same, I made the following experiments to determine what effect the condensing of the covering would have, the quantity of matter of which it was composed remaining the same, but the thickness being decreased.

As I had already, by means of the foregoing experiments, determined the warmth of coverings of raw silk, wool, cotton, and linen when taking 16 grains of each substance, and making thereof, about the bulb of a thermometer, a globular covering half an inch thick, I now took 16 grains of moderately coarse threads of each of these four substances, and with them I made four new experiments.

Instead of filling with these threads the entire space between the bulb of the thermometer and the inner surface of the globe, in the middle of which was the bulb, I wound it around the bulb of the thermometer, so that the latter looked exactly like a little ball.

I now introduced, as before, the thermometer bulb

thus enveloped, into the middle of a glass globe an inch and a half in diameter; to this globe was attached a *neck* ten inches long, and of such a width as to allow of the insertion of the bulb of the thermometer wrapped up as described above, together with the attached scale.

The results of these four experiments may be seen in the following table; and that they may the more easily be compared with those made with the same quantity of the substances, but differently disposed, I have placed side by side the results of the comparative experiments.

| Loss of Heat. | The Bulb of the Thermometer was covered with 16 grains of one of the following substances. | | | | | | | |
|---|---|---|---|---|---|---|---|---|
| | Silk. | | Wool | | Cotton. | | Linen. | |
| | Raw. | In threads. | Raw. | In threads. | Raw. | In threads. | Raw. | In threads. |
| From 70° to 60° | 94″ | 46″ | 79″ | 46″ | 83″ | 45″ | 80″ | 46″ |
| 60 50 | 110 | 62 | 95 | 63 | 95 | 60 | 93 | 62 |
| 50 40 | 133 | 85 | 118 | 89 | 117 | 83 | 115 | 83 |
| 40 30 | 185 | 121 | 162 | 126 | 152 | 115 | 150 | 117 |
| 30 20 | 273 | 191 | 238 | 200 | 221 | 179 | 218 | 180 |
| 20 10 | 489 | 399 | 426 | 410 | 378 | 370 | 376 | 385 |
| From 70° to 10° | 1284 | 904 | 1118 | 934 | 1046 | 852 | 1032 | 873 |

It would carry me too far if I brought forward in detail all the experimental results obtained in my researches undertaken to investigate the manner in which heat propagates itself through the various coverings. In my printed memoirs I have said all upon this subject that can with reason be said. For the present I have indicated clearly, not only the course upon which I entered at the very beginning of my researches, but also the object I had in view. Philosophers may decide whether this course was the right one, and whether I pursued it with zeal and perseverance.

The few remarks and observations which follow were occasioned by my researches made at that time.*

All the different substances which I had yet made use of for covering the bulb of the thermometer (which was contained within a glass globe an inch and a half in diameter) had in a greater or less degree confined the heat and prevented it from passing into or out of the bulb of the thermometer as rapidly as it would otherwise have done. Here then arose the important, and as yet unanswered question, how and by what mechanical operation had the coverings in question produced these effects?

This much is certain, that the slowness of the cooling of the bulb of the thermometer cannot by any possibility be a result of the non-conducting powers of those substances of which the coverings consisted, considered simply as having hindered the passage of the heat, for if, instead of regarding them merely as bad conductors of heat, we were to suppose them to have been totally impervious to heat, still their volumes — that is, the sum of all their solid parts or fibres — would be so inconsiderable in proportion to the space they occupied, that they would either have produced no effect on the air filling their interstices, or this air would have been sufficient of and for itself to have conducted all the heat communicated in less time than was actually taken up in the experiments. Here is the proof of this statement.

The diameter of the glass globe being 1.6 inches, its contents amounted to 2.14466 cubic inches. The di-

* See my eighth Essay.[3]

ameter of the thermometer bulb was 0.55 of an inch, and its contents 0.08711 of a cubic inch. Taking now from the contents of the globe (2.14466 cubic inches) the contents of the thermometer bulb (0.08711 of a cubic inch), there remain 2.05755 cubic inches as the measure of the space occupied by the substances by which the bulb of the thermometer was surrounded.

Although the above-mentioned substances *occupied* this space, they were very far from *filling* it, as will be observed without my calling attention to the fact; on the contrary, this space contained a large quantity of air, which occupied and filled the small interstices of the substances in question.

For example, in one of the experiments the bulb was covered with 16 grains of raw silk. As I had already learned from experiment that the specific gravity of the silk was to that of water as 1734 to 1000, it follows that the volume of 16 grains of silk was equal to the volume of 9.4422 grains of water. Further, as 1 cubic inch of water weighs 253.185 grains, it follows incontrovertibly that the space occupied by 9.4422 grains of water can be reckoned at the highest at 0.037294 of a cubic inch, and this amount of water (9.4422 grains) corresponds in volume to 16 grains of silk.

We know, however, that the space which this small quantity of silk (0.037294 of a cubic inch) occupies is 2.05755 cubic inches; hence it appears that, since 0.037294 is to 2.05755 as 1 is to 54, the silk which I used in the experiment in question could not fill more than $\frac{1}{55}$ of the space in which it was confined.

The longer we meditate upon these investigations, the more we are struck by the importance of the results that follow from them. I have never been able to explain them without rejecting altogether that hypothesis according to which it is supposed that the heat which may be in the air is communicated directly from one particle of this fluid to another.

My researches on the propagation of heat in liquids are sufficiently well known.* From them it has probably been seen how and in what manner I was compelled by the results of my numerous experiments to adopt the opinion with regard to this subject which I have developed in my various writings.

I have examined with the greatest care the objections which have been offered to the deductions which I have drawn from my experiments, and I can assert with truth — and to say this is a duty I owe to myself — that neither in these objections nor in the result of any new experiment, as far as my knowledge extends, has the least thing occurred which could serve as a reason for altering my opinion in regard to this subject. In a paper which I sent last year to the Royal Society at London,† I think that I have proved that water is really a non-conductor of heat, as I suspected six years ago.

I have now only a few words to say in addition, about the various experiments which I made at different times, to enable me (if it were in any way possible) to answer decisively that important and much contested question as to the materiality of heat, about which philosophers have striven for so long a time.

Those who regard heat as a substance must, of necessity, assume that it possesses weight. If now the

* A detailed description of my investigations in regard to this interesting subject is contained in my seventh Essay, which appeared in London in the year 1797, in two parts, together 188 octavo pages.[2]

† See p. 274.

heating of a body is caused by the accumulation of this substance in the body, it follows naturally that the body must be heavier when it is warm than when it is cold. Some natural philosophers have sought to determine this point; I feel confident, however, that no one has made more decisive experiments in this direction than myself.*

I was provided with excellent instruments, and spared neither trouble nor expense to arrive, by means of my experiments, at a certain and convincing result. The results obtained are, in few words, as follows.

I had a ball of very fine gold made, and weighed it when perfectly cold, and again after heating it to such a temperature that it was on the point of melting. Further, I weighed a considerable amount of water, which I had sealed hermetically in a flask, first in its liquid state, then at the temperature of melting ice, then as actual ice, and then again at its original temperature. All these experiments convinced me that the weight of a body is not changed in the least by heat.

Now although, as a consequence of the results of these experiments, I was only still more strengthened in those doubts which a number of other natural phenomena had raised in my mind with regard to the existence of caloric, still I saw at the same time only too well that the essential point of the controversy was far from being decided thereby. The defenders of caloric would still object (as they have actually done) that this substance is far too subtile to be weighed upon our ordinary balances.

* A paper in which are described in detail all my experiments upon this subject may be found in the Philosophical Transactions for 1799.[12]

After I had long meditated upon a way of putting this interesting problem entirely out of doubt by a perfectly conclusive experiment, I thought finally that I had discovered it, and I think so still.

I argued that if the existence of caloric was a fact, it must be absolutely impossible for a body or for several individual bodies, which together made one whole, to communicate this substance continuously to various other bodies by which they were surrounded, without this substance gradually being entirely exhausted.

A sponge filled with water, and hung by a thread in the middle of a room filled with dry air, communicates its moisture to the air, it is true, but soon the water evaporates and the sponge can no longer give out moisture. On the contrary, a bell sounds without interruption when it is struck, and gives out its sound as often as we please without the slightest perceptible loss. Moisture is a substance; sound is not.

It is well known that two hard bodies, if rubbed together, produce much heat. Can they continue to produce it without finally becoming exhausted? Let the result of experiment decide this question.

It would be too tedious to describe here in detail all the experiments which I undertook with a view of answering in a decisive manner this important and disputed question. They may be found in my memoir On the Source of Heat excited by Friction. I have had it printed in the Philosophical Transactions for the year 1798; still these experiments bear too close a relation to my later researches on heat for me to omit attempting at least to give the reader a clear idea of the experiments and of their results.

The apparatus which I used in these investigations

is too complicated to be represented in this place; still it will not be difficult for the reader, with the help of the accompanying figure (see Plate V.), to form a conception of the principal experiments and their results.

Let A be the vertical section of a brass rod which is an inch in diameter and is fastened in an upright position on a stout block, B; it is provided at its upper end with a massive hemisphere of the same metal, three and a half inches in diameter. C is a similar rod, likewise vertical, to the lower end of which is fastened a similar hemisphere. Both hemispheres must fit each other in such a way that both the rods stand in a perfectly straight vertical line.

D is the vertical section of a globular metallic vessel twelve inches in diameter, which is provided with a cylindrical neck three inches long and three and three-quarters inches in diameter. The rod A goes through a hole in the bottom of the vessel, is soldered into the vessel, and serves as a support to keep it in its proper position.

The centre of the ball, made up of the two hemispheres which lie the one upon the other, is in the centre of the globular vessel, so that, if the vessel is filled with water, the water covers the ball as well as a part of each of the brass rods.

If now the hemispheres be pressed strongly together, and at the same time the rod C be turned, by some means or other, about its axis, a very considerable quantity of heat is generated by means of the friction which takes place between the flat surfaces of the two hemispheres.

*Plate V.*

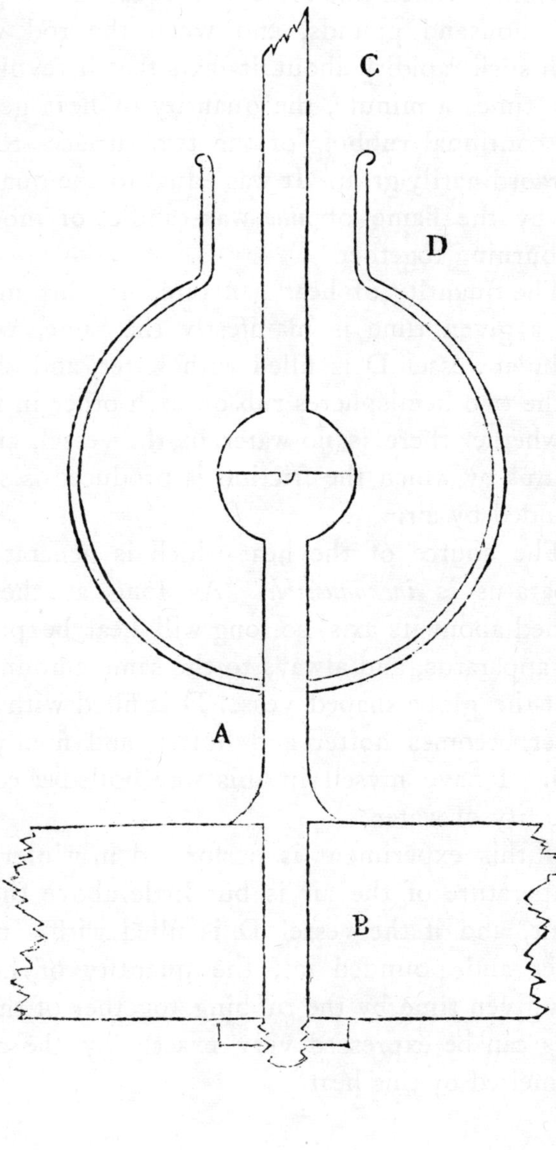

The quantity of the heat excited in this manner is exactly proportional to the force with which the two surfaces are pressed together, and to the rapidity of the friction. When this force was equal to the pressure of ten thousand pounds, and when the rod was turned with such rapidity about its axis that it revolved thirty-two times a minute, the quantity of heat generated by the continual rubbing of the two surfaces together was extraordinarily great. It was equal to the quantity given off by the flame of *nine* wax-candles of moderate size all burning together.

The quantity of heat generated in this manner during a given time is manifestly the same, whether the globular vessel D is filled with water, and the surfaces of the two hemispheres rub on each other in this liquid, or whether there is no water in the vessel, and the apparatus by which the friction is produced is simply surrounded by air.

The source of the heat which is generated by this apparatus is *inexhaustible*. As long as the rod C is turned about its axis, so long will heat be produced by the apparatus, and always to the same amount.

If the globe-shaped vessel D is filled with water, this water becomes hotter and hotter, and finally begins to boil. I have myself in this way boiled a considerable quantity of water.

If this experiment is performed in winter when the temperature of the air is but little above the freezing-point, and if the vessel D is filled with a mixture of water and pounded ice, the quantity of heat caused in a given time by the rubbing together of the two surfaces can be expressed very exactly by the amount of ice melted by this heat.

Since the apparatus affords heat continuously, and always to the same amount, we can melt in this way as much ice as we please.

But whence comes this heat? This is the contested point, to determine which was the real aim of the experiment.

It is certain that it comes neither from the decomposition of the water nor from the decomposition of the air Various experiments on this point, which I have described at length in my memoir in the Philosophical Transactions, are more than sufficient to establish this fact beyond doubt.

Just as little does it come from a change in the capacity for heat brought about by friction in the metal of which the hemispheres are composed. This is shown, first, by the continuance and uniformity of the production of the heat; and, secondly, by an experiment bearing directly on this point, by which I am convinced that not the slightest change had taken place in the capacity of the metal for heat.

Just as little does it come from the rods which are attached to the hemispheres, for these rods were always warm, the hemispheres communicating heat to them.

Much less could this heat come from the air or the water immediately surrounding the hemispheres, for the apparatus communicated heat to both these fluids without cessation.

Whence, then, came this heat? and what is heat actually?

I must confess that it has always been impossible for me to explain the results of such experiments except by taking refuge in the very old doctrine which rests on the supposition that heat is nothing but a vibratory motion taking place among the particles of bodies.

A bell, on being struck, immediately gives forth a sound, and the oscillations of the air produced by these vibrations forthwith cause a quivering motion in those bodies with which they come in contact. On the other hand, a sponge filled with water cannot give off its moisture to the bodies in its vicinity for any length of time without itself losing moisture.

A very illustrious philosopher, for whom I have always entertained the greatest respect, and whom, moreover, I have the good fortune to count among my most intimate friends, M. Bertholet, has, in his admirable *Essai de Statique Chimique*, attempted to explain the results of this investigation, and to reconcile them with that theory of heat which is founded upon the hypothesis of caloric.

If a man as learned, as honest, as worthy, and as renowned as is M. Bertholet, spares no pains in opposing the errors of a natural philosopher or chemist, one cannot and dare not keep silence unless he wishes to acknowledge himself vanquished. If, however, one can produce proofs — a fortunate thing for all those who find themselves driven to similar self-vindication — that the objections of M. Bertholet have no foundation, he has done very much towards establishing beyond doubt the opinions and facts in question.

I will now endeavour to answer the objections which M. Bertholet has offered to my explanation of the above-mentioned experiments; and, that the reader may be in a position to give to these objections their just value, I will insert them here in the writer's own words.

"Count Rumford has made a curious experiment with regard to the heat which may be excited by friction. He causes a blunt borer to revolve very rapidly (*this borer revolved about its axis only thirty-two times a minute*) in a brass cylinder weighing thirteen pounds, English weight (*the cylinder weighed one hundred and thirteen pounds and somewhat more*), and says that he observed that this borer in the course of two (*one ana a half*) hours, and under a pressure equal to 100 cwt., reduced to powder 4145 grains (8½ ounces Troy) of brass, and that an amount of heat was generated during this operation sufficient to bring to boil 26.38 pounds of water, previously cooled to the freezing-point. He asserts that he did not discover the slightest difference between the specific heat of the metallic dust and that of the brass which had not experienced the friction. Hence he supposes that the heat was excited by the pressure alone, and was not at all due to caloric, as is the opinion of most chemists.

"I will for the present satisfy myself with simply inquiring whether it necessarily follows from this experiment that we must renounce entirely the received theory of caloric, according to which it is regarded as a substance which enters into combination with bodies, or whether this result cannot be explained in a satisfactory manner by applying to the case in question those laws of nature in accordance with which the operations of heat are manifested under other conditions.

"If the evolution of heat be regarded as a consequence of the decrease of volume caused by the pressure, then not only the metallic powder but also all the rest of the brass cylinder must have contributed, though not in an equal manner, to this evolution, by the powerful expansive effort of that portion which experienced the greatest pressure, and consequently acquired the greatest temperature, without being able to assume the dimensions proper to this same temperature on account of the less heated and less expanded parts; consequently there must have arisen, necessarily, a certain condensation of the metal in respect of its natural dimensions, which condensation gradually decreased from the point where the pressure was greatest to the surface. We may suppose that this operation took place in a similar manner in all parts of the cylinder.

"As a consequence of this decrease of volume, an amount of caloric was given out equal to that which would have caused a similar increase of volume, on the supposition, that is, that the specific heat of the metal does not change through this range of the scale of the thermometer, and that the expansions are equal; and this, considering the range of temperatures and the consequent expansions, is probably not far from the truth. The entire amount of heat disengaged would have raised the cylinder to about 180° of Reaumur's scale; and if the expansion of brass by heat is equal to that of iron, which has been found to be $\frac{1}{75000}$ for each degree of the thermometer, the 180 degrees would have caused an expansion of $\frac{18}{7500}$ in each direction, and the decrease of volume must have brought about the same degree of heat if we suppose that the pressure stood in equal relation to this expansion.

"Now there is a change, and sometimes a very considerable one, wrought in the specific gravity of a metal, by percussion, by the action of a fly-wheel, or by the compression of a wire-drawing machine. It appears, for example, that the specific gravity of platina and of iron, on being forged, is thus increased by a twentieth part.

"Hence it appears that the experiment of Count Rumford is far from explaining satisfactorily a property which is well known, and called in question by no one.

"It is easy, it is true, to arrange side by side in an imposing manner the phenomena of heat; if, however, you were to say to one who has little or no knowledge of chemical speculations, 'Count Rumford's cylinder has, in the course of two hours, by means of a violent friction, afforded all the heat required to dissolve in water, without changing its temperature, 15 kilogrammes of ice, or as much as 2 hectogrammes ($6\frac{1}{2}$ ounces) of oxygen would require [*sic*] in its combination with phosphorus,' I do not know at which of these phenomena he would be most astonished.

"The slight changes which can take place in the amount of combined caloric have so inconsiderable an influence on the capacity for work of the caloric within the narrow limits of the thermometric scale, that it cannot be computed. Moreover, we have not, as yet, adequate data for determining the nature of the changes in this respect which take place in a solid body in consequence of the particular condition of condensation into which it has been brought by means of a certain mechanical force, and by degrees of heat differing greatly from each other.

"Besides, Rumford, in the experiment to determine the specific heat of the filings of bell-metal thus obtained, heated them to the temperature of boiling water. But this extremely elastic metal would very naturally as soon as left to itself, and especially during the operation just mentioned, resume that state of expansion and that capacity for heat which is proper to it at a given temperature, so that the effect of the pressure to which it has been subjected partly disappears again, just as a piece of metal which has been hammered resumes its natural properties on being annealed."

In reply to these remarks, I will call to mind what follows.

1st. The discovery which I made, that no considerable change had taken place in the specific heat of the metallic dust produced by the friction, led me in no way to the supposition that the heat excited in the experiment could not come from the caloric set free. I only found that the source of this heat was inexhaustible. To explain this phenomenon, which has never yet been explained, is the point now in question, and I do not see how it can be explained except by giving up altogether the hypothesis adopted in regard to caloric.

2d. If we actually suppose (and it is far from having been proved) that the simple pressing together of a metal is sufficient to expel the caloric contained in it, still the explanation of such a natural phenomenon would be advanced little or none; for since the action of the force which causes the pressure is continuous, the condensation of the metal brought about by this force would in a short time reach its maximum; and if really in this operation ever so much caloric had been disengaged from the metal, still it would very soon disperse. The rubbing surfaces, on the contrary, continue to give forth heat, and that always to the same amount.

3d. In regard to the objection made to the experiment which was undertaken with a view of determining whether a change had taken place in the capacity of the metallic dust for heat, this can very readily be answered, and in such a way that nothing, it seems to me, can be said against it. If the temperature of boiling water were really sufficient to give to these small, forcibly condensed particles of metal the quantity of heat necessary to bring them back to their original condition as far as their capacity for heat is concerned, then, as the water by which the apparatus was surrounded finally began to boil, they must, without doubt, have taken the necessary amount of heat from this water. If, now, these particles of metal received finally from the water the caloric which in the beginning they imparted to it, the question arises, whence came the caloric which served to heat, not only the water, but also the metal and the objects immediately surrounding it?

I am far from desiring to deceive any one by an imposing arrangement of facts; but the facts in my experiments were so very striking that it was altogether impossible for me to help instituting comparisons and making calculations with regard to them which would make them clear, especially to those not yet sufficiently acquainted with such investigations.

I will now close my remarks with an entirely new computation. I will show whether it is probable that the metal could supply all the heat which was produced by friction in the experiment in question. If we are to make this supposition, we must, in the first place, allow that all the heat came directly from the particles of metal which were separated from the solid mass of metal by the friction; for, since the mass remained in the same condition throughout the entire experiment, it is evident that it could contribute in no measure to the effect produced.

We will now inquire how much heat would have been developed if the experiment had been carried on without cessation, until the whole mass of metal had been reduced to powder by the friction.

After the experiment had lasted an hour and a half, there were 4145 grains (Troy) of the metallic dust, and during that time an amount of heat was produced by the friction sufficient to raise 26.58 pounds of ice-cold water to the boiling-point.

Since the mass of metal weighed 113.13 pounds, or 791,910 grains, all this metal would have been reduced to powder if the experiment had lasted uninterruptedly, day and night, for 477½ hours, or for 19 days 21½ hours, and during this time an amount of heat would have been produced sufficient to have raised 5078 pounds of water to the boiling-point.

Since the metal used in this experiment showed a capacity for heat which was to that of water as 0.11 to 1, it is evident that this amount of heat would have been sufficient to raise a mass of the same metal 46,165 pounds in weight through 180 degrees of Fahrenheit's scale, or from the temperature of melting ice to that of boiling water.

This amount of heat would be sufficient to melt a mass of metal sixteen times heavier than that which I used in the experiment.*

* Brass melts at a temperature of 3807° Fahrenheit; copper at 4587°; bell-metal melts more easily than copper; if, however, we suppose that it requires the same heat for fusion, we find by a very simple calculation, that the amount of heat necessary to raise the temperature of 46,165 pounds bell-metal through 180 degrees would be sufficient to raise the temperature of 1811½ pounds through 4587

Is it at all conceivable that such an enormous quantity of caloric could really be present in this body? But even this supposition would be by no means sufficient for the explanation of the fact in question, as I have shown by a decisive experiment that the capacity of the metal for heat has not sensibly altered.

Whence, then, came the caloric which the apparatus furnished in such abundance?

I leave this question to be answered by those persons who believe in the actual existence of caloric.

In my opinion, I have made it sufficiently evident that it was impossible for it to come from the metallic bodies which were rubbed together, and I am absolutely unable to imagine how it can have come from any other object in the neighbourhood of the apparatus, for all these objects received their heat constantly from the apparatus itself.

I will now proceed to give an account of my further investigations on the subject of Heat.

In the summer of the year 1800, I visited Scotland, and on this occasion spent some months in Edinburgh.

It is well known that the University at that place stands in high repute on account of the eminent scholars occupying chairs there for more than fifty years in uninterrupted succession.

One day I found myself in the company of Professor Hope (the successor of the celebrated Black), Professors Playfair and Stewart, and several other persons. We repeated the experiment which Pictet undertook with a view to determine the condensation and contraction

degrees, or to bring this number of pounds to the melting-point. From this calculation it appears that a quantity of bell-metal, the temperature of which is at the melting-point of ice, on being reduced by friction to the state of powder, gives out sixteen times as much heat as would be necessary to melt it.

of air by the cooling influence of cold bodies. It now happened that for the first time my opinion on the subject of heat was publicly announced.

Two metallic mirrors fifteen inches in diameter, with a focal distance of fifteen inches, were placed opposite each other, sixteen feet apart. When a cold body (for example, a glass bulb filled with water and pounded ice) as was the case on this occasion, was placed in the focus of one of the mirrors, and a very sensitive air-thermometer was placed in the focus of the other mirror, the latter thermometer began immediately to fall. If, instead of being placed directly in the focus, the thermometer was removed a short distance from it to one side, the cooling power which in the former case the cold body had exerted upon it was no longer perceptible.

The matter was, however, not allowed to rest with merely repeating the experiment of Pictet just as he describes it, but I was allowed, in addition, to make various changes, that I might lay aside every doubt, and elucidate in the most convincing manner the fact in question.

I expressed my opinion on the results of these experiments in the following words: —

"It is not possible that caloric has an actual existence. The communication of heat and the communication of sound seem to be completely analogous. The cold body in one focus compels the warm body (the thermometer) in the other focus to *change its note.*"

It is owing to a peculiar circumstance, the further discussion of which would be neither appropriate nor useful in this place, that I here introduce word for word the expression which I used on this occasion.

A considerable time before, I had already projected

a series of experiments on the subject of radiant heat, and in my sixth Essay, which treats Of the Management of Fire and the Economy of Fuel, published at London in 1797, I had openly announced my purpose of taking the work in hand as soon as possible.

The experiments I have just mentioned as being performed in my presence by Professor Hope determined me not to put off this intention of mine a moment longer.

As soon as I returned to London, I began immediately to make all preparations for my researches. I therefore communicated my intentions to Sir Joseph Banks, at that time President of the Royal Society, also to Mr. Cavendish, because both these gentlemen (as well as myself) were managers of the Royal Institution. As I wished to carry out my experiments in the most decisive manner, and consequently with the apparatus as perfect as possible, — which to all appearance would require a considerable outlay, — I was, at my request, authorized by the managers of the Royal Institution to procure the new instruments needed at the expense of the Institution, with the condition, however, that these instruments should remain at the Institution as its property, and be kept in its cabinet.

As the principal object in this investigation was to establish beyond doubt the cooling emanations from cold bodies, I desired to accumulate the emanations and concentrate them as much as possible, in order that their action might be so much the more sensible.

Pictet took for his experiment, as is well known, two metallic reflectors, and placed a cold body in the focus of one of them, and a thermometer in the focus of the other. That the cooling influence which the cold object exerted on the thermometer might be doubled, I proposed in my experiment to have two cold bodies and one reflector, and, in order to increase so much the more the cooling effect on the thermometer, I intended to place it in the upper part of an open cylindrical vessel, the two cold bodies, however, being placed somewhat lower.

In Pictet's experiments both reflectors were in a horizontal line, and the thermometer on which the cooling influences were exerted was continually heated by the vertical current from the air above, which was caused necessarily by the cooling of the stratum of air immediately surrounding the thermometer; as a consequence, the frigorific influence of the cold body was lessened by the calorific influence of this current to such an extent that an equilibrium resulted. Still I expected that in my case I should, in all probability, be able to carry the cooling of the thermometer still farther, as I hoped by the arrangement of my apparatus to prevent this current, and at the same time to double the cooling effect.

After long delay on the part of the workmen, the necessary mirrors, four in number, were finally completed. They are now in the physical cabinet of the Royal Institution at London, and are used in the annual lectures on physics. If I am not mistaken, there are several other instruments kept in the same place, which I had expected to use in the projected experiments on the radiation of bodies; but most of the instruments designed for this investigation (made by Mr. Fraser, New Bond Street) were made at my own expense and are still in my possession.

It is only necessary to see this apparatus, which I had made in the summer of 1801, to be immediately convinced that I pursued my researches on the subject of heat zealously and connectedly.

In the beginning of the year 1802 I was recalled to Bavaria. I was, therefore, obliged to leave London in the early part of the month of May, after I had actually begun on a very few only of the experiments which I had planned with so much pains. But as I was firmly resolved to devote myself to them again, as soon as I could obtain any leisure, however little, I took back to Germany with me the greater part of this apparatus which I had procured during my stay in England.

During my journey, I remained three months at Paris, so that I did not reach Munich before the end of August. In the early part of the month of October, however, I began my experiments.

As I had not been able to bring with me from London the four large reflectors belonging to the Royal Institution, and as I could not procure similar ones in Bavaria, I was obliged to change the plan of my investigations, and to try whether it might not be possible to discover the radiation from bodies in some other way, and to make the effects of these radiations manifest without the aid of the concentration brought about by means of the metallic reflectors.

In the first experiments which I undertook, I had this object in view, to determine whether the invisible heating rays which a warm body (a heated stove, for example) gives out are not of the same character as those coming from the sun. For this purpose I procured three cylindrical boxes of very thin, soft wood, precisely alike, four and a half inches in diameter, three inches high, and open above. In each of these boxes, an inch and a quarter from the bottom, I put a circular metallic disk, a quarter of a line in thickness, and of the same diameter as the inside of the box. This disk, which formed a sort of optical screen in the inside of the box, was fastened in its place by a number of very short wooden pegs, which went through the side of the box.

In the middle of the bottom of the box was a circular aperture, three quarters of an inch in diameter, closed by a cork stopper.

In this stopper was a hole of three lines' diameter, into which fitted a small mercurial thermometer provided with an oval reservoir. The divisions of the scale were engraved upon the tube itself.* By means of this stopper the thermometer was introduced into the inside of the box in such a way that its bulb was situated in the axis of the box, and in the middle of the space between the bottom and the metallic disk. This space, which was designed to serve as a reservoir of heat, was filled with a certain quantity of flat silver threads, which had been picked out of old silver lace.

In one box the metallic disk or reflector was brass; in the second, tinned iron; and in the third, ordinary sheet-iron.

* I had four such thermometers made for me in England, and they did me good service throughout the whole course of my experiments on the subject of heat. Their tubes were made of very hard glass, three lines in diameter, and were polished down on one side so as to present a flat surface, on which the divisions of the scale were etched with fluoric acid. The tubes are six or seven inches long, and the bulbs for the mercury are pear-shaped, and consequently not so liable to get broken as cylindrical ones. In the pointed or lower part of the pear, the glass can be quite thick without any disadvantage.

The accompanying figure represents the vertical section of one of these boxes in a horizontal position. The stopper is also shown by diagonal lines, and a part of the thermometer in its proper place.

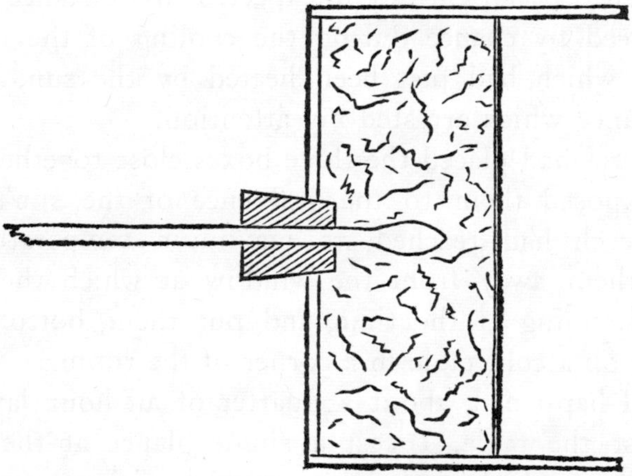

In order to diminish the loss of heat which might take place through the bottom and the sides of the box, each one was covered inside and outside with well-sized paper, then coated three times with copal varnish, and, in addition to this, they were covered during the experiment with an envelope of fur.

When one of the boxes was placed for a certain length of time in the sun, so that its rays fell vertically upon the metallic disk, there was a certain amount of heat excited in the same; and, as this heat was evenly distributed within the box by means of the metallic threads, it was possible to observe very exactly the various degrees of heat by means of the thermometer; if all three boxes were placed at the same time in the sun, it was possible to determine with certainty the relative amounts of heat excited at the surface of the three different metals used in the experiment.

I was not at all surprised to find that the rays of the sun excited more heat in a given time on the black and unpolished iron disk than on the other two disks, which were bright and polished; I was, however, all the more astonished by an entirely unexpected circumstance which I noticed by chance during the cooling of the instruments which had just been heated by the sun, a circumstance which arrested my attention.

After I had placed the three boxes close together, and had exposed them to the influence of the sun's rays until each had reached its maximum temperature, I took them away from the window at which they had been standing at the time, and put them, bottom upwards, on a cold table in a corner of the room.

As I happened, about a quarter of an hour later, to go past the table, I cast a single glance at the thermometers, which now, in a vertical position, projected from the reversed boxes. To my no slight astonishment, I saw that the box which before contained the most heat (the one which had the iron disk) was now the coldest of all.

This phenomenon surprised me so much the more, as I was convinced that this rapid cooling could not be due to the fact that this box did not have sufficient room to take up just as much heat as the others. For, as I very well knew how much in these experiments depended upon the boxes being precisely alike as regards their contents, I had taken the greatest pains by similar distribution of the silver threads to arrange them alike before I began my experiments.

I cannot allow myself here to give in detail all the conjectures and projected experiments of which this discovery was the cause. I will, therefore, say nothing further except that it made a firm and lasting impression on my mind, and afterwards exerted much influence on the manner in which I carried on my inquiries.

Meanwhile I did not allow this occurrence to hinder me in the least from carrying out to the end the experiments for which I had devised my apparatus. I had, therefore, a cylindrical iron stove put into the middle of a large room, and having surrounded it with fire-screens, I caused all the windows in the room to be opened. When the stove was sufficiently heated, I found that no sensible change had taken place in the mean temperature of the room. I now removed the screens which surrounded the stove, and placed all three of the boxes at the same time in the same position, that is, twenty-four inches from the stove.

The box containing the iron disk, which previously had contained the most heat after standing in the rays of the sun, was also now the warmest after being subjected to the influence of the rays which proceeded, although invisibly, from the stove.

In order to become more closely acquainted with these rays, I had several new instruments constructed; among others, four large air thermometers, and three other thermometers of the same size filled with spirit of wine. The bulbs of these thermometers were an inch and three quarters in diameter, and contained either various substances mixed with air or else simply spirit of wine. The bulb of the first thermometer was filled with air alone; in the second was a mixture of air and eider-down; in the third a mixture of air and very thin flat silver threads; the fourth contained air, eider-down, and, at the same time, flat silver threads.

In the bulb of the first of the thermometers filled

with spirit of wine, there was nothing besides this fluid, in the second there was a mixture of spirit of wine and eider-down, and in the bulb of the third thermometer there was a mixture of spirit of wine and flattened silver threads.

If, then, I exposed these thermometers in turn, now to the influence of the rays of the sun, and again to the influence of rays coming from bodies warmed by the fire, I could, from the rapid or gradual heating of the different thermometers, arrive at a sufficiently just conclusion with regard to the identity of the radiations or to the difference between them.

It would be too tedious if I were to describe these experiments here in detail. From some of them I obtained many, in certain respects, very remarkable results,* which allowed me to draw such conclusions as pointed out clearly enough the way in which I must proceed towards the chief object of my researches.

I afterwards procured other thermometers of very large size. Their bulbs are round, and are made of copper; they are four inches in diameter. Their tubes, which are glass, are thirty inches long, and are filled with linseed oil. I use them in experiments designed to determine the relative rapidity with which a warm body (the thermometer itself) cools in different liquids having the same temperature. This instrument is the

* I noticed, among other things, that the thermometer whose bulb contained a mixture of spirit of wine and flat silver threads was much more sensitive in general, and especially to very slight changes of temperature, than another thermometer of the same size, the bulb of which contained only spirit of wine. It would, perhaps, be of advantage to procure similar thermometers to use, if not ordinarily, at least on certain occasions. I am firmly convinced that a thermometer whose bulb is filled with mercury and platina cut into threads will be much more sensitive, that is, will indicate the temperature much more quickly, than a thermometer of the same size filled, as is usual, with mercury alone.

only one that I could ever devise for such experiments without fearing important objections on account of the apparatus employed.

I possess, also, various other thermometers, intended simply to receive and collect within themselves the calorific or frigorific rays which fall upon their surfaces. The reservoir of each consists of two cones of very thin sheet brass, which lie one within the other, and are fastened to each other, on the under side, in such a way that there is an empty space, not quite a line in width, between the inner surface of the outer cone and the outer surface of the inner cone. The inner cone is four inches in diameter, four inches high, and ends in a point above. The diameter of the outer cone is four and a quarter inches, and it ends above in a cylindrical tube three quarters of an inch in diameter and four inches long. In this cylinder is fixed a glass thermometer tube, and if the space between the two cones be filled with linseed oil or coloured spirit of wine, this instrument answers the same purpose as an ordinary thermometer. The scale of this thermometer is fastened firmly to this glass tube.

The outer wall of the instrument is shielded and protected from the calorific or frigorific influences of the surrounding air by means of a cylindrical box of dry wood, thickly coated with varnish, and filled with eider-down; into this the body of the instrument fits. This cover is four and a half inches high, and is, on the inside, of the same diameter as the lower part of the outer cone; and the tube of the instrument, with its attached scale, goes through a hole made in the bottom of the box.

If, now, the outer blackened surface of the inner

cone be held in the neighbourhood of and towards an object which is giving off calorific (or frigorific) rays, the heat (or cold) caused by these rays is communicated to the fluid contained in the space between the two cones, and this change of temperature brings about a corresponding change in the level of the liquid in the upper part of the tube; by this means the amount of heat (or cold) communicated can be estimated and measured.

An instrument of this description, which I procured in the year 1801, during my stay in England, is at present in the physical cabinet of the Royal Institution at London. Two similar ones, fitted up in Bavaria, are still kept in my cabinet at Munich. I have described this instrument thus minutely, simply because I am convinced that it is of very great service in experiments on the calorific and frigorific radiations from various bodies, and because it has been my earnest desire to induce natural philosophers to devote their attention to this subject, so worthy of investigation.

It only remains for me to say a few words in regard to the experiments which I have described very fully in the memoir read before the Royal Society on the 3d of February, 1804, which has been translated into French by Professor Pictet.[13]

I performed these experiments in Munich, in 1803, during the months of January, February, and March. According as the results seemed of importance, I immediately acquainted my friends in England and France with them. Among others, I communicated to Sir Joseph Banks, then President of the Royal Society of London, the very striking results of an experiment which I made, on the 11th of March, with two metallic vessels, both of which — one being naked, and the other having a covering of linen — I allowed to cool, exposed to the air, after having first filled them with warm water. In addition to this, I wrote him that I had made several experiments with various vessels *blackened* and *covered with repeated coatings of varnish*, and I announced the results obtained. I also informed him of the discovery which I made, with the help of my *thermoscope*, that different bodies of the same temperature give out very different quantities of calorific rays, and that frigorific rays have just as real an existence as the calorific rays from warm bodies.

Since Sir Joseph showed my letters to various persons, and since I did not keep my experiments or their results a secret from him or from any one else, my discovery was publicly mentioned in London even as early as the spring of the past year. As an incontrovertible proof of this fact, I can bring forward a letter from a friend of mine (to whom I had not mentioned my new discovery in any way), in which he congratulates me on the success of my researches, and informed me at the same time that he had learned what he knew with regard to my discoveries from Mr. Davy, a Professor in the Royal Institution, who had spoken publicly of them in his lectures on chemistry.

The memoir in which I gave an account of my investigations was finished early in May (1803); in the early part of June I left Munich for a journey into Switzerland. As I intended to proceed from Geneva to Paris, I took with me my memoir and some of my newly invented instruments, and among others the thermoscope.

On reaching Geneva, in August, I read my memoir in the presence of Professor Pictet, De Saussure, and various other persons, and at the same time repeated some of my experiments with the thermoscope.

As soon as I reached Paris, in the latter part of October, I had my memoir copied (by Mr. Cadel, of Glasgow, who was then in Paris), and sent it, in the middle of December, to London by the younger Mr. Livingston, who was kind enough to deliver it in person to Sir Joseph Banks on the 23d of December. As the Christmas recess of the Royal Society begins just after this time, my memoir could not be read in a public meeting of the Society until the 3d of February.[13]

The 6th of June there were sent to me from London (through the elder Mr. Livingston, Minister Plenipotentiary of the United States of North America at Paris) two copies of my memoir, published by order of the Royal Society. At the same time I received a letter from Mr. Davy, Professor of Chemistry at the Royal Institution, in which he informed me that Mr. Leslie had, a short time previously, published a memoir on heat, and that in it he had described various experiments which bore a resemblance to some which I had performed.

The 2d of June I received at Paris, from M. Bertholet, Mr. Leslie's book, which was sent to me by Sir Joseph Banks. M. Bertholet had at the same time received from England a copy of the work, which was sent to him by one of his friends there.

As I had, only a short time before, occupied the attention of the National Institute with an account of my recent researches and discoveries,* the appear-

* Between the 19th of March and the 7th of May, 1804, I presented to the Na-

ance of a book coming from England, and containing a description of a number of experiments and discoveries in many respects not dissimilar to my own, could not fail to create a certain feeling of surprise among the philosophers of Paris, as I could plainly enough perceive. I find myself, therefore, compelled, although against my will, to explain as far as possible an occurrence which it is highly important for me should appear in its true light.

I am far from intending to assert that Mr. Leslie had any knowledge of those experiments of mine which bore a resemblance to those which he announced publicly in print. It is, however, equally certain that I did not know, and could not have known, the least thing about his. It will not be difficult for me to prove this.

It might, perhaps, be just as easy for Mr. Leslie to bring forward proofs that he knew absolutely nothing about my experiments. This would be all the more readily believed as he (in the course of certain remarks made in a note with regard to the observations which I offered in explanation of the propagation of heat in liquids) speaks of me as of a man *already dead** at the time when he made these remarks.

It is certain that we are perfect strangers to each other, that we do not know each other even by sight, and that we never had any sort of correspondence with each other.

As regards the *priority of the public announcement* of our discoveries, this point can be easily made clear by

tional Institute five different memoirs on this subject. They will probably be printed in the "Mémoires de l'Institut."[14]

* See the thirty-ninth note at the end of his work, beginning with the following words : "*A late* ingenious experimenter."

the statement of certain facts which do not admit of doubt.

It is true that I cannot determine with any great accuracy the time when Mr. Leslie's book first saw the light; it cannot, however, possibly have been published before the middle of May of this year, for the dedication is dated at Largo, in Fifeshire (Scotland), the 20th of May, 1804. This would be, consequently, nearly a year after the time when the most remarkable results of my investigations were known in London; it would be nine months from the time when, in Geneva, I read the memoir containing the circumstantial and detailed account of these investigations in the presence of a number of celebrated philosophers; it would be five months later than the time at which this memoir was placed in the hands of the President of the Royal Society of London; and it would be more than a quarter of a year from the time at which it was read publicly before this Society.

Still the *priority* in question, considered in and by itself, is of such slight importance that I should not have mentioned it at all, were it not that the facts which go to establish it tend at the same time to strengthen a far more important assertion, namely, that I am actually the *discoverer* of what I announced as discoveries.

If Mr. Leslie and myself, the one in Scotland, the other in Bavaria, each for himself and at about the same time, did actually make the same discoveries, this is a condition of things which has already happened more than once before our time; and then, as far as the interpretation of these phenomena is concerned, we differ from each other in our mode of explanation to such an extent that there can no question arise between us in regard to the ownership of our opinions. Nothing is more certain than that, in this respect, we have not borrowed one from the other in the slightest degree.

Besides, I have every reason for believing that even if I had not described so particularly the facts which I have brought forward, still all those who will take the trouble to consider impartially the numerous experiments on the subject of heat which I have made during more than twenty years, will be convinced that I must have been led to the investigations and discoveries in question by the entirely natural connection of ideas caused by my opinions on the subject, without needing to borrow, in the slightest degree, from any person whomsoever.

To close this historical review of my various researches on the subject of heat, I will give a very brief account of my labours in this connection, from my arrival in Paris, until the close of the month of October of the last year (1803).

As I had brought with me two thermoscopes, I had them adjusted, by Dumontier, with all possible care; I also sent to Munich for several other instruments which I had used, the year before, in my experiments on heat.

I also procured several new instruments, in order to make new experiments; among others, an apparatus which I intended to use to determine the progress of heat in a massive bar of metal, in glass, and in other solid substances. All these instruments I showed to several members of the National Institute, namely, to MM. Laplace, Delambre, Prony, and Biot.

To these philosophers, and at the same time to M. Bertholet as well, I proposed to perform the now well-known experiment of the cold body and the speaking-tube (and this was before the instrument necessary for the purpose had been invented), and thus to terminate our (in all respects very friendly) controversy on the reality of caloric.

This experiment was afterwards performed in the physical cabinet of the National Institute in the presence of Laplace, Bertholet, and Charles. The result was precisely as I had predicted.

On the 28th Ventose of the year 12 (the 19th of March, 1804) I presented to the Mathematical and Physical Class of the National Institute my first memoir, in which I described my thermoscope and a few of the discoveries that I had made with the help of this instrument.[15]

On the 5th Germinal (26th of March, 1804) I presented to the same Class a second memoir, in which I sought to develop my ideas on the nature of heat, as well as on the manner in which it is excited and communicated. At the same time I gave the results of certain experiments which I had made on the cooling of warm bodies in the air.[3]

On the 19th Germinal (9th of April, 1804) I presented to the Class a third memoir, which treated of an experiment, which I had recently made in Paris, on the nature of heat; by this experiment the influence of the rays emanating from cooling bodies was rendered manifest in a manner entirely new.[16]

On the 10th Floréal (30th of April, 1804) my fourth memoir was presented to the Class.[17] In this memoir I described an experiment which I performed with two flasks of equal size. One was made of glass, the other of tinned iron. Both were filled with boiling water, and exposed at the same time to the air, in which they were allowed to cool. The water contained in the glass flask cooled twice as fast as that in the one made of tinned iron, although the walls of the latter were much thinner than those of the glass flask. This memoir ends with some considerations on the comparison which has been instituted between a warm body and a sponge filled with water, and on the influence of radiation during the warming and cooling of bodies.

On the 17th Floréal (7th of May, 1804) I laid before the Class my fifth memoir,[18] in which I gave an account of an entirely new series of experiments, which I had made in Paris, on the manner in which heat is propagated in a massive bar of metal, six inches long and an inch and a half in diameter. This bar was heated at one end by boiling water, and cooled at the other end sometimes with a mixture of pounded ice and water, and sometimes simply with water of the temperature of the air.

M. Biot, member of the Institute, made, at about the same time with myself, several successive experiments on the propagation of heat in metallic bars and other solid bodies. He, however, used for this purpose bars of a different length from mine, and higher temperatures. Otherwise we obtained the same results from our experiments.

He hit upon the fortunate idea of employing similar experiments for measuring very high degrees of temperature; such, for example, as is necessary in the preparation of porcelain, or for melting metals not readily fusible.

As I was invited to prepare a condensed description of my recent experiments on heat, to be read at the public sitting of the National Institute on the 6th Messidor (June 26, 1804), I presented the memoir which follows. As it has already been printed (in the *Moniteur* of the 10th Messidor, or the 29th of June, 1804), I may be allowed to introduce it into this collection. The case is otherwise with the five other memoirs which I presented to the first class of the Institute; for as they will be embodied in the Mémoires of the Class it would not be proper for me to publish them earlier.

To complete this historical review, I must, in addition, say a word or two on my attempts to perfect the application of heat to the arts and to all sorts of domestic purposes. Among the fifteen Essays which I have published in three octavo volumes are no less than eight which treat of the use of heat. They are as follows: —

Essay IV. Of Chimney Fireplaces; VI. On the Management of Fire and Economy of Fuel; X. Of Kitchen Fireplaces; XII. Of the Salubrity of Warm Rooms in Winter; XIII. Of the Salubrity of Warm Baths, and the Mode of their Preparation; XIV. Of the Management of Fire in Closed Fireplaces; XV. Of the Use of Steam as a Vehicle for transporting Heat.

# REFERENCES TO RUMFORD'S OWN WORK

1. "An Account of some Experiments made to Determine the Quantities of Moisture absorbed from the Atmosphere by various Substances." Read before the Royal Society, March 22, 1787.
2. The paper which appeared in 1786 is Part I of "The Propagation of Heat in various Substances," and the paper which appeared in 1792 is Part II of that essay.
3. Part I of "The Propagation of Heat in various Substances."
4. This refers to Part I of "The Propagation of Heat in Fluids," which can be found in Rumford's *Essays, Political, Economical and Philosophical*, Vol. II.
5. "The Propagation of Heat in Fluids," Part II.
6. Essay VI was originally published by T. C. Cadell and W. Davies in London in 1797.
7. Part II of "The Propagation of Heat in various Substances" was published in the *Philosophical Transactions* of 1792.
8. "The Propagation of Heat in Fluids."
9. Essay VII was originally published by T. C. Cadell and W. Davies in London in 1797.
10. Essay VIII was originally published by T. C. Cadell and W. Davies in London in 1798.
11. "Essay on the Force of Fired Gunpowder."
12. "An Inquiry concerning the Weight ascribed to Heat."
13. "An Inquiry concerning the Nature of Heat, and the Mode of its Communication."
14. Five essays on researches on heat were read in 1804 before the French Institute, on: 19 March, 26 March, 9 April, 30 April, and 7 May.
15. "Description of a New Instrument for Physics."
16. "Short Account of a new Experiment on Heat."
17. "Experiments on Cooling Bodies."
18. "Heat is Communicated through Solid Bodies."

# FACTS OF PUBLICATION

## AN EXPERIMENTAL INQUIRY CONCERNING THE SOURCE OF THE HEAT WHICH IS EXCITED BY FRICTION

Read before the Royal Society, January 25, 1798.

*Philosophical Transactions of the Royal Society of London*, *88* (London, 1798), 80–102.

*Bibliothèque Britannique* (*Science et Arts*), edited by Auguste Pictet, Charles Pictet, and F. G. Maurice (Geneva, 1798), VIII, 3–34.

Sir Benjamin Thompson, Count of Rumford, *Essays*, *Political*, *Economical and Philosophical* (London: T. Cadell, jr. and W. Davies, 1798), II, 469–496.

*Allgemeines Journal der Chemie*, edited by Alexander Nicolaus Scherer (Berlin: H. Frolich, 1798), I, 9–31; remarks by the editor, 31–37.

*The Complete Works of Count Rumford* (Boston: American Academy of Arts and Sciences), I (1870), 469–493.

## AN INQUIRY CONCERNING THE WEIGHT ASCRIBED TO HEAT

Read before the Royal Society, May 2, 1799.

*Philosophical Transactions of the Royal Society of London*, *89* (London, 1799), 179–194.

Sir Benjamin Thompson, Count of Rumford, *Philosophical*

*Papers: Being a collection of memoirs, dissertations and experimental investigations relating to various branches of natural philosophy and mechanics, together with letters to several persons on subjects connected with science and useful improvements* (London: T. Cadell, jr. and W. Davies, 1802), I, 366–383. This contains a Supplement, included in the American Academy edition of Rumford's *Works*, II, 17–22.

*Bibliothèque Britannique* (*Science et Arts*), edited by Auguste Pictet, Charles Pictet, and F. G. Maurice (Geneva, 1800), XIII, 217–238.

*A Journal of Natural Philosophy, Chemistry and the Arts: Illustrated with Engravings*, edited by William Nicholson (London, 1799–1800), III, 381–390.

*The Complete Works of Count Rumford* (Boston: American Academy of Arst and Sciences), II (1873), 1–22.

## OF THE PROPAGATION OF HEAT IN VARIOUS SUBSTANCES

London: T. Cadell, jr. and W. Davies, 1798.

*Philosophical Transactions of the Royal Society of London*, *76* (London, 1786), 273–304; *82* (1792), 48–80.

*Bibliothèque Britannique* (*Science et Arts*), edited by Auguste Pictet, Charles Pictet, and F. G. Maurice (Geneva, 1796), I, 11–45.

Sir Benjamin Thompson, Count of Rumford, *Essays, Political, Economical and Philosophical* (London: T. Cadell, jr. and W. Davies, (1798), II, 391–468.

*The Complete Works of Count Rumford* (Boston: American Academy of Arts and Sciences), I (1870), 401–468.

## THE PROPAGATION OF HEAT IN FLUIDS

Part I.

London: T. Cadell, jr. and W. Davies, 1797.

Sir Benjamin Thompson, Count of Rumford, *Essays, Political, Economical and Philosophical* (London: T. Cadell, jr. and W. Davies, 1798), II, 199–313.

*Bibliothèque Britannique* (*Science et Arts*), edited by Auguste Pictet, Charles Pictet, and F. G. Maurice, with remarks by Auguste Pictet (Geneva 1797), V, 90 (announcement), 97–200; also published separately.

*Annalen der Physik*, angefangen von F. A. C. Gren, fortgesetzt von L. W. Gilbert (Halle, 1799), I, 214–241, 323–351, 436–463.

*Neues Journal der Physik*, edited by D. Friedrich Ulbrecht (Leipzig: Carl Gren, 1797), IV, 418–450.

*Chemische Annalen für die Freunde der Naturlehre*, *Arzney*, *Gelahrtheit*, *Haushaltungskunst und Manufacturen*, edited by D. Lorenz Crell (Helmstädt: C. G. Fleckeisen, 1797), 78–104, 149–170, 233–246, 342–358, 446–464, 488–502.

Part II.

Published with a second edition of Part I (London: T. Cadell, jr. and W. Davies, 1798).

*Bibliothèque Britannique* (*Science et Arts*), edited by Auguste Pictet, Charles Pictet, and F. G. Maurice (Geneva, 1798), VIII, 85–121, 201–339.

Sir Benjamin Thompson, Count of Rumford, *Essays*, *Political*, *Economical and Philosophical* (London: T. Cadell, jr. and W. Davies, 1798), II, 313–386.

*The Complete Works of Count Rumford* (Boston: American Academy of Arts and Sciences), I (1870), 237–400.

## SHORT ACCOUNT OF A NEW EXPERIMENT ON HEAT; NOTICE D'UNE NOUVELLE EXPÉRIENCE SUR LA CHALEUR

Read at the Institut de France, 19 Germinal, An 12 (April 9, 1804).

*Mémoires de la classe des Sciences, Mathématiques et Physiques de l'Institut de France* (Paris: Baudouin, Imprimeur de l'Institut de France, January 1806), VI, 88–96.

*A Journal of Natural Philosophy*, *Chemistry and the Arts: Illustrated with Engravings*, edited by William Nicholson (London, 1805), XII, 65–70.

*The Complete Works of Count Rumford* (Boston: American Academy of Arts and Sciences), II (1873), 131–137.

## THE HEAT PRODUCED IN A BODY BY A GIVEN QUANTITY OF SOLAR LIGHT IS THE SAME WHETHER THE RAYS BE DENSER OR RARER, CONVERGENT, PARALLEL, OR DIVERGENT; RECHERCHES SUR LA CHALEUR EXCITÉE PAR LES RAYONS SOLAIRES

Read at the National Institut de France, 11 Germinal, An 13 (April 2, 1805).

*Mémoires de la classe des Sciences, Mathématiques et Physiques de l'Institut de France* (Paris: Baudouin, Imprimeur de l'Institut de France, January 1806), VI, 123–133.

*A Journal of Natural Philosophy, Chemistry and the Arts: Illustrated with Engravings*, edited by William Nicholson (London, 1805), XII, 164–171.

*Annalen der Physik*, angefangen von F. A. C. Gren, fortgesetzt von L. W. Gilbert (Halle, 1805), XX, 177–186.

*Journal de Physique, de Chimie, d'Histoire Naturelle et des Arts, avec des planches en taille-douce*, par J. C. Delamétherie (Paris: Chez Courcier, 1805), LXI, 32–39.

*The Complete Works of Count Rumford* (Boston: American Academy of Arts and Sciences), II (1873), 158–165.

## REFLECTIONS ON HEAT; MÉMOIRE SUR LA CHALEUR

Read at a public session of the National Institute of France, 6 Messidor, An 12 (June 25, 1804).

*Le Moniteur Universel ou Gazette Nationale* (Paris, 1804), 10 Messidor, An 12 (June 29, 1804).

Le Comte de Rumford, *Mémoires sur la Chaleur* (Paris: Chez Firmin Didot, 1804), 129–156; published in German in Rumford, *Kleine Schriften* (Weimar, 1805), vol. IV.

*The Complete Works of Count Rumford* (Boston: American Academy of Arts and Sciences), II (1873), 166–187.

AN INQUIRY CONCERNING THE NATURE OF HEAT, AND THE MODE OF ITS COMMUNICATION

Read before the Royal Society, February 2, 1804.
London: W. Bulmer and Co., 1804.

*Philosophical Transactions of the Royal Society of London*, *94* (London, 1804), 77–182.

Le Comte de Rumford, *Mémoires sur la Chaleur* (Paris: Chez Firmin Didot, 1804), 1–128 (translated by Auguste Pictet); published in German in Rumford, *Kleine Schriften* (Weimar, 1805), vol. IV.

*The Complete Works of Count Rumford* (Boston: American Academy of Arts and Sciences), II (1873), 23–130.

RESEARCH ON HEAT, SECOND MEMOIR

Read 5 Germinal, An 12 (March 26, 1804).

*Mémoires de la classe des Sciences, Mathématiques et Physiques de l'Institut de France* (Paris: Baudouin, Imprimeur de l'Institut de France, 1806), VI, 79–87.

HISTORICAL REVIEW OF THE VARIOUS EXPERIMENTS OF THE AUTHOR ON THE SUBJECT OF HEAT

Le Comte de Rumford, *Mémoires sur la Chaleur* (Paris: Chez Firmin Didot, 1804), vii–lxviij, published in German in Rumford, *Kleine Schriften* (Weimar, 1805), vol. IV.

*The Complete Works of Count Rumford* (Boston: American Academy of Arts and Sciences), II (1873), 188–240.

# INDEX

(VOLUME I)